Introduce to
Advanced Optical Fiber Telecommunications
and Quantum communications

Theory and Design

Han-xiong Lian

2016.10.01.

INTRODUCE TO ADVANCED OPTICAL FIBER TELECOMMUNICATIONS AND QUANTUM COMMUNICATIONS
THEORY AND DESIGN

iUniverse books may be ordered through booksellers or by contacting:

iUniverse
1663 Liberty Drive
Bloomington, IN 47403
www.iuniverse.com
1-800-Authors (1-800-288-4677)

Because of the dynamic nature of the Internet, any web addresses or links contained in this book may have changed since publication and may no longer be valid. The views expressed in this work are solely those of the author and do not necessarily reflect the views of the publisher, and the publisher hereby disclaims any responsibility for them.

Any people depicted in stock imagery provided by Thinkstock are models,
and such images are being used for illustrative purposes only.
Certain stock imagery © Thinkstock.

ISBN: 978-1-5320-1290-7 (sc)
ISBN: 978-1-5320-1291-4 (e)

Print information available on the last page.

iUniverse rev. date: 12/05/2016

Preface

Nowadays, human is in the era of knowledge exploring. The key technologies for this era are the computer and the optical fiber communications.

After optical fiber invented by Dr. Gao Kun in 70s, the single mode fiber displays an important role to provide super wide bandwidth for the computer data transmission. Since then human has been experience five generations of optical fiber communications systems.

- The first generation: After a period of reach starting from 1975, the first commercial fiber-optic communications system operated at $0.8\mu m$ and at a bit rate of $45Mb/s$ and used GaAs semiconductor lasers, and the repeater spacing is $10km$.

- The second generation: The second gencration used in 1980s operated at $1.3\mu m$, and used InGaAsP semiconductor lasers and the multi-mode fiber, and in 1981 the single mode fiber was revealed to greatly improve system performance. Until 1987, these systems were operating at bit rate of up to $1.7Gb/s$ with repeater spacing up to $50km$.

- The third generation: The third-generation operated at $1.55\mu m$ and had single mode fiber losses about $0.2dB/km$. Scientists overcome the pulse-spreading at this wavelength by using dispersion-shifted fiber to have minimal dispersion at $1.55\mu m$ by limiting the laser spectrum to a single longitudinal mode. These improvements eventually allowed third-generation system to operate commercially at $2.5Gb/s$ with repeater spacing in excess of $100km$.

- The fourth generation: The fourth generation starting 1992, used optical amplifiers to reduce the number of repeaters and used the wavelength-division multiplexing to increase data capacity. Those caused a revolution that results in the doubling of system capacity every 6 months until a bit rate of $10Tb/s$ was researched by 2001. In 2006, a bit rate of $14Tb/s$ was researched over a single $160km$ using optical amplifiers. Especially, 10 Gbps optical fiber transponder designed and developed by authors professor Han-xiong Lian and senior engineer Ying-xiu Ye in Multiplexer Inc. New Jersey, USA by 2002 become a new favorite, which has been chosen as the optical repeater and digital adder/droper in the long haul optical fiber transmission links first in

USA, Canada, and then in China and India, ⋯ **to form a Global long-haul 10Gbps optical fiber transmission networks** as shown in Figure 0-1.

- The fifth generation: Based on the 10Gbps optical fiber transponder the fifth generation is to expend the wavelength range by using a wavelength division multiplexing (**WDM**) system, in which each wavelength can be used as the 10Gbps transponder transmission system. Now, 162 channels of DWDM has been developed to provider 162×10 Gbps data stream transmission system. Principally, the transmission capability of the single mode fiber is no limit since it has a window promising of the range to 1.30 - 1.65μm.

All of those research come to three milestones as follows.

- **The first milestone — 10Gbps optical fiber transponder**

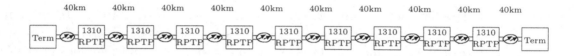

Figure 0-1   Conventional TDM Transmission - 10Gbps

Obviously, *10Gbps optical fiber transponder in fourth generation is the 1st milestone* in optical fiber communications networks. The main advantage of which is the simplest solution for 10Gbps long haul optical fiber transmission networks, since the optical modulation is an interior amplitude modulation of laser compared with the external optical modulation. However, the compensation for which is that the laser is a narrow line-width single longitudinal mode laser at $+2dB$ output power and the photo-detector is supper sensitivity with the sensitivity at $-17dBm$. The function of the 10Gbps optical fiber transponder is an optical repeater, with optical amplify, O/E converter (in photo-detector), E/O converter (in laser) and digital adder/dropper. And then it can drop down data-stream to local station and add up data-stream from the local station as shown in Figure 0-1. Where "RPTP" represents 10Gbps optical fiber transponder, "RP" is the receiver side and "TP" is the transmitter side. The spacing between two transponder is $40km$ with a pair of optical single mode fiber as the transmission medium.

The research and development of 10Gbps optical fiber transponder has to face the following challenge:

a. Supper narrow line-width single longitudinal mode laser,

b. Super high sensitivity of photo-detector and preamplifier.

c. Digital multiplexing, digital de-multiplexing and digital adding-dropping multiplexing.

Japanese company Hitachi first provided the single longitudinal mode laser diode with output power more than $+2dBm$. And Multiplex Inc. New Jersey, USA, based on the Lucent company technology in United State, provided the supper high sensitivity of photo-detector diode, reached to a supper high photoelectric sensitivity of $-17dBm$, which means that, as an optical amplifier, 10Gbps optical fiber transponder can support transmission link distance up to 40km and beyond. Both of them gave multiplex Inc. in USA a opportunity to design and develop a *10Gbps optical fiber transponder*. Based on these achievements above, the authors Professor Han-xiong Lian and Senior Engineer Ying-xiu Ye obtained an opportunity to design and develope the 10Gbps optical fiber transponder and got successful in Multiplexer Inc. New Jersey, USA by 2002.

- **The second milestone — All optical fiber transmission networks**

Based on the 10Gbps optical fiber transponder, using wavelength division multiplexing (WDM) to form all optical fiber transmission network is a smart choice, which combines 128 channels of 10Gbps optical fiber transponders to form a super wide bandwidth Metropolitan Area Networks **to increase data capacity more than 120 times**. *Which is another milestone in the developing of optical fiber communications systems.* An example of which is shown in Figure 0-2.

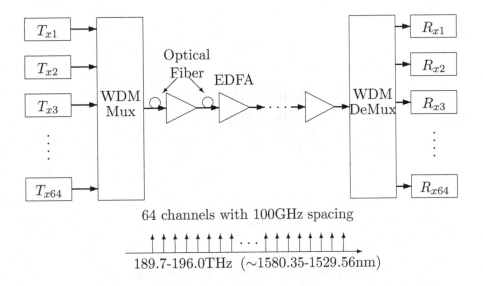

64 channels with 100GHz spacing

189.7-196.0THz ($\sim$1580.35-1529.56nm)

Figure 0-2   All optical fiber DWDM communications networks with 64 channels

Table 0-1 62 channels in ITU-T DWDM Grid (spacing=100MHz)

Channel Code	λ (nm)	Channel Code	λ (nm)	Channel Code	λ (nm)	Channel Code	λ (nm)
18	1563.05	30	1553.30	42	1543.73	54	1534.25
19	1562.23	31	1552.53	43	1542.94	55	1533.47
20	1561.42	32	1551.72	44	1542.14	56	1532.68
21	1560.61	33	1550.92	45	1541.35	57	1531.90
22	1559.80	34	1550.12	46	1540.56	58	1531.12
23	1558.98	35	1549.32	47	1539.77	59	1530.33
24	1558.17	36	1548.52	48	1538.98	60	1529.55
25	1557.36	37	1547.72	49	1538.19	61	1528.77
26	1556.56	38	1546.92	50	1537.40	62	1527.99
27	1555.75	39	1546.12	51	1536.61		
28	1554.94	40	1545.32	52	1535.82		
29	1554.13	41	1544.53	53	1535.04		

For a 64 channels dense wavelength division multiplexing (DWDM) as shown in Figure 0-3 and Table 0-1, the spacing between two adjacent channels is $100MHz$. Thus, the design and implementation of optical de-multiplexing from one optical data-stream into 64 optical data-streams or optical multiplexing from 64 optical data-streams into one optical data-stream are all not a easy duty.

In an all optical fiber DWDM communications network, all of the components are optical fiber components (such as how to design a fiber Bragg grating filter) or thin film components (such as how to design a Mach-Zehnder filter). Which will be discussed in detail in Chapter 3.

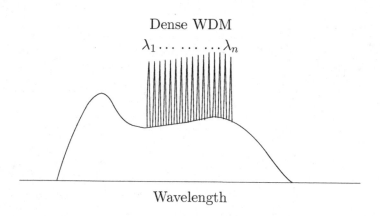

Figure 0-3   ITU DWDM Wavelengths

- **New milestone — Quantum communication**

For the conventional optical communications, the information is carried by the intensity of photons, where light "on" is "1" and light "off" is "0". However, in quantum communications, the information is carried by the quantum state, or, the information carrier is the polarization state of photon as shown in Table 0-1,

Table 0-1

Basis	0	1
+	↑	→
x	↗	↘

Where, in basis "+",

| ↑> is a note to represent the vertical polarization state of the photon,

| →> is a note to represent the horizontal polarization state of the photon,

In basis "x", | ↗> is a note to represent the 45^0 polarization state of the photon,

| →> is a note to represent the -45^0 polarization state of the photon.

The quantum communications is based on the entangled photons a pair of photons with entanglement or correlations. Which means that

a. In basis "+", if the state of the first one is | ↑> the state of the second one is | →> definitely, vice verse. In other words, if the first one is "0", the second one is "1" definitely, vice verse.

b. In basis "x", if the state of the first one is | ↗>, the state of the second one is | ↘> definitely. Or say, if the first one is "0", the second one is "1" definitely, vice verse.

When a pair of entangled photons are separated a distance, the entanglement properties occurs any where and any time, no matter how far the separated distance is.

The polarization state of the photon in entangled photons is a random state, and then the measured results of the state is a probability. Who knows what the state of each one is to be. If you want to know the state of each one, you have to measure the state, and the state you measured is not the original state, since it is in random state. However, we knows that if the measured state of the first one is "0", the state of the second one is "1" definitely and the changing of the second state take no time if you measure the state of the first one. Which is so called the "action at a distance". The "action at a distance" take no time means that the transmission speed from the first one to the second one is 10000 times of light speed in an entangled photons.

Now, the technology allows one to distribute sequence of random bit, (or quantum state), whose randomness and secrecy are guaranteed by the laws of quantum physics. These sequences can then be used as secret keys with conventional cryptography technologies to guarantee the confidentiality of data transmissions. In which Alice (sender) and Bob (receiver) can shared a random and secret keys for encryption and decryption. If the eavesdropper (Eve) want to eavesdrop, he has to measure the state of the photon, and then Alice(sender) and Bob(receiver) will aware his action. This is based on the principle of the quantum mechanics: any measurement on the quantum system will interference the quantum system.

After public communication between Alice and Bob, both of them can make decision to abandon the states that Eve measured and remain the others as their secret keys to do quantum key distribution (QKD). However, the conventional cryptography technologies can no aware the eavesdrop from Eve.

Until now, the conception of the entangled photons is unacceptable for many physicists, even the great physicist Einstein, "since the light speed is the speed limit in the unverse", Einstein thinks. However, all of the experimental results can not support the idea from Einstein. And the implementation of the quantum communications is underway in China, America, Australia, Europe countries and Japan. Therefore, we encourage young researchers participate the development of the quantum communications and quantum computer. This is extremely important project in information technology for the future.

Now, the implementation of quantum communication is underway in two options:
a. Festival, the quantum communication can be implemented on the existing optical fiber infrastructures. However, the quantum state can not pass the optical amplifier in the existing optical fiber infrastructures.
b. Therefore, most of scientists consider that quantum satellite communication is a new option to extend the transmission distance of quantum communication to form a Global long-haul quantum communications links. In which, we need to set up quantum repeaters over long distance.
In Switzerland, the Qurep Quantum Repeater technology is concentrated around quantum light-matter interactions at the quantum level in ensembles of rare earth ions frozen in a crystal that store quantum information by coherent control of the quantum degrees of freedom.

The storage of quantum state allows us to wait for successful transmission of a photon over an extended distance, thus overcoming the problem of fiber loss.

In 1993, the American scientist C.H. Bennett proposed the conception of quantum teleportation, which means that the quantum communication is to transmit the quantum state, which carry information.
After than, 6 scientists from different country proposed a scheme to realize "Teleportation" by using quantum channel with classical channel. In this case, an unknown quantum state at A (Alice) is transmitted to another place B (Bob), then make another particle at Bob onto the prepared quantum state. And the original particle still remain at A (Alice). Which means that the "transmission signal" from Alice to Bob is "quantum state" only but not the photon it-self. However, the information has been transmitted from Alice to Bob by the preparation of another particle to the quantum state at B (Bob). Which is so called "Teleportation". This is an important idea, according to this idea, a large information can be transmitted from Alice to Bob with three features:

1. Security:
a. First, the quantum cryptography key is random and used only once. Although

it has been eavesdropped, the eavesdropper, Eve, is unable to get the correct key, and is unable to break the information.

b. Secondary, in the process of quantum communication between Alice and Bob, two entangled articles, one in Alice hand and another one in Bob hand, any observation and interference any particle will cause the decay of the article. Therefore, the part of infirmation obtained by Eve is a "broken" information, but not the original information.

2. High efficiency:

The unknown quantum state is in entanglement before testing, which represents two state. For instance, one quantum state represents two number "0" and "1". Seven quantum states represents $2^7 = 128$ states and "2×128" number. Therefore, the quantum communication efficiency is two times of classical communication efficiency or more.

3. Instant transmission:

In quantum communication, the information transmission is instant transmission since the original information photon is still in Alice hand, and Bob get the information is by the preparation of another particle to the quantum state. Which can be implemented instantly. Which is equal to say that the speed of the information transmission is almost 10,000 times of light velocity "c".

This book try to cover the three milestones knowledge:

a. 10Gbps optical fiber transponder and Global long-haul optical fiber TDM links,

b. Wavelength division multiplexing and DWDM super wide bandwidth Metropolitan Area Networks,

c. Quantum communication and Global quantum communication networks, including principle, technology and network to the deep that can help you to design the components and networks. Which requires a new mix of competencies, from telecommunication engineering to theoretical physics.

To cover all of the knowledge, we need 8 Chapters including:

- Introduction in Chapter 1.

- The all optical fiber communications network in Chapter 2.

- The all optical fiber communications technologies in Chapter 3.

- The optical fiber as the transmission medium in Chapter 4.

- The longitudinal mode laser as the transmitter in Chapter 5.

- The photo-detector preamplifier as the receiver in Chapter 6.

- The 10Gbps optical fiber transponder in Chapter 7.

- The quantum communications in Chapter 8.

The author Han-xiong Lian has solid background in optical fiber communications, microwave communications and physics. He has been involved in the research and development of optical fiber communications for almost 30 years.

From 1984, his research area was nonlinearity of the single mode fiber and Raman Scattering Amplifier supported by the Ministry of Posts and Telecommunications in Chinese government.

Another project is Erbium-doped single mode fiber and Erbium-doped single mode amplifier supported by the Ministry of Posts and Telecommunications in Chinese government.

The third project is single mode fiber sensor supported by Electrical Power Ministry in Chinese government.

In 1984, as a professor in optical fiber communications, he has joined a Government Investigation Group to discuss with **Dr. Gao Kung** (the inventor of the optical fiber) in STD Inc. in Germany about a project to manufacture the optical fiber in China.

In 1989, his university invited **Dr. Tang yi Lee** (the leading scientist of the photo detector reacher group in Lucent Inc. USA), he got a change to know his research in Lucent USA.

From 1995 to 1998, both of Han-xiong Lian and Senior engineer Ms. Ying-xiu Ye had to face a challenge to develop a broadband millimeter wave transmitter and receiver to set up a system to transmit 400 channels calor TV in **BNI**, *Northern Telecom*. Canada and then to design and develop the millimeter transmitter and receiver for the space satellite communications in **FEI USA**.

From 2000 to 2002, Both of Han-xiong Lian and Ying-xiu Ye have jointed *Atrium Network Inc.* to develop *40Gbit/s optical fiber transponder*. All of those challenge forced Han-xiong Lian and Ying-xiu Ye to prepare all of the knowledge to face a new challenge to design and develop the *10Gbps optical fiber transponder* in **Multiplexer Inc. New Jersey USA** and got successful in 2002. And the boss of Multiplexer Inc. **Dr. Wang** was one of the leading scientists of the photo detector research group in Lucent with **Dr. Tang Yi Lee**.

The third author Ms. Wenlei Lian is Working in this field for Keysight Technology (formerly Agilent Technologies) for 15 years, which gave her most updated information on what her customers really need. She gave those feedbacks to Mr. Han-xiong Lian and Ms. Yingxiu Ye. We discussed a book topic that will allow us to share our knowledge and at the same time, to meet the market's needs.

From 2003 till now, the writing group of Han-xiong Lian, Ying-xiu Ye and Wenlei Lian has collect their knowledge, and two books have been published:

a. Han-xiong Lian, "Analytical Technology in Electromagnetic Field Theory in RF, Wireless and Optical Fiber Communications" Published by iUniversal Inc. New York Bloomington, 2009.

b. Han-xiong Lian, Ying-xiu Ye and Wenlei Lian, "Supper low noise PLL Oscillator and Low Jitter Synthesizer", Published by iUniversal Inc. New York Bloomington, 2014.

In which, the main writer is Han-xiong Lian.

Han Xiong Lian

2016.10.01

Contents

Preface   iii

1   Introduce to the Optical Fiber Networks   1
   1.1   Introduction 1
   1.2   High bit rate digital transmission network with TDM 2
   1.3   All-optical transmission network with DWDM 5

2   All Optical Fiber DWDM Communication Networks   9
   2.1   The Global Networks Hierarchy 9
      2.1.1   TDM and SONET 13
      2.1.2   SONET with DWDM 17
      2.1.3   Additional Drivers In Metropolitan Area Networks 25
      2.1.4   Value of DWDM in the Metropolitan Area 27
      2.1.5   Technologies in the Metropolitan Area Networks 28
      2.1.6   Topologies and Protection Schemes for DWDM 34
   2.2   Introduce to all optical fiber DWDM communication system 42
      2.2.1   Optical WDM multiplexing and de-multiplexing 43
      2.2.2   All optical fiber DWDM communications networks 45

3   All Optical Fiber DWDM Communications Technology   47
   3.1   Optical Mach-Zehnder filter 47
      3.1.1   Introduce to Optical Mach-Zehnder filter 47
      3.1.2   Single Coupler Mach-Zehnder Filter 50
      3.1.3   Cascaded CMZ filter with weighted coupler distribution . . . 55
   3.2   Optical Fiber Bragg Grating 65
      3.2.1   Introduce to fiber Bragg grating (FBG) 65
      3.2.2   Single fiber Brag grating of (FBG) 68
      3.2.3   Cascaded FBG filter with index change 80
   3.3   Optical (de)multiplexer 88
      3.3.1   Optical (de)multiplexer using Mash-Zhnder filter 88
      3.3.2   Array Waveguide Grating (AWG) 89
   3.4   Optical ADD/DROP cross-connector 102
   3.5   Optical amplifier 104
      3.5.1   Nonlinearity in single mode fiber 104

      3.5.2   Erbium-Doped Fiber Amplifier 106

      3.5.3   Raman amplifier 111

      3.5.4   Brillouin amplifier 126

4 Signals in Optical Fiber   **135**

  4.1   Introduction 135

      4.1.1   Critical angle θ_c 136

      4.1.2   Numerical aperture NA 137

  4.2   Modes in optical fiber 138

      4.2.1   Mode analysis of multi-mode optical fiber 141

      4.2.2   Mode analysis of single mode optical fiber 149

      4.2.3   The conception of LP_{mn} modes (Linearly Polarized modes) . 151

      4.2.4   The characteristics of single mode fiber 152

      4.2.5   Band loss of single mode fiber 152

  4.3   Attenuation in optical fiber 153

  4.4   Group velocity and dispersion in optical fiber 157

      4.4.1   Phase velocity v_p and group velocity v_g 157

      4.4.2   Dispersion in optical fiber 160

      4.4.3   Control of dispersion in single mode fiber 165

  4.5   Polarization effects in single mode fiber 167

  4.6   Nonlinearity in single mode fiber 169

      4.6.1   Stimulated Raman Scattering 170

      4.6.2   Stimulated Brillouin scattering 182

      4.6.3   Nonlinear Kerr Effect and Soliton 183

      4.6.4   Four-wave mixing 185

  4.7   Basic knowledge 187

      4.7.1   Maxwell equation 187

      4.7.2   Material equations 188

      4.7.3   Wave equation 189

      4.7.4   Plane wave 189

      4.7.5   Reflection and refraction of plane wave 197

5 Optical source - Semiconductor Laser   **205**

  5.1   The fundamental of a Laser 205

      5.1.1   Introduction 205

      5.1.2   The energies model of atoms 205

      5.1.3   Boltzmann distributions and thermal equilibrium 207

      5.1.4   The interaction of electrons and photons 208

      5.1.5   Optical amplification 213

      5.1.6   Pumping 216

      5.1.7   Three-level lasers 217

      5.1.8   Four-level lasers 218

      5.1.9   The energy levels of electrons in a crystal 220

      5.1.10 Energy band model of direct semiconductor 221

5.2   The interaction of electrons and photons in semiconductor   228
    5.2.1   The spontaneous emission in semiconductor   228
    5.2.2   The stimulated emission in semiconductor   230
    5.2.3   The absorption in semiconductor   231
    5.2.4   Condition of optical amplification in semiconductor   232

5.3 $p - n$ junction in semiconductor   235
    5.3.1   Laser diode structure   235
    5.3.2   The double heterstructure of the semiconductor laser diode .   236
    5.3.3 $p - n$ junction in semiconductor   239
    5.3.4   Optical gain and the net rate of stimulated emission   241
    5.3.5   Maximum Gain   243
    5.3.6   Linewidth enhancement factor and transverse confinement factor   244

5.4   The direct modulation of semiconductor laser   253
    5.4.1   The direct modulation   253
    5.4.2   Carrier density rate equation   253
    5.4.3   The steady-state solution of the carrier density rate equation   256
    5.4.4   The frequency response of the direct modulation of semiconductor laser   257

5.5   Fabry-Perot semiconductor laser amplifier   259
    5.5.1   Original Fabry-Perot Etalon   259
    5.5.2   Fabry-Perot semiconductor laser amplifier   260
    5.5.3   Anti-reflected coating and the output   266

5.6   The performance of the Fabry-Perot semiconductor laser   272
    5.6.1   Threshold condition   273
    5.6.2   The longitudinal modes   275
    5.6.3   The saturation problem   276

5.7   Single Longitudinal Mode Semiconductor Laser   276
    5.7.1   Passive periodic waveguide - Bragg Grating   278
    5.7.2   Active periodic waveguide   286
    5.7.3   Introduce to the single longitudinal mode laser(1)   290
    5.7.4   The performance of the single longitudinal mode laser   294
    5.7.5   Distributed Bragg Reflector (DBR) semiconductor laser . . .   299
    5.7.6   Distributed Feedback (DFB) semiconductor laser   305
    5.7.7   Appendix   310

6   Optical Photo-detector   315
6.1   Introduction   315
6.2   PIN photo-detector diode   315
    6.2.1   p-n photodiode   315
    6.2.2   PIN photodiode   316
    6.2.3   The frequency response of a PIN photodiode   318
    6.2.4   Detection sensitivity of photodiode   321
    6.2.5   Photodiode preamplifier   323

6.3   Avalanche photodiode 324

7   10 Gbps Optical Fiber Transponder   329
7.1   Introduction to 10 Gbps optical fiber transponder 329
       7.1.1   The block diagram of the 10Gbps optical fiber transponder . 330
       7.1.2   The 10Gbps optical transponder functions 331
       7.1.3   Digital ADD/DROP Technology 336
7.2   Digital technology 340
       7.2.1   16:1 Multiplexer 340
       7.2.2   1:16 De-multiplexer 355
       7.2.3   The block diagram of the transponder and FIFO (First-In-
               First-Out) 367
       7.2.4   Register 375
       7.2.5   Clock and Synthesizer 383
       7.2.6   CDR- Clock and Data Recovery 385
7.3   Layout of 10 Gb/s optical fiber transponder 391
       7.3.1   Schematic 391
       7.3.2   Layout 394

8   Introduce to Quantum communication   397
8.1   Introduction 397
       8.1.1   The conception of the quantum communications 397
       8.1.2   The conception of Quantum Cryptography 398
8.2   The principle of quantum entanglement 400
       8.2.1   The phenomena of quantum entanglement 400
       8.2.2   EPR Paradox and Bell's inequality 405
8.3   Quantum Key Distribution — QKD 415
       8.3.1   BB84 protocol and QKD 415
8.4   Quantum optical fiber telecommunications 420
       8.4.1   Quantum key distribution (QKD) based on the optical fiber
               infrastructure 422
       8.4.2   The conception of Quantum Teleportation 426
       8.4.3   The implement of Quantum Teleportation 428
       8.4.4   Argument 435
8.5   Quantum satellite communications 437
8.6   Quantum communications networks 440
       8.6.1   The development of quantum communications in western . . 441
       8.6.2   Quantum communications networks 442

Chapter 1

Introduce to the Optical Fiber Networks

1.1 Introduction

The traditional optical fiber transmission networks in the years about 2002 based on the SONET/SDH format using 10Gbps optical fiber transponder are already installed in the world wide area such as USA, Canada, China, India etc. This transport network has a layered structure as defined by the International Standardization Sector (ITU-T) with OC-48 and OC-192. It consist of a circuit (optical fiber transponder), a transmission medium (installed single mode fiber) and path layer.

In 21st century, there are explosive growth in the amount of information being transmitted by digital services such as electronic commerce, software distribution, and digital/music distribution services. The capability required to handle all of those information will be provided by using a new communication technologies. IP/ATM and all-optical network (or photonic network) are key-enablers for realizing terabit capabilities and effective and reliable use of networks.

The maturity of OC-48, OC-192, OC-678 and extremely dense WDM force us to rethink the strategy for cost-efficiency bandwidth management using these layers. The introduction of an optical path layer with high bit rate TDM pipes (using 10Gbps optical transponders) multiplexing by DWDM which can be managed by OADM (Optical Add Drop Multiplexer) and OXC (Optical Cross-connector) will be effective for overall network efficiency.

Summery

From discussion above, we may simple say: that

a. The traditional optical transmission network in the years about 2002 is a high bit rate (10Gbps) digital transmission network using **TDM** (time division multiplexing) along with installed single mode fiber as the transmission medium.

b. New optical transmission network is an *all-optical transmission network* using high bit rate TDM pipes (for instance, 10Gbit optical transponders) multiplexing by DWDM based on the installed single mode fiber as the transmission medium.

Which means that each wavelength can be used to transmission 10Gbps data-stream (information) on an installed optical fiber. And more than 64 channels (or 64 wavelengths) can be multiplexed in an all-optical network based on an installed optical fiber. Therefore, which equals to increase the number of installed optical fiber to more than 64 but on one installed fiber.

The information being transmission in an all optical network are still the digital signal. Which is high quality signal for customers.

Thus, we need to introduce the high bit rate TDM pipes (or 10Gbps optical transponder) as the source along with the DWDM as the all-optical networks.

1.2   High bit rate digital transmission network with TDM

High bit rate digital transmission network is based on the 10Gbps optical fiber transponder as shown in Figure 1-1.

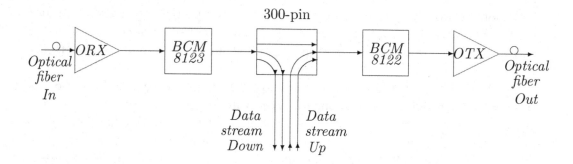

Figure 1-1   A 10Gbps optical fiber transponder

As shown in Figure 1-1, a 10Gb/s optical fiber transponder is consist of

- A 10Gb/s photo-receiver with an integrated limiting amplifier (ORX):

- An optical transmitter (OTX).

- A 300-pin with MSA (Multi Source Agreement) compliant as the data-stream inter-cooperated with the local station:

Where

- The 10Gb/s photo-receiver with an integrated limiting amplifier (ORX) is to transfer the input 10Gb/s optical data-stream into 10Gb/s electric data-stream. The 10Gb/s series electric data steam then de-multiplexed into 16 channels of parallel electric data-stream at 622Mb/s by BCM8123.

- At the optical transmitter (OTX), 16 channels parallel 622Mb/s electric data-stream are multiplexed into 10Gb/s series electric data-stream by BCM8122 and then send to the laser to transfer the series electric 10Gb/s data-stream into 10Gb/s series optical data-stream.

- The 300-pin with MSA (Multi Source Agreement) compliant is used as the data-stream inter-cooperated with the local station:

  - The 300-pin MSA is used to down load part of the electric data-stream from the receiver side into the local station.

  - Meanwhile, the 300-pin MSA is used to pick up part of the electric data-stream from the local station into the transmitter side. Which combine with part of 16 channels of electric data-stream from the receiver side to send into the BCM8122 to be multiplexed into 10Gb/s series electric data-stream.

Thus, the 10Gb/s optical transponder functions are:

- As an optical amplifier: The sensitivity of a 10Gb/s photo-receiver (ORX) is, for instance, -17.5dBm, which is far better than the MAS's requirements. The output of the optical transmitter is , for instance, +2dBm, which meets the MAS's requirement as well. So as the 10Gb/s optical fiber transponder supports transmission link distance up to 40km and beyond, with exceptionally good eye diagram and jitter performance.

- As an electric data-stream inter-cooperation with the local station through 300-pin with MSA compliant.

- To recover the data clock by using a clock recovery circuit within the de-multiplexer BCM8123.

10Gbit/s data steam is a very high speed data. that means the pulse of the data is narrow to $100ps$ (or $100\mu\mu s$). For instance, any $100ps$ time delay will make a big bit error. In this case,

- The dispersion of the optical fiber will make time delay different in a broad-band data-stream.

- The line-width of the laser spectrum must be narrow to less than 100ps.

- The design of the 10Gbit/s optical fiber transponder has to meet the critical requirement.

All of those will be discussed in detail in Chapter 7.

Technologies in 10Gbps optical fiber transponder

Thus, there are two key technologies in the transponder:

- electric data-stream de-multiplexing and multiplexing.

- electric data-stream adding and dropping.

Both of them are implemented in the electric domain using TDM (time-division multiplexing) as shown in Figure 1-2. Where, the 10Gb/s series electric data steam de-multiplexed into 16 channels of parallel electric data-stream at 622Mb/s by BCM8123, and 16 channels parallel 622Mb/s electric data-stream are multiplexed into 10Gb/s series electric data-stream by BCM8122. And electric data-stream adding and dropping is implemented by 300-pin MSA.

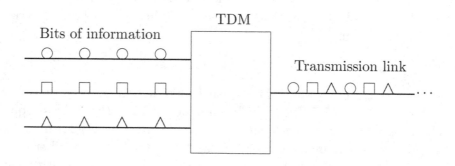

Figure 1-2   TDM conception

Thus, all of the functions of de-multiplexing, multiplexing , adding and dropping require that all of the data-stream should have the same format and same protocol. This is the one limitation of the high bit rate digital transmission network.

Another limitation of the high bit rate digital transmission network is the bit rate. As we mentioned before, this network is built up by using 10Gbps optical fiber transponder. And the TDM is realized by bias modulation of the laser. Experimental tell us that the bit rate limitation of the such laser modulation is 10Gbps. Thus the another limitation of the high bit rate digital transmission network is that the maximum bit rate is 10Gbps. This is way we need a new all-optical fiber network to overcome the limitation of the traditional network.

The all-optical fiber DWDM network is the best choice for the new network.

a. Several high bit rate digital piles are independently transmission on an installed single mode fiber without considering the same format and same protocol issue.

b. An installed optical fiber can provide very wide bandwidth with low-loss range around $\lambda = 1,550nm$ or $\lambda = 1,300nm$. Any one of DWDM network can contain more than 64 (or 120) channels wavelengths for the all optical fiber DWDM networks. Which can be seen from the WDM conception as shown in Figure 1-3. And each wavelength can transmit 10Gbps optical data-stream. Modern system can handle up to 160 channels wavelengths and can thus expand a basic 10Gbps system over a single mode fiber to over 1.6Tbps.

1.3   All-optical transmission network with DWDM

As we mentioned above (Figure 1-1), the traditional optical fiber networks is a system with O/E (Optical-to-electric) and E/O (electric-to-optical) at the very edge both side of the system, and then does de-multiplexing, adding and dropping, then multiplexing digital signals (TDM) in electrical.

However, all optical fiber networks is a system, which does de-multiplexing, adding and dropping , then multiplexing optical signals (WDM) in optical.

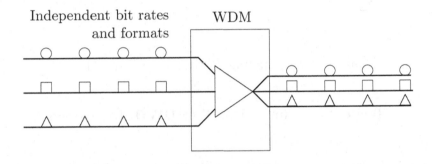

Figure 1-3   Increasing Capability with WDM

Figure 1-3 shows the conception of the WDM (wavelength division multiplexing) in optical. In which, the WDM (wavelength division multiplexing) is a technology, which multiplexes a number of optical data-stream signal onto a single mode fiber by using different wavelength of the laser light. Which means that each input optical data-stream signal is in individual wavelength of laser light. WDM technology multiplexes a number of different wavelength signal onto a single mode fiber. Thus, in optical fiber WDM system:

a. The format of each optical data-stream signal is independent without considering format and protocol issue.

b. The bit rate of each optical data-stream signal is independent as well. Thus each input optical signal can be OC-48 (2.5Gbps), or OC-192 (10Gbps).

c. The number of the input optical data-stream signals can be 64 (or 120) channels or 160 channels as long as each optical data-stream signal can be separated by the optical filter. The low-loss bandwidth of the single mode fiber at $1,550nm$

or $1,300nm$ are both wide enough. Thus the capability of the DWDM system is almost no up limit.

d. The traditional high bit rate optical networks (SONET) can be updated to 64 (or 120) channels (or 160 channels) input based on installed optical single mode fiber as shown in Figure 1-4. Which is a realizing terabit capabilities and effective and reliable networks.

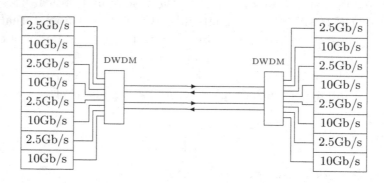

Figure 1-4   Upgrading SONET with DWDM

Technologies in optical fiber DWDM networks

As shown in Figure 1-5, all optical fiber networks are enable the construction of high-capability and flexible optical communication systems for the data-center.

The key technologies of the new optical networks are:

a. Optical add drop multiplexers (OADMs), and

b. Optical cross connectors (OXCs)

c. Along with already mature DWDM (density wavelength division multiplexer) systems.

The wavelength multiplexing and de-multiplexing may be implemented by cascaded Mach-Zehnder filter or cascaded fiber Bragg grating filter. Which will be discuss in Chapter 3. Cascaded Mach-Zehnder filter or cascaded fiber Bragg grating filter may also be used to implement the optical add drop multiplexers (OADMs). And the optical cross connectors (OXCs) is more complicated, one option to implement the OXCs is the array waveguide grating (AWG). All of those technologies will be discussed in Chapter 3.

Nowadays, the all optical communication DWDM system provides a very wide, say almost no limited data-transmission systems over the world. This transmission systems provide customer almost no limited service, which include:

No limited channels HDTV (high density television),

No limited data-dream of internet,

No limited cellula phone and ipad iphone etc... as long as you would like to pay for them.

The all optical communication system is very complicated, however, we may simply say, it is a combination of TDM (time division multiplexing) and WDM (wavelength division multiplexing).

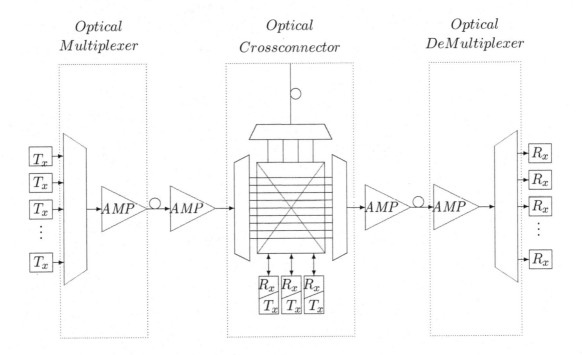

Figure 1-5   All optical fiber communication networks [39]

Chapter 2

All Optical Fiber DWDM Communication Networks

2.1 The Global Networks Hierarchy

Optical fiber communication based on a single mode fiber is a strong transmission medium for the data and analog signals in the world. After the 10Gbps optical fiber transponder been installed as the long haul transmission system in the world wide area, the demand of high speed data, such as high density TV, Internet, cell phone, still is the strong driver to develop the world wide optical fiber communications system in the world. First, the demand drives the devices to be developed. And then, the optical fiber communication transmission system forms a Global Network Hierarchy as shown in Figure 2-1, which is based on a pair of single mode fibers being installed in the world wide area already. From Figure 2-1, we will learn the process of the data transmission in the world area clearly.

The **Global Networks Hierarchy** includes three parts:

- **The long-haul networks**, which includes the long-haul networks with the 10Gbps optical fiber transponders every 40km distance in the world.

Long-haul networks are the core of the global networks. Dominated by a small group to handel the large transnational and global carriers, long-haul networks connect the MANs (Metropolitan Area Networks). Their application is transport and their primary concern is the transmission capacity. Meanwhile, these networks adopts SONET (Synchronous Optical Network) for the Northern America area and SDH (Synchronous Digital Hierarchy) for the other area. In many cases, technologies are experiencing fiber exhaust (mainly in dispersion issue for optical fiber side and the diverse protocols issue in the system)) because of being using one wavelength only in a single mode fiber. However, the potential transmission capability of a single mode fiber is much much larger than now on if we use WDM (wavelength division multiplexer) in a single mode fiber. Which means that every wavelength independently transmission optical data in a single mode fiber transmission system.

Therefore, the transmission capability of an installed single mode fiber will increase 4 times even to 126 times if we use WDM technology.

† Which will overcome the diverse protocols issue since each wavelength will be transmission independently.

† Also overcome the dispersion issue in a single mode fiber as long as the bit rate of each wavelength transmission is less than 10Gbps.

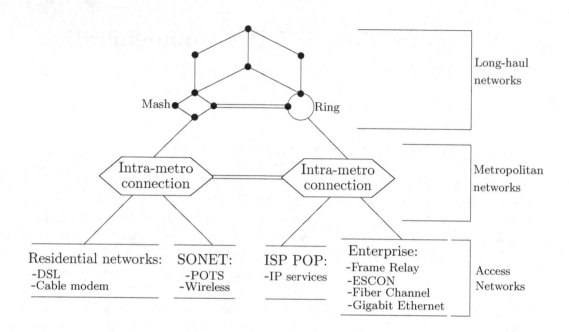

Figure 2-1   Global Network Hierarchy [1]

- **The access networks**.

At the other end of the spectrum is **the access networks**. These networks are the closest to the end users, at the end of the MAN (metropolitan area networks). The access networks includes:

† *The residential networks*:

a. Installing the DSL for the home phone service over Bell home phone line. Digital subscriber line (DSL) is a family of technologies that are used to transmit digital data over telephone lines.

b. Installing the cable modem for home TV with Rogers cable TV system.

† *The SONET (Synchronous Optical Networks)*:

a. Provides wireless services for cellular phone.

b. Provides POTS services. POTS (Plain old telephone service) is voice-grade telephone service employing analog signal transmission over copper loops.

† *The ISP-POP (Internet Service Provider-Point of Presence)*: Provides the IP services for personal Internet services.

† *The enterprise*: such as IBM's Enterprise System Connection (ESCON). They are characterized by the diverse protocols and infrastructures, and span a broad spectrum of rates. Customers range from residents Internet users to large corporations and institutions. The predominance of IP (Internet Protocol) traffic, with its inherently burst, asymmetric, and unpredictable nature, presents many challenges, especially with new real-time applications. At the same time, these networks are required to continue to support legacy traffic and protocols, such as IBM's Enterprise System Connection (ESCON).

- **Metropolitan Area Networks.**

Between these two large and different networking domain above is the **Metropolitan Area Networks** (simply called MANs). In these networks, channel traffic within the metropolitan domain (among business, offices, and metropolitan areas) and traffic between the large long-haul points of presence (POPs).

† The Metropolitan Area Networks (MANs) have many of the same characteristics as the access networks, such as diverse protocol and channels speeds. Metropolitan Area Networks (MANs) have been traditionally SONET/SDH based, using point to point or ring topologies with add/drop multiplexers (ADMs).

† On the other hand, the MANs lies at a critical juncture, it must meet the needs created by the dynamics of the ever-increasing bandwidth available in long-haul transport networks.

† Meanwhile, the MANs must address the growing connectivity requirement and access technologies that are resulting in demand for the high speed, customized data services.

† There is a nature tendency to regard the MANs as a scala-down version of the long-haul networks. Networks serving the metropolitan area encompass short distance than in long-haul transport networks.

† Network shape is more stable in long-haul networks, while topologies change frequently in the metropolitan area networks (MANs).

† Many more types of services and traffic types must be supported in MANs, such as traditional voice and leased line services to new applications, and video.

† Metropolitan networks today differ from trunk-oriented long-haul networks is that they encompass a collection of low bit-rate asynchronous and synchronous transmission equipment, short loops, small cross-section, and variety of users with varying bandwidth demands. These fundamental differences between metropolitan networks and the long-haul networks have powerful implications for the requirements in the metropolitan domain. Protocol and speed transparency, scalability, and dynamic provisioning are at least as important as capacity, which rules in the long-haul market.

An Alternative view

In reality, the lines between three domains are not always so clear-cut. Long-haul and metropolitan networks are sometimes not clearly delineated; the same situation holds true for the access and metropolitan networks. Other views of the global network exist. One, for example, define the access network as part of, rather than separate from, the MAN, while also including enterprise connectivity in the metropolitan domain. In this view, the metropolitan market breaks down as follows:

- Core — These are essentially scaled-down long-haul system. They are considered the core of the MAN, because they interconnect carrier points of presence (POPs) and do not directly interface with end users.

- Metropolitan access — This is the segment between carrier points of presence (POPs) and access facilities, which could be equipment at customer premises or at an aggregation point.

- Enterprise — This the part of the network dedicated to serving the needs of enterprises. Using owned or leased fiber (or leased fiber capacity), connectivity is provided between geographically disparate enterprise sites and for new applications, such as storage area networks (SANs).

Options for increasing Carrier Bandwidth

Carriers have two options to face the challenge of dramatically increasing capacity while constraining the cost:

† Install new optical fiber:

Install new fiber means to expend their networks. However, install new fiber is a costly proposition. It is estimated at about 70,000 dollars per mile, most of which is the cost of permits and construction rather than the fiber itself. Installing new fiber may make sense when it is desirable to expand the embedded base.

† Increase the effective bandwidth of existing optical fiber:

Increase the effective bandwidth of existing optical fiber can be accomplished in two ways:

a. Increase the bit rate of existing system. Using TDM (Time-Division Multiplexing), data is routinely transmitted at 2.5Gbps (OC-48), 10Gbps (OC-192), 40Gbps (OC-678) as indicated in Table 2-1.

Table 2-1

Optical Carrier	Max. Bit Rate	Limitation of Distance (by chromatic dispersion)
OC-48	2.5Gb/s	
OC-192	10Gb/s	40km - 50 km
OC-678	40Gb/s	8km - 10km

However, The electronic circuit is complex and costly, both to purchase and maintain. The transmission at 10Gbps OC-192 over single mode fiber is 16 times more affected by chromatic dispersion than 2.5Gbps OC-48. The greater transmission power required by the higher bit rates also introduces nonlinear effect that can affect waveform quality. Finally, polarization mode dispersion, another that limit the distance a light pulse can travel without degradation, is also as issue.

b. Increase the number of wavelengths on a existing fiber. In this approach, many wavelengths are combined onto a single mode fiber. Using wavelength division multiplexing (WDM) technology several wavelengths, or light colors, can simultaneously multiplex signal of 2.5Gbps to 40Gbps each over a stand of fiber. Without having to install new fiber, the effective capability of existing fiber plant can routinely be increased by a factor of 4 to 162 with high density on the horizon. The specific limits of this technology are not yet known.

2.1.1   TDM and SONET

Now we will discuss TDM (Time Division Multiplexing) and SONET (Synchronous Optical Networks).

TDM conception:

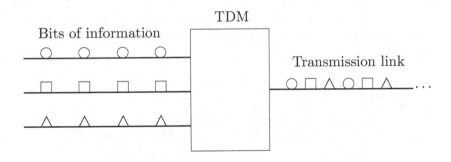

Figure 2-2   TDM conception

TDM can be explained by the bit traffic over a link. To transmission all the bit traffic from four providers to another city, you can send all the bit traffic on one single mode fiber, providing the feeding providers are fairly serviced and the bit traffic is synchronized. When sending bits over a link, TDM increases the capability of the transmission link by slicing time into smaller intervals so that the bits from multiple input sources can be carried on the link, effectively increase the number of bits transmitted per second as shown in Figure 2-2.

So if each of the four feeds put bits of a smaller interval onto the link every four bits time slot, then the link would get a bits of a time slot at the rate of one each bits time slot. As long as the speed of all the bits is synchronized, there would be no collision. At the destination the bits of a time slot can be taken off the link and feed to the local network by the same synchronous mechanism, in reverse. This is the principle used in synchronous TDM when sending bits over a link. Though fair, this method results in inefficiency, because each time slot is reserved even when there is no data to send. This problem is mitigated by the statistical multiplexing used in **Asynchronous Transfer Mode** (ATM).

Although ATM offers better bandwidth utilization, there are practical limits to the speed that can be achieved due to the electronics require for segmentation and reassembly (SAR) of ATM cells that carry packet data.

SONET/SDH TDM

The telecommunications industry adopted the Synchronous Optical Network (SONET) or Synchronous Digital Hierarchy (SDH) stand for optical transport of TDM data. Here, SONET, used in North-America, and SDH, used elsewhere, are two closely rated stands that specify interface parameters, rates, framing formats, multiplexing methods, and management for synchronous TDM over optical fiber.

SONET/SDH takes n bits streams multiplexes them, and optically modulates the signal, sending it out using a light emitting device (laser) over optical fiber with a bit rate equal to "incoming bit rate x n". Thus traffic arriving at the SONET multiplexer from four places at 2.5Gbps will go out as a single stream at "4 X 2.5Gbps", or 10Gbps. This principle is illustrated in Figure 2-3, which shows an increase in the bit rate by a factor of four in time slot T.

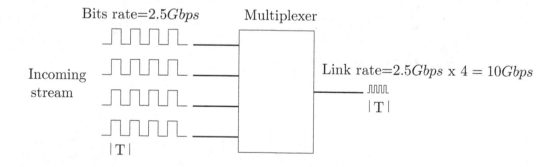

Figure 2-3   SONET TDM

The original unit used in multiplexing telephone call is 64*kbps*, which represents one telephone call. 24 (in North America) or 32 (outside North America) of these units are multiplexed using TDM into higher bit-rate signal with an aggregate speed of 1.544*Mbps* (24 channels) or 2.048*Mbps* (32 channels) for transmission over T1 or E1 lines, respectively. The hierarchy for multiplexing telephone calls is indicated in Table 2-2.

Table 2-2 Telco Multiplexing Hierarchy

Signal	Bit rate	Voice Slot
DS0	64*kbps*	1 DS0
DS1	1.544*Mbps*	24 DS0s
DS2	6.312*Mbps*	96 DS0s
DS3	44.736*Mbps*	128 DS0s

These are the basic building block used by SONET/SDH to multiplex into a standard hierarchy of speeds, from STS-1 at 51.84*Mbps* to STS-192/STM-64 at 10*Gbps*. Table 2-3 shows the relationship between the telco signal rates and the most commonly used levels of the SONET/SDH hierarchy.

Table 2-3

Optical Carrier	SONET/SDH Signal	Bit Rate	Capacity
OC-1	STS-1	51.84*Mbps*	28DS1s or 1 DS3
OC-3	STS-3/STM-1	155.52*Mbps*	84 DS1s or 3 DS3s
OC-12	STS-12/STM-4	622.08*Mbps*	336 DS1s or 12 DS3s
OC-48	STS-48/STM-16	2488.32*Mbps*	1344 DS1s or 48 DS3s
OC-192	STS-192/STM-64	9953.28*Mbps*	5379 DS1s or 192 DS3s

This multiplexing and aggregation hierarchy is depicted in Figure 2-4.

Again, originally, the SONET is designed for the telephone transmission over the single mode fiber. Where in Figure 2-4:

- Each 24 channels or 32 channels telephone is multiplexed into DS1 $\cdots$.

- Using a standard called **virtual tributaries (VT)** for mapping low-speed channels into STC-1, the 28 DS1 signal can be mapped into the STS-1 payload.

- Or, the 28 DS1 signals can be multiplexed to DS3 with an M13 multiplexer and fit directly into the STS-1.

- Also, using packet over SONET (POS), ATM and layer 3 traffic can feed into the SONET terminal from switches equipped with SONET interfaces.

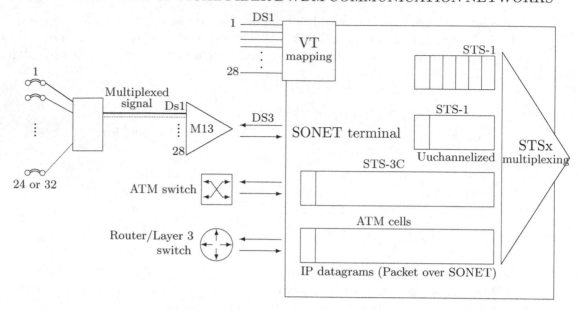

Figure 2-4   TDM and SONET Aggregation [2]

The drawbacks of SONET/SDH:

SONET/SDH does have some drawbacks:

- The notions of priority or congestion do not exist in SONET or SDH.

- Meanwhile, the multiplexing hierarchy is a rigid one. When more capacity is need, a leap to the next multiple must be made, same like which will results in an outlay for more capacity than is initially need. For example, the next incremental step from 10Gbps (STS-192) is 40Gbps (STS-768).

- Also, since the hierarchy is optimized for voice traffic, there are inherent inefficiencies when carrying data traffic with SONET frames. Some of these inefficiencies are indicated in Table 2-4.

- DWDM, by contrast, can transport any protocol, including SONET, without special encapsulation.

Table 2-4

Ethernet	SONET/SDH signal	Bit Rate	Wasted Bandwidth
10BASE-T (10Mbps)	STS-1	51.8540 Mbps	80.709%
100BASE-T (100Mbps)	STS-3/STM-1	155.5.520 Mbps	35.699%
1000BASE-T (1000Mbps)	STS-48/STM-16	2488.32 Mbps	59.812%

Summery

To summarize, the demand placed on the transport infrastructure with a single optical wavelength application and the explosive growth of the Internet has exceed the limits of traditional TDM. Single mode fiber, which once promised, seems like being unlimit bandwidth, is been exhausted, and expense, complexity. And the scalability limitations of SONET infrastructure, all of those are be coming increasingly problematic.

2.1.2 SONET with DWDM

Except Time Division Multiplexing (TDM) we have another option, Wavelength Division Multiplexing (WDM).

Wavelength Division Multiplexing and DWDM:

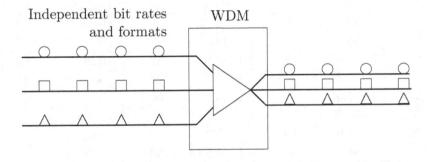

Figure 2-5   Increasing Capability with WDM

WDM increases the capacity of the single mode fiber using complete different method from TDM. WDM assigns incoming optical signals to special wavelength within a certain frequency band. This multiplexing resembles different wavelength without interfering each other as shown in Figure 2-5 since each channel is transmitted at a different frequency (or different wavelength). And then, we can select from them using a tuner. Another way to think about the WDM is that each channel is a different color of light, several channels then make up a "rainbow". In this case, each channel in WDM is independent not only in bit rate but also in format.

In a WDM system, each of the wavelengths is launched into the fiber, and the signals are demultiplexed at the receiving end. Like the TDM, the resulting capacity is an aggregate of the input signals , however, WDM carriers each input signal independently with the others. This means that each channel has it own dedicated bandwidth and independent format; all signals arriver at the same time, rather than being broken and carried in time slots.

The different between WDM and dense wavelength division multiplexing (DWDM) is fundamentally one of only degree. DWDM spaces the wavelengths more closely than does WDM, and there has a greater overall capacity. The limits of the spacing are not precisely known, and have been reached 128 channels on one single mode fiber, though the optical fiber transmission system are available in mid-year 2000.

In commercial, the spectral in ITU-T DWDM Grid wavelengths is shown in Figure 2-6, which can be used for the DWDM system with 62 channels listed in Table 2-5.

Table 2-5 62 channels in ITU-T DWDM Grid (spacing=100GHz)

Channel Code	λ (nm)	Channel Code	λ (nm)	Channel Code	λ (nm)	Channel Code	λ (nm)
18	1563.05	30	1553.30	42	1543.73	54	1534.25
19	1562.23	31	1552.53	43	1542.94	55	1533.47
20	1561.42	32	1551.72	44	1542.14	56	1532.68
21	1560.61	33	1550.92	45	1541.35	57	1531.90
22	1559.80	34	1550.12	46	1540.56	58	1531.12
23	1558.98	35	1549.32	47	1539.77	59	1530.33
24	1558.17	36	1548.52	48	1538.98	60	1529.55
25	1557.36	37	1547.72	49	1538.19	61	1528.77
26	1556.56	38	1546.92	50	1537.40	62	1527.99
27	1555.75	39	1546.12	51	1536.61		
28	1554.94	40	1545.32	52	1535.82		
29	1554.13	41	1544.53	53	1535.04		

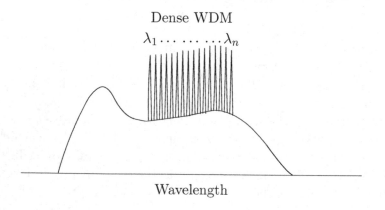

Dense WDM

$\lambda_1 \ldots \ldots \ldots \lambda_n$

Wavelength

Figure 2-6   ITU DWDM Wavelengths

The spacing between two channels is 100Gbps in ITU-T. Which is supported by three technologies:

- One is the supper narrow bandwidth spectrum of the laser as shown in Figure 2-6, in which each of them represents a supper narrow spectrum from laser. And the different spectrum in Figure 2-6 can be obtained by a temperature tuner (slow) or by switching the injection current of the laser.

- Another one is the optical bandpass filter with narrow bandwidth. Which can be used to filter out each channel from the optical de-multiplexer in the receiving end.

- DWDM refers originally to optical signals multiplexed within the 1550nm band so as leverage the capability (and cost) of erbium doped fiber amplifiers (EDFAs), which are effective for wavelengths between approximately 1525 - 1565nm (C band), or 1570 - 1610nm(L band). EDFAs ware originally developed to replace SONET/SDH optical-electrical-optical (OEO) regenerators, which they have made practically obsolete. EDFAs amplifies any optical signal in their operating range, regardless of the modulated bit rate. In terms of multi-wavelength signals, so as EDFA has enough pump energy available to it, it can amplify as many as optical signals as can be multiplexed into its amplification band (though signal densities are limited by choice of modulation format). EDFAs therefore allow a signal-channel optical link to be upgraded in bit rate by replacing only equipment at the ends of the links, while the existing EDFA or series of EDFAs through a long-haul router. Furthermore, single-wavelength links using EDFAs can similarly to be upgraded to WDM links as reasonable cost. The EDFA's cost is thus leveraged across as many channels as can be multiplexed into 1500nm band.

The introduction of the ITU-T G.694.1 [3] frequency grid in 2002 has made it easier to integrate WDM with older but more standard SONET/SDH systems. WDM wavelengths are position in a grid having exactly 100Ghz (about 0.8nm) spacing in optical frequency, with a frequency fixed at 193.10THz(1552.25nm) [4]. The main grid is placed inside the optical fiber amplifier bandwidth, but can be extended to wider bandwidths. Today's DWDM system use 50GHz or even 25GHz channel spacing for up to 160 channels operation [5].

TDM and WDM Interface:

As shown in Figure 2-7(a), SONET TDM takes synchronous and asynchronous signals and multiplexer them to a single higher bit rate for transmission as a single wavelength over fiber. In which, source signal have to be converted from electrical to optical in a laser, or from optical to electrical in a photodetector and then back to optical from another laser before multiplexed, then WDM takes multiple optical signals, maps them to individual wavelengths, and multiplexes the wavelengths over a single mode fiber.

However, as shown in Figure 2-7(b), compered with TDM, WDM can carry multiple protocols without a common signal format. while SONET TDM cannot. Some of the key differences between TDM and WDM are graphically illustrated in Figure 2-7.

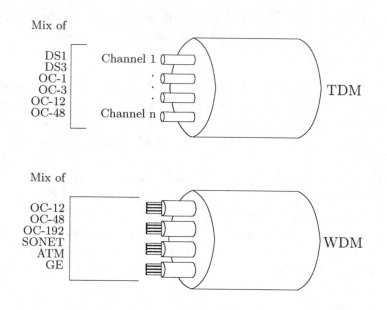

Figure 2-7   TDM and WDM interface [6]

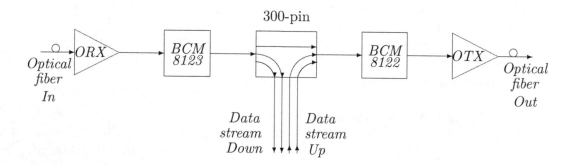

Figure 2-8   A 10Gbps optical fiber transponder

- SONET TDM is the early products, which is to map the low bit rate data channels into a high bit rate data channels for SONET TDM to transmission the high bit rate optical data over a single mode fiber. One way to explain the mapping is illustrated in Figure 2-8. Which is a 10Gbps optical fiber transponder used for SONET TDM system. In which,

  - The 9953.28 Mbps optical data over a single mode fiber is input into the transponder. This optical data is converted from optical to electric by a photo-detector. The electric data are de-multiplexed inside BCM8123. After de-multiplexing, the electric data are sent to Drop/Add (300 pin).

  - The Drop/Add (300 pin) is the interface of the SONET TDM, which can drop data to the local networks and pick up data (or add data) from the local networks. Which means that it can map the low bit rate electric data from DS1, DS3, OC-1, OC-3, OC-12, OC-48 into a high bit rate electric data by multiplexer BCM8122. **The multiplexing in SONET TDM is synchronous or asynchronous, which is taken in the electric domain. In synchronous or asynchronous, all data have to use a common protocol.**

  - The high bite rate electric data are then converted into optical data by a laser modulation. Then transmission over the single mode fiber.

  - The limitation of the bite rate, 10Gbps more, in SONET TWD is caused by the modulation of the laser. Which means that in SONET TDM the bit rate is 9953.28 Mbps.

- However, WDM is used to speed up the bit rate using the infrastructure of the single mode fiber, so as to satisfy the demand of the customers.

  - The WDM is wavelength multiplexing in optical domain using optical Drop/Add, but not in electric domain.

  - Each optical data channels in WDM is independent since it use its own wavelength. then multiplexing them to wavelengths over one single mode fiber.

  - Therefore, the protocol and bit rate of each wavelength in WDM is independent. In other words, **WDM can carry multiple protocols without a common signal format. while SONET TDM cannot.**

SONET with DWDM

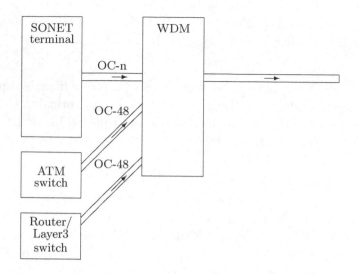

Figure 2-9   Direct SONET Interfaces from Switch to DWDM [7]

Using DWDM, existing SONET equipments can be preserved.

BY using DWDM (Density Wavelength division Multiplexing) as a transport for TDM, existing SONET equipments can be preserved. New implementations can be eliminated layers of equipment. For example, SONET multiplexing equipment (ADM) can be avoid altogether by interface directly to DWDM equipment from ATM (Asynchronous Transfer Mode) and packet switches, Router/Later 3 switch, where OC-48 interfaces are common (see Figure 2-9). Additionally upgrades do not have to conform to specific bit rate interface, as with SONET, where aggregation of tributes is locked to special values.

From SONET TDM to SONET with DWDM

Figure 2-10(a) shows the infrastructure required to transmit 10Gbps (4xOC-48 SR interface) across a span of 360km using SONET equipment. Here, each 1310/RPTP is a regenerator to regenerate 2.5Gbps optical data by using SONET TDM OC-48, In which, O/E converter is used to covert optical signal into electric signal and E/O converter is used to convert electric signal into optical signal.. Meanwhile, there are three process between O/E converter and E/O converter. Which are:

a. digital de-multiplexing,
b. digital add/drop, and

c. digital multiplexing.

All of those means that 1310/RPTP represents a regenerator.

DWDM eliminates regenerators

The long-haul conventional TDM transmission, SONET OC-192 (10Gbps transponder), must be periodically regenerated in core networks since optical signals become attenuated as they travel through the optical fiber. In SONET/SDH optical networks prior to the introduction of DWDM, each separate fiber carrying a single optical signal, typically at 2.5Gbps (OC-48), required a separate electric regenerator every 60 to 100 km as shown in Figure 2-10(a). As additional fibers were "turned up" in a core network, the total cost of the facilities to house and power them, had to be considered. The need to add regenerators also increased the time required to light new fibers.

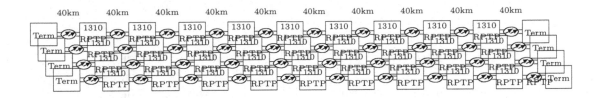

(a)   Conventional TDM Transmission - 10Gbps

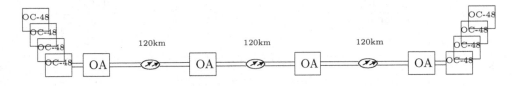

(b)   DWDM Transmission - 10Gbps

Figure 2-10   DWDM Eliminates Regenerators

Now, Figure 2-10(b) shows the infrastructure required for the same capacity using DWDM. While optical amplifiers could be used in the SONET case to extended the distance of spans before having to boost signal power, there would still need to be an amplifier for each fiber. Because with DWDM all four signals can be transported on a single fiber pair (versus four), fewer pieces of equipment are required. Eliminating the expense of regenerators (RPTP) required for each fiber results in considerable saving.

A single optical amplifier can re-amplify the channels on a DWDM fiber without de-multiplexing and processing them individually, with a cost approaching that of a single regenerator. The optical amplifier signals may still need to be regenerated periodically. But depending on the system design, signals can now be transmitted anywhere from 600 to thousands of kilometers without regeneration.

Upgrading SONET with DWDM

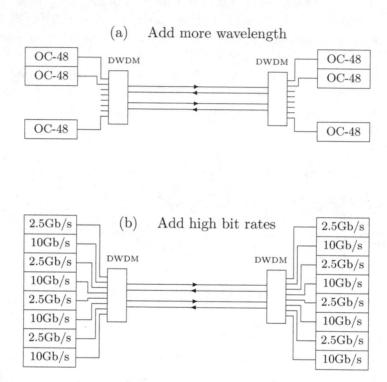

Figure 2-11   Upgrading SONET with DWDM

In addition to dramatically reducing the cost of regenerators, DWDM systems greatly simplify the expansion of network capacity. The only equipment is to install additional or higher bit-rate interfaces in the DWDM systems at either end of the fiber. In some cases it will only be necessary to increase the numbers on the fiber by deploying interfaces, as shown in Figure 2-11(a). The existing optical amplifiers amplifies the new channel without additional regenerators. In the case of adding higher bit-rate interfaces, as shown in Figure 2-11(b), fiber type can become a consideration. Although amplifiers are of great benefit in long-haul transport, they are often unnecessary in metropolitan networks. Where distances between network elements are relatively short, signal strength and integrity can be adequate without amplification. But with MAN expending in deeper into long-haul reaches, amplifiers will become useful.

2.1.3   Additional Drivers In Metropolitan Area Networks

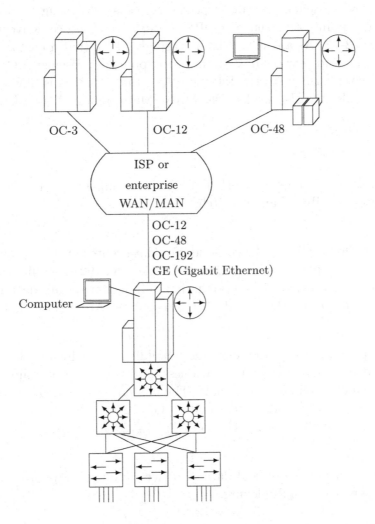

Figure 2-12   High-Speed Enterprise WAN Bandwidth Migration [8]

Bandwidth, the chief driver in the long-haul market, is also a big driver in metropolitan area, access, and large enterprise networks as shown in Figure 2-12. In these types of networks, additional applications driving demand for bandwidth include storage area networks (SANs), which make possible the serverless office, consolidation of data center, and real-time transaction processing backup.

There is also rapidly increasing demand on access networks, which function primarily to connect end users over low-speed connections, such as dial-up lines, digital subscriber line (DSL), cable, and wireless, to a local point of presence (POP). These connections are typically aggregated and carried over a SONET ring, which at some point attaches to a local point of presence (POP) that serves as an Internet gateway for long-hauls. And the networks is enterprise **WAN** (wide Area Networks). Here, ISP is Internet Service Provider.

Now, the growing demand for high-speed services is prompting service providers to transform the point of presence (POP) into dynamic service-delivery center as shown in Figure 2-12. As a result, it is increasingly likely that a customer now obtains many high-speed services directly from the point of presence (POP), without ever using the core segment to the Internet. And the networks is now extended to include WAN (wide Area Networks) and MAN (Metropolitan Networks).

As shown in Figure 2-12,

- The bottom is the customer's dail-up line, conncted to the consolidation data center via fiber channel switch.

- The middle building is the consolidation data center. Which provides different optical transponder (OC-3, OC-12, OC-48, OC-192) for different buildings on the top. Meanwhile the consolidation data center obtain data from WAN (wide area networks) then to MAN (metropolitan Area Networks) vis optical fiber.

- The top buildings are the customers buildings, in which each computer is connected to a router via wire or wireless, and the router is connected to the modem, which combines the signals. The modem is provided by the information provider and is connected to OC-3/OC-12/OC-48. The function of the transponders OC-3/OC-12/OC-48 OC-3/OC-12/OC-48 are:

  - To convert the optical data into electric data via photodetector then are sent to de-multiplexing.

  - The electric data are then sent to customers via Add/Drop. Meanwhile the data from customers are pick-up via Add/Drop, then are sent to multiplexer.

  - Then the data from multiplexer are convert into optical data via laser, then send to the optical fiber.

Note:

† The consolidation data center obtains the data from the WAN (wide Area Networks) via SONET infrastructure (long-haul transmission link) and send out to the MAN (Metropolitan Area Networks) vis optical fiber with DWDM.

† The cable modem to each customer is a shared connection. The line that runs

to your home is a combining signal which also provides to your neighborhood. That means that if the bandwidth of the combining signals is not wide enough, it may become congested causing decreased speed.

2.1.4 Value of DWDM in the Metropolitan Area

DWDM is the clear winner in the backbone. It was first developed on long-haul routers in a time of fiber scarcity. Then the equipment savings made it the solution of choice for new long-haul routers even when ample fiber was available. While DWDM can relieve fiber exhaust in the metropolitan area, its value in this market extends beyond this single advantage. Alternatives for capability enhancement exist, such as pulling new cable and SONET overlays, but DWDM can do more. The additional value in the metropolitan area is:

- The addition of DWDM is fast.

- DWDM is flexible protocol and bit rate.

- DWDM can offer new and higher-speed serves at less cost.

The need to provision services of varying types in a rapid and efficient manner in response to the changing of customers is distinguishing characteristic of the metropolitan networks.

- With SONET, which is the foundation of the vest majority of existing metropolitan area networks's (MAN's), service provisioning is a lengthy and complex process. Which includes "Network planning and analysis", "administrative provisioning", "Digital Cross-connect System (DCS) reconfiguration", "path and circuit verification", and "service creation", all of though can take serval weeks.

- By contrast, with DWDM equipment in place provisioning new service can be simple as tuning on another light wave in an existing fiber pair.

Potential providers of DWDM-based services in metropolitan areas, where abundant fiber plant already exists or is being built, include incumbent local exchange carrier (ILECs), competitive local exchange carriers (CLECs), inter-change carriers (IXCs), Internet service providers (ISPs), cable companies, private network operators, and utility companies. Such carriers can often offer new services for less cost than older ones. Much of the cost saving in due to the reducing unnecessary layers of equipment, which also lowers operational costs and simplifies the network architecture.

Carriers can create revenue today by providing protocol-transparent, light-speed

Local Area Networks (LAN) and Storage Area Networks (SAN) services to large organizations, as well as a mixture of lower-speed services (Token Ring, FDDI, Ethernet) to smaller organizations. Implementing an optical networks, they are ensuring that they can play in the competitive field of the future.

2.1.5 Technologies in the Metropolitan Area Networks

Numerous technologies for transport and encapsulation of data have been advocated in the metropolitan market. A characteristic of these networks is that they are called upon to support a variety of older and new traffic types and rates. Overall, however, there is a trend toward using a common optical layer for transporting digital data. As shown in Figure 2-13,

DSL vs Cable:

DSL (Digital Subscriber Line) provides you with Internet signal over the telephone lines which run to your home. Traditional telephone service and Dial-up Internet use a very small portion of the available frequency ranges (less than 3Ghz) which could pass over a telephone wire.

† DSL requires an active phone line. This means that you either need active POTs service with telephone provider.

† DSL is a dedicated line. Your line is yours and only yours. The speed you get is the speed you will always get.

† The quality of a DSL line depends on your distance from the CO/Remote that you are connect to. If your home is too far away from this point, you may end up with lower speed than you expected or you simply not quality for service.

Cable: Cable Internet provides you with an Internet signal over a coaxial cable lines which run to your home. Cable Internet makes use of unused channels (predefined frequency ranges) to pass your data instead of television programming.

† Cable is a shared connection. The line runs to your home is a branch of a larger trunk which also feeds your neighbor's homes. This means that if trunk is not wide enough to feed your neighborhood it may become congested causing decreased speed.

SONET/SDH:

SONET/SDH has been the foundation for MANs over one decade from 1990, serving as the fundamental transport layer for both TDM-based circuit networks and most over lay data networks. While SONET/SDH has evolved into a very resilient technology, it remains fairly expensive to implement. Inherent inefficiencies in adapting data services to the voice optimized hierarchy and an inflexible multiplexing hierarchy remain problematic. More importantly, capability scaling limitations – OC-768 may be the practical limit of SONET/SDH caused by the dispersion in the fiber, and unresponsiveness to burst IP traffic make any TDM-based technology a poor choice for the further.

ATM:

Many service provider favor ATM (Asynchrony Time Multiplexing) because it can encapsulate protocols and traffic types into a common format for transmission over a SONET infrastructure. Meanwhile the data networks world, which is overwhelmingly IP-oriented, favors packet over SONET (POP), which obtains the costly ATM intermediate layer. Advancements in IP, combined with the scaling capacity of gigabit and multigigabit router, makes it possible to envisage as IP-based network that is well suitable for carrying primarily data traffic, and secondarily voice.

Nevertheless ATM remains strong in the metropolitan area. It can accommodate higher line interfaces and provide managed virtual circuit services while offering traffic management capabilities. Thus ATM edge devices are commonly used to terminate traffic, including VoIP, DSL, and Frame Relay.

Gigabit Ethernet:

Gigabit Ethernet (GE) is a proven technology for easy migration from integration into traditional Ethernet. It is relatively inexpensive compared to other technology that offer the same transmission rate, but does not provide quality of serve (QOS) or fault tolerance on its own. When confined to point-to-point topologies, collisions and carrier sense multiple access (CSMA) are not of concern, resulting in more effective use of full bandwidth. Because the optical physical layer can support much longer distance that traditional Category 5 cable, Gigabit Ethernet over fiber (1000BASE-LX, for example) can be extended into the wide-area realm using DWDM.

The latest advancement in Ethernet technology, 10Gbps Ethernet, is being driving by a need to interconnect Ethernet LANs operating at 10,000, or 1000Mbps. Ten Gigabit Ethernet can be used for aggregating slower access links, in the backbone networks, and for WAN access. Using 1550nm serial laser, distances of 40 to 80km are possible with 10Gbps Ethernet over standard single mode fiber. With such technology, service providers can build simple Ethernet networks over dark fiber without SONET or ATM and provision high speed 10/100/1000Mbits services at very low cost. In addition, a very short reach (VSR) OC-192 interface can be used to connect 10Gbps Ethernet to DWDM equipment over multi-mode fiber.

Ethernet offers the technical advantages of a proven, adaptable, reliable, and uncomplicated technology. Implementation are standard and inter-operable, and cost in much less than SONET or ATM.

Architecturally, Ethernet's advantage is its emerging to serve as a scalable, end-to-end solution. Network management can also be improved by using Ethernet across the MAN and WAN.

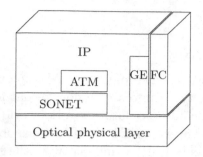

Figure 2-13   Data Link and Network Protocols Over the Optical Layer

IP:

Clearly, as traditional circuit-switched services migrate to IP networks and data grows, networks must evolve to accomudate the traffic. However, IP may need to become as complex as ATM to replace its functionally. Thus, both ATM and IP are candidates for transport directly over DWDM. In either case, the result is simplified networks infrastructure, lower cost due to fewer network elements and less fiber, open interface, increased inflexibility, and stability. The question is, in which format will IP travel over an optical network: IP over ATM over SONET. IP over SONET (as POS), or IP over Gigabit Ethernet or 10Gbps Ethernet? (see Figure 2-13). Here FC is Fiber Channel.

Fiber Channel:

Fiber Channel is the predominant data link technology used in Storage Area Networks (SANs), which will be discuss after then. Fiber Channel is an economic replacement for the Small Computer System Interface (SCSI) protocol as a high-speed interface for applications such as data backup, recovery, and mirroring. Fiber Channel interface are available at 16Gbps, such as Dell QLogic 2662 Dual port low profile Host bus adapter.

Fiber Channel comes without the very short distance limitations of SCSI; it also avoids the termination restrictions of SCSI because each note acts as an optical repeater. Fiber Channel can be implemented in a point-to-point, arbitrated loops, or mesh topology using a switch. As shown in Figure 2-13, Fiber Channel, like other protocols, can be carried directly over the optical layer using DWDM.

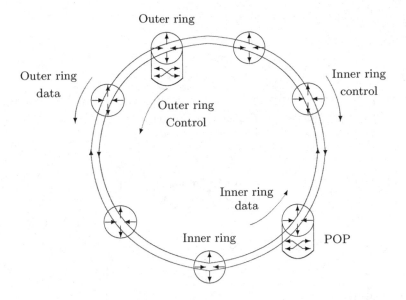

Figure 2-14   Dynamic Packet Transport Ring Architecture

Dynamic Packet Transport

Dynamic Packet Transport (DPT) is a Cisco protocol that provides an alternative to SONET for more efficient transport of data in ring architectures. DPT support basic packet processing, fairness, multi-casting, Intelligent Protection Switching (IPS), topology discovery, Address Resolution Protocol (ARP), routing, and network management. DPT (dynamic Packet Transport) can run over dark fiber, SONET, or WDM.

DPT's chief advantage over SONET is its ability to reuse bandwidth that would have otherwise been lost. Bandwidth is consumed only on traversed segments, and multiple notes can transmit concurrently.

DPT is based on bi-directional counter rotating rings (see Figure 2-14). Packets are transported on both rings in concatenated payload, which control massages are carried in the opposite direction from data.

Storage Area Networks:

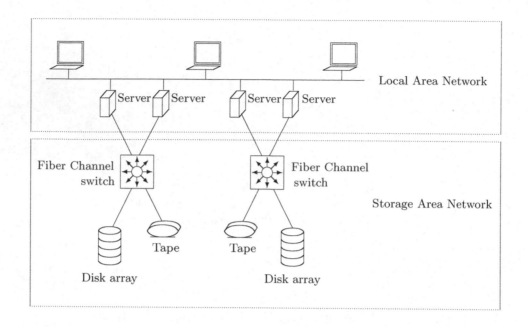

Figure 2-15   Storage Area Networks Architecture [9]

Storage Area Networks (SANs) represent the last stage in the evolution of mass data storage for enterprises and other large institutions. In host-center environments, storage, as well as applications was centralized and centrally managed. With the advent of client/server environments, information that was previously centralized became distributed across the networks. The management problem created by this decentralization are addressed in two principle ways: networks attached storage (NAS), where storage devices are directly attached to the LAN (local Area Networks), and SANs (Storage Area Networks).

Composed of servers, storage devices (tapes, disk arrays), and network devices (multiplexers, hubs, routers, switches, and so on), a SAN (Storage Area Networks) constitutes as entirely separate networks from the LAN (see Figure 2-15). As a separate network, the SAN can relieve bottlenecks in the LAN by providing the resources for applications such as data mirroring, transaction processing, and backup and restoration. A number of types of interfaces have been used to connect servers to devices in a SAN. the most prevalent is IBM's Enterprise System Connection (ESCON), a 17-Mbps half-duplex protocol over fiber.

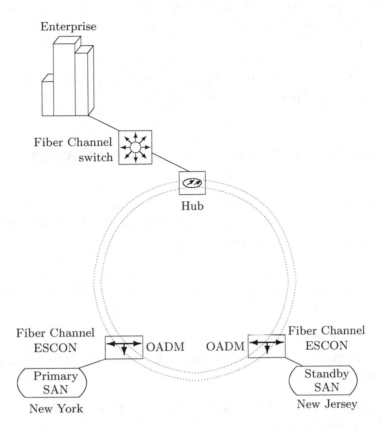

Figure 2-16   SAN Access over Optical Layer [10]

Fiber Channel, on which IBM's FICON (Fiber Connection) is based, is also frequently employed in SANs and has a much higher capability than ESCON (see the **Fiber Channel**). Both technologies, however, significant distance limitations. For example, the standard maximum distance without repeaters is around 3km for half duplex ESCON and around 10km for full duplex 100-Mbps Fiber Channel. There is performance degradation as distance increase beyond these numbers.

This distance limitation can be overcome by transporting data between one or more enterprise locations and one or more SANs over the optical layer using DWDM. In Figure 2-16, the distance separating the enterprise location and then SAN sites can be greatly extended. Access to the ring is by way of OADMs (Optical ADD/DROP Multiplexer) with Fiber Channel or ESCON interfaces at each SAN location (one of these Link interfaces, used in IBM environment for distributing loads across the numbers of a server complex).

In addition to overcome distance limitations, DWDM can also reduce fiber requirements in SANs. Both ESCON and FICON require a pair of fibers for every channel. By multiplexing these channels over DWDM transport, significant savings be realized.

Finally, in Figure 2-16, at Hub node traffic originates, terminated and managed, and connectivity with other networks is established.

2.1.6 Topologies and Protection Schemes for DWDM

Network architectures are based on many factors, including types of application and protocols, distances, usage and access patterns, and legacy network topologies. In metropolitan market, for example:

- **Point-to-Point** topologies might be used for connecting enterprise locations,

- **Ring** topologies for connecting inter-office facilities (IOFs) and for residential access,

- **Mesh** topologies might be used for inter-POP connection and connections to the long-haul backbone.

In effect, the optical layer must be capable of supporting many topologies and, because of unpredictable developments in this area, those topologies must be flexible.

Today, the main topologies in development are point-to-point and ring. With point-to-point links over DWDM between large enterprise sites, there needs only to be a customer premise device for converting application traffic to specific wavelengths and multiplexing. Carriers with linear-ring topologies can evolve toward full rings based on OADMs. As configurable optical cross-connects and switches become more common, there point-to-point and ring networks will be interconnected into meshes, transforming optical meterpolitan networks into full flexible platforms.

Point-to-point Topologies

Point-to point topologies can be implemented with or without OADM. These networks are characterized by *ultra-high channel speed* (10 to 40 Gbps), *high signal integrity and reliability*, and *fast path restoration*. In long-haul networks, the distance between transmitter and receiver can be several hundred kilometers, and the number of amplifiers required between endpoints is less than 10. In MAN, amplifiers are often not need.

Protection in point-to-point topologies can be provided in a couple of ways. In first generation equipment, redundancy is at system level. Parallel links connect redundant system at either end. Switch over in case of failure is the responsibility of client equipment (a switch or router, for example), while the DWDM systems themselves just provide capacity.

In second generation equipment, redundancy is at the card level. Parallel links connect signal systems at either end that contain redundant transponders, multiplexers, and CPUs. Hear protection has migrated to the DWDM equipment, with switching decisions under local control. One type of implementation, for example, uses a 1+1 protection scheme based on SONET Automatic Protection Switching (APS) as shown in Figure 2-17.

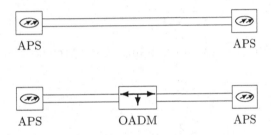

Figure 2-17   Point-to-Point Architecture

Ring Topologies:

Rings are the most common architecture found in metropolitan areas and span tens of kilometers. The fiber ring might contain as few as wavelength channels, and typically fewer nodes than channels. Bit rate is in the ring of 622 Mbps to 10 Gbps per channel.

Ring configuration can be deployed with one or more DWDM systems, supporting any-to-any traffic or they can have a hub station and one or more OADM nodes, or satellites (see Figure 2-18). At the hub node traffic originates, terminated and managed, and connectivity with other networks is established. At the OADM notes, selected wavelengths are dropped and added, while the others pass through transparently (express channels). In this way, ring architectures allow nodes on the ring to provide access to network elements such as routers, switches, or servers by adding or dropping wavelength channels in the optical domain. With increase in number of OADMs, however, the signal is subject to loss and amplification can be required.

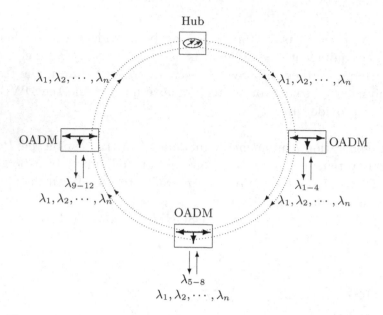

Figure 2-18   DWDM Hub and Satellite Ring Architecture

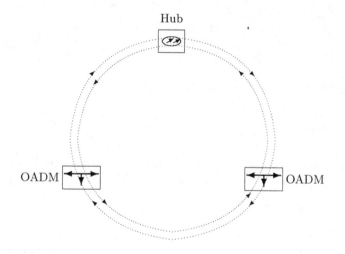

Figure 2-19   UPSR Protection on a DWDM Ring

Candidate networks for DWDM application in the metropolitan area are often already based on SONET ring structure with 1+1 fiber protection. This scheme, such as Unidirectional Path Switched Ring (UPSR) or Bi-direction Line Switched Ring (BLSR), can be reused for DWDM implementations. Figure 2-19 shows a UPSR (unidirectional Path Switched Ring) scheme with two fibers. Hire, hub and nodes sends on two counter-rotating rings, but the same fiber is normally being used by all equipment to receive the signal; hence the name unidirectional. If the working ring should fair, the receiving equipment switches to the other pair. Although this provides full redundancy to the path, no bandwidth reuse is possible, as the redundant fiber must always be ready to carry the working traffic. This scheme is most commonly used in access networks.

Other schemes, such as Bidirectional Line Switched Ring (BLSR), allow traffic to travel from sending to the receiving nodes by the most direct route. Because of this, BLSR is considered preferable for core SONET networks, especially when implemented with four fibers, which offers complete redundancy.

Mesh Topologies

Mesh topology has been successfully used in cell phone, which mention us to consider the mesh topologies to be used for the all-optical networks.

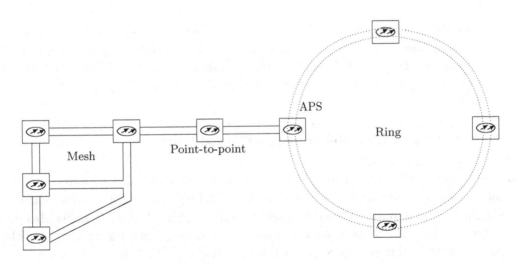

Figure 2-20   Mesh, Point-to-Point, and Ring Architectures

As networks evolve, rings and point-to-point architecture will still have a place, but mesh promises to be the most robust topology. This development will be enabled by the introduction of configurable optical cross-connects and switches that will in some cases replace and in other cases supplement fixed DWDM devices.

From a design standpoint, there is a graceful evolutionary path available from point-to-point to mesh topologies. By beginning with point-to-point links, equipped with OADM nodes at the outset for flexibility, and subsequently interconnecting them, the network can evolve into a mesh without a complete redesign. Additionally, mesh and ring topologies can be jointed by point-to-point links (see Figure 2-20).

DWDM mesh networks, consisting of interconnected all-optical nodes, will require the next generation of protection. Where previous protection schemes relied upon redundancy at the system, card, or fiber level, redundancy will now migrate to the wavelength level. This means that because of a fault a data channel might change wavelength as it makes its wave through the network. The situation is analogous to that of a virtual circuit through at ATM could, which can experience changes in its virtual path identifier (VPI)/virtual channel identifier (VCI) value at switching points. In optical networks, this concept is sometimes called a light path.

Mesh networks will therefore require a high degree of intelligence to perform the functions of protection and bandwidth management, including fiber and wavelength switching. The benefits in flexibility and efficiency, however, are potentially great. Fiber usage, which can be low in ring solutions because of the requirement for protection fibers on each ring, can be improved in a mesh design. Protection and restoration can be based on shared paths thereby requiring fewer fiber pairs for the same amount of traffic and not wasting unused wavelengths.

Finally, mesh networks will be highly dependent upon software for management. A protocol based on Multiprotocol Label Switching (MPLS) in under development to support routed paths through an all-optical network. In addition, network management will require an as-yet unstandardized channel to carry messages among the network elements.

New Generation Metropolitan Optical Network

DWDM will continue to provide the bandwidth for large amount of data. In fact, the capability of systems will grow as technologies advance that allow closer spacing, and therefore higher numbers of wavelengths. But DWDM is also moving beyond transport to become the basis of all-optical networking with wavelength provisioning and mesh-based protection. Switching as the photonic layer will enable this evolution, as will the routing protocols that allow light path to traverse the network in much the same way as virtual circuits do today.

These and other advances are converging such that an all-optical infrastructure, using mesh, ring, and point-to point topologies at optical layer to support the needs of enterprise, metropolitan access, and metropolitan core networks as shown in Figure 2-21.

Migration from SONET/SDH to DWDM

As a transport technology, SONET is an "agnostic" protocol that can transport all traffic types, which providing interoperability, protection schemes, network management, and support for a TDM hierarchy. Although SONET may continue to be the interface standard and transport protocol of choice of the foreseeable future, upgrading it is expensive, as line-rate specific network elements are required at each point of traffic ingress or egress.

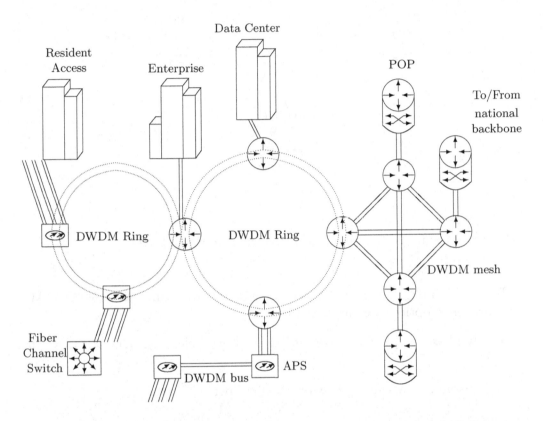

Figure 2-21   New Generation Metropolitan Optical Network [11]

Using DWDM to increase the capability of embedded fiber, while preserving SONET infrastructure, offers an alternative to expensive SONET upgrades. Migration from SONET to DWDM may in fact be the most important application in the near term. In general, this migration by replacing backbones with DWDM, then moves toward the edges of the network.

In one common situation, bandwidth on a SONET ring can be increased greatly by replacing SONET ADMs with DWDM equipment. Where:

† SONET ADMs means that the input and output of ADMs are the optical data based on an existing fiber. And the add/drop of the ADMs can dropping down electrical data to or pick up electrical data from the node of the ring.

† DWDM means wavelength division multiplexing, each OC-48 or OC-192 or · · ·

is independent transmission on its own wavelength based on the exiting fiber.

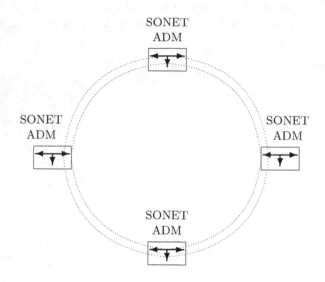

Figure 2-22   A traditional ring with SONET ADMs

In the example shown in Figure 2-22, a traditional ring with SONET ADMs, there are three options for upgrating the ring:

- Replace or upgrade the equipment, for example, from OC-48 to OC-192.

- Install a new ring on new or existing fiber.

- Install one or more new rings by deploying DWDM.

As shown in Figure 2-23, the third option can be realized by using DWDM to increase the capability of the existing ring, so as one fiber can essentially act as many fibers.

For the second option, DWDM can be used to remove an entire class of equipment, the SONET ADMs. This change, which might constitute a second phase of SONET migration, allows routers and other devices to bypass SONET equipment and interface directly to DWDM, which simplifying traffic from IP/ATM/SONET to POS (Packet Over SONET) to eventually over the optical layer as shown in Figure 2-24. In this phase of migration, end user sites are served by OADMs rather than SONET ADMs.

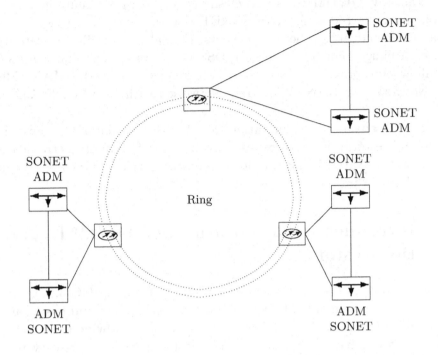

Figure 2-23   Migrating the SONET ring to DWDM - First stage

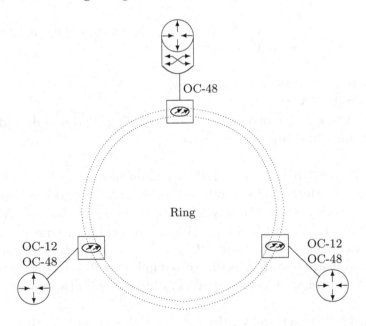

Figure 2-24   Migrating the SONET ring to DWDM - Second stage

In the way, DWDM rings and mesh networks can eliminate increased cost and complexity of introducing more SONET elements into the networks to meet the demand. The advantage here for carriers is the ability to offer bit-rate-independent services, feeling them from the DS1/DS3/OC-n framework. Such a scheme would also allow enterprise LAN access to be extended across the MAN (Metropolitan Area Networks) or WAN (Wide Area Networks) without a SONET infrastructure.

A further advantage in migrating from SONET to the optical layer is that protection and restoration becomes less susceptible to failure of electronic components; a common survivability platform for all network services is created, including those without built-in protection.

2.2   Introduce to all optical fiber DWDM communication system

10Gb/s optical fiber transponder (OC-192), including OC-3, OC-12, OC-48, is the fourth generation optical fiber communications system and has been installed in long haul optical fiber communications link in the worldwide area such as in USA, Canada, China, India, and European in last ten years. These systems presents the forth generation of optical fiber communication system. Saying those are the traditional TDM transmission system using SONET ADM (add-drop-multiplexing) technology.

However, the demend from customers is still the driver to rapidly develop the optical fiber transmission system, which includes

a. High density **Video** such as blue ray,
b. High density **TV** channels,
c. Internet and cellular phone networks, including iPad 6 and iPad pro,
c. **Cloud data** in computer networks.

Now, 40Gb/s, 80Gb/s, ... and 120Gb/s transmission systems are available in commercial for the Metropolitan Area Networks (MAN). And the bast way to construct these network is using the wavelength division multiplex (WDM) technology based on SONET TDM, which includes 10Gb/s optical fiber transponder (OC-192) or 2.5Gbps optical fiber transponder (OC-48) .... Which is the fifth generation optical fiber communications system as illustrated in Figure 2-25. This networks is an all optical fiber WDM communication network, which includes two special technologies:

a. Optical Multiplexer and Optical De-multiplexer: The Optical multiplexer is used to combine several wavelengths (λ_1, λ_2,...,λ_n) and then output to a single optical fiber. And the Optical De-multiplexer is used to divide the wavelengths ($\lambda_1 + \lambda_2 +..., +\lambda_n$) into individual wavelengths (λ_1, λ_2,...,λ_n).

b. Optical Add-Drop-multiplexing (OADM): The optical **Dropper** is used to drop several wavelengths (such as λ_1, λ_2) from the networks into the local area. The **Adder** is used to add several wavelengths (such as λ_3, λ_4) from the local area into the networks.

In which, each wavelength is for one 10Gb/s optical fiber transponder. Where, T_x is the transmitter part of the transponder, R_x is the receiver part of the transponder. And **AMP** is an optical amplifier.

The Optical-Cross connector is just the OADM (Optical-Add/Drop-Multiplexer) in Figure 2-25, which is complicated, we will discuss in detail in the following section.

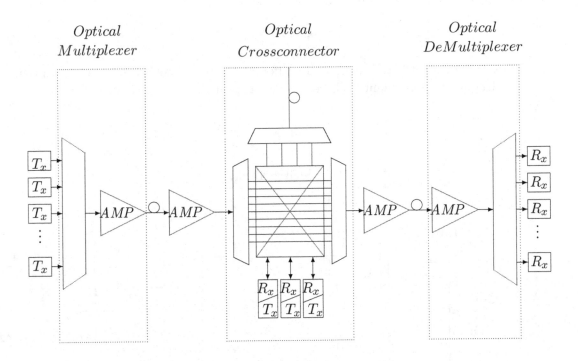

Figure 2-25   Fifth general optical fiber communication networks [39]

2.2.1   Optical WDM multiplexing and de-multiplexing

The optical WDM multiplexing and optical WDM de-multiplexing can be implemented by using

a. Prisms.

b. Optical thin film filter such as the optical Mach-Zehnder (M-Z) filter. We will discuss in Chapter 3.

c. Optical interference filter such as the optical fiber Bragg Grating (FBG). We will discuss in Chapter 3.

Using Prisms

As shown in Figure 2-26, the prisms can be used to **combine** serval wavelengths into a single wavelength output or to **split** a group of wavelengths (or colors) into several individual wavelengths (or colors). Which is based on the refracted law as

$$n_1 \sin \theta_i = n_2 \sin \theta_r \tag{2.1}$$

Which gives the refracted angle θ_r as

$$\theta_r = (n_1/n_2)\theta_i \tag{2.2}$$

Means that the refracted angle θ_r is different if the refractive index of the prism, n_2, is frequency dependent. Where $n_1 = n_0$ is in air.

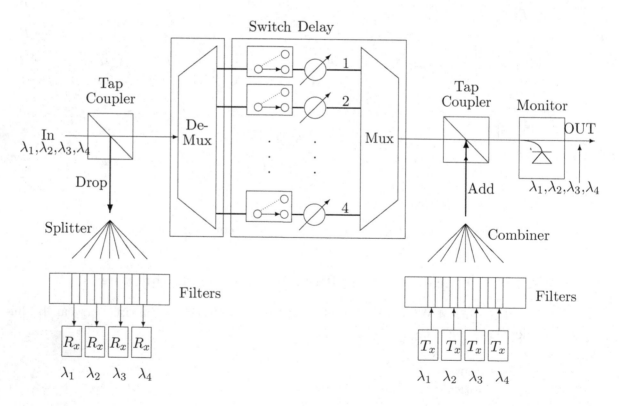

Figure 2-26   All optical fiber communications network by using prisms [12]

2.2.2 All optical fiber DWDM communications networks

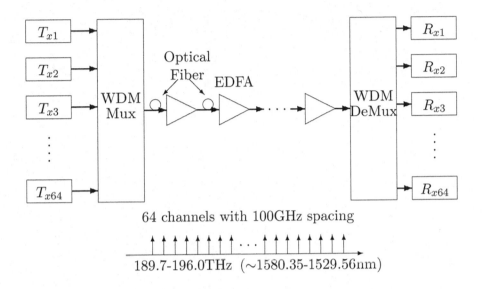

Figure 2-27   All optical fiber DWDM communication networks with 64 channels [1]

Figure 2-27 is an all optical fiber DWDM communication networks with 64 channels. Where "R_{xn}" is the receiver side of the optical fiber transponder and "T_{xn}" is the transmitter side of the transponder. The optical multiplexer and de-multiplexer can be implemented by optical fiber Bragg grating filter or cascaded Mach-Zehnder filter.

This is so called the fifth generation optical fiber communications network or all optical fiber communications network.

In forth generation optical fiber communications network, the key technologies are digital multiplexing, digital de-multiplexing, digital adding dropping, or simple say the digital add drop multiplexing (DADM) technology.

In fifth generation optical fiber communications network, the key technologies are optical multiplexing, de-multiplexing, adding dropping, or simple say optical add drop multiplexing (OADM) technology. Thus the fifth generation optical fiber communications network is all optical fiber communications network.

The input optical DWDM multiplexer is used to multiplexing 64 wavelengths from the transmitters of the transponders.

The 64 wavelengths is transmitted simultaneously and independently on an installed single mode fiber and amplified simultaneously by optical Er-doped fiber amplifiers (EDFA).

Then the 64 wavelengths data-stream are de-multiplexing into individual 64 wavelengths by optical DWDM de-multiplexer, and then sent to 64 receivers of the transponder.

Chapter 3

All Optical Fiber DWDM Communications Technology

The optical DWDM multiplexing and optical DWDM de-multiplexing can be implemented by using

    a. Optical thin film filter such as the optical Mach-Zehnder (M-Z) filter

    b. Optical interference filter such as the optical fiber Bragg Grating (FBG).

3.1 Optical Mach-Zehnder filter

3.1.1 Introduce to Optical Mach-Zehnder filter

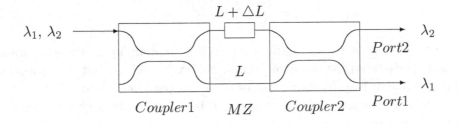

Figure 3-1   Single state Mach-Zehnder filter

A wavelength-Division-Multiplexed (WDM) optical system [13] require filters that select individual wavelength channels from the WDM signal stream [14]. Full multiplexers/demultiplexers [15], [16] direct all individual channels of WDM signal into physically separate outputs. In contrast, channel adding/dropping filters [17], [18], [19], [20], [21], [22], select certain (one or more) wavelength channels in one physical output, while leaving the other channels undistributed in the second

47

output. Optical channel dropping filter have been implemented using a number of different approaches: Mach-Zehnder interferometer with an output directional coupler [17]; grating-resonator coupled wavelength filter [18]; grating assisted co-directional coupler filter [19], [20]; or the meander coupler [21], [22].

The Mach-Zehnder filter is based on the interference of two coherent monochromatic source that are based on the length difference ($\triangle L$), thus the phase difference of two path ($\triangle \phi = \beta \triangle L$), thus contributing positive output or negative output depending on different wavelength.

In fiber-optical systems, a phase difference between two optical paths may be artificially induced. As shown in Figure 3-1, consider an input fiber with two wavelengths λ_1 and λ_2 is equally split with directional coupler 1, and each half is coupled into a waveguide, one of which ($L + \triangle L$) is longer than the other (L). The two half power arrive at a second directional coupler or combiner at different phases and based on the phase variation and the position of the output fiber, each wavelength interferes constrictively on one of the two output fibers and destructively on the other. That is, wavelength λ_1 interferes constructively on the first fiber (port 1) and wavelength λ_2 on the second (port 2)(Figure 3-1).

This arrangement is used to construct an integrated device that functions as filter or as wavelength separator, known as a *Mach-Zehnder filter.* according to which two frequencies at its input port are separated and appear at two output ports, port 1 and port 2.

A mix of two wavelengths ($\lambda 1$ and $\lambda 2$) arrives at coupler 1. Coupler 1 equally distributes the power of wavelengths λ_1 and λ_2 into two waveguides having an optical path difference ($\triangle L$). Because of the path difference, the two waves arrive at coupler 2 with a phase difference

$$\triangle \phi = \beta \triangle L = \frac{2\pi}{\lambda}\triangle L = \frac{2\pi f(\triangle L)n}{c} \tag{3.1}$$

Where n is the refractive index of the waveguide. At coupler 2, the two waves recombine and are directed to two output ports. However, each output port supports the one of the two wavelengths that satisfies a certain phase condition.

Wavelength λ_1 is obtained at output port 1 if the phase difference at the end of L satisfies the condition

$$\triangle \phi_1 = (2m-1)\pi \tag{3.2}$$

Where $\triangle \phi_1 = (2\pi/\lambda_1)\triangle L$, and then condition (3.2) becomes

$$\triangle L = (2m-1)\frac{\lambda_1}{2} \tag{3.3}$$

In this case, λ_1 contributes maximally at Port 1. Wavelength λ_2 is obtained at port 2 if the phase difference satisfies the condition

$$\triangle \phi_2 = 2m\pi \tag{3.4}$$

Where $\triangle\phi_2 = (2\pi/\lambda_2)\triangle L$, and then condition (3.4) becomes

$$\triangle L = m\lambda_2 \qquad (3.5)$$

In this case, λ_2 contributes maximally at port 2; m is a positive integer.

Thus, this filter exhibits periodic pass bands as shown in Figure 3-2. Meanwhile, we need a mathematic analysis to explain the behavior of the Mach-Zehnder filter. This analysis will be given in the next sub-section.

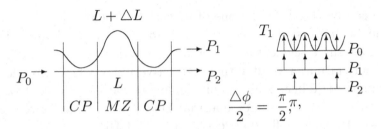

Figure 3-2   Single state Mach-Zehnder filter

Where P_0 is the input power of two wavelengths (λ_1, λ_2), which is equally split by directional coupler 1 (CP) and each half is coupled into a waveguide (Mach-Zehnder,MZ), one of which is longer than the other L by ($\triangle L$). The two waves arrive at directional coupler 2 (CP) with a phase different phase $\triangle\phi$. Each wavelength interference constructively on one of the two output fibers and destructively on the other. For instance, λ_1 interferences constructively on port 1, and destructively on port 2. λ_2 interferences on port 2 and destructively on port 1 as shown in Figure 3-2. Thus, port 1 exhibits periodic pass band centered at λ_1, and port 2 exhibits periodic pass band centered at λ_2. Where

$$T_1 = \frac{P_1}{P_0} \qquad (3.6)$$

Now form (3.1),(3.2) and (3.3), we have

$$\frac{2\pi f_1 \triangle L n}{c} = (2m - 1)\pi \qquad (3.7)$$

$$\frac{2\pi f_2 \triangle L n}{c} = 2m\pi \qquad (3.8)$$

From last two relations, the optical channel spacing $\triangle f$ is

$$\triangle f = \frac{c}{2n(\triangle L)} \qquad (3.9)$$

Here, channel spacing $\triangle f$ is the separative interval between two channels λ_1 and λ_2.

Tunability of the Mach-Zehnder filter

If the quantity $\triangle L$ can be adjusted at will, it is clear that the Mach-Zehnder filter can be tuned. The purpose of the quantity is to introduce the desired phase shift at entry point of the directional coupler 2. Thus, the phase shift is controlled by controlling the propagation delay of path $L + \triangle L$ with respect to the path L. This accomplished either the refractive index or the path (and thus the effective optical path), by altering its physical length or by both means.

The phase can be controlled by one of several methods:
† Mechanical compression, by means of a piezoelectric crystal, alters the physical length of the waveguide segment and its refractive index.
† Certain optical material alter their refractive index when exposed to beat; a thin-film thermoelectric heater placed on the longer path would control the refractive index of the path. A polymer material known to change its refractive index when exposed to heat is per-fluoro-cycle-butane (PFCBP).

Thus by controlling the refractive index of the path, the phase on the effective optical path $L + \triangle L$ is controlled and the wavelength selectability of the devise is accomplished, making the Mach-Zehnder filter a tunable *optical frequency discriminator* (OFD).

3.1.2   Single Coupler Mach-Zehnder Filter

Single Coupler Mach-Zehnder Filter

Now we will describe the design and characteristics of cascaded coupler Mach-Zehnder (CMZ) filter with weighted coupling that function as channel dropping filters with very low Side-lobe levels.

A Mach-Zehnder interferometer with input and output directional couplers [17] acts as a channel dropping filter, as illustrated in Figure 3-1. For 3dB input and output couplers, the power transmission is as a function of frequency in sinusoidal

$$T_1 = P_1/P_0 = cos^2(\triangle\phi/2) \tag{3.10}$$

Here

$$\triangle\phi = \beta(\triangle L) \tag{3.11}$$

Where β is the propagation constant of the waveguides, $L + \triangle L$ and L are the lengths of the two arms.

The filter response is periodic in frequency with the period, from (3.9),

$$\triangle f_p = \frac{c}{2n(\triangle L)} \tag{3.12}$$

Where the period $\triangle f_p = 2\triangle f$ is indicated in Figure 3-2. The period is the interval between two maximums, and the channel spacing is the interval between λ_1 and λ_2.

Because of power conservation, the sum of two outputs must be equal to the input power, namely, $P1 + P2 = P_0$. Thus, we have

$$T_2 = P_2/P_0 = 1 - T_1 \tag{3.13}$$

Thus, a WDM input stream P_0 with channel separation $\triangle f = \triangle f_p/2$ will be divided between two outputs, with alternate channels going to the outputs P_1 and P_2.

Such an optical filter can be implemented either with discrete optical fibers/couplers or by an integrated optical devices [15], [16]. Using the InP integrated optical technology [16], a Mach-Zehnder filter can be made with a period $\triangle f_p = 400GHz$ ($3.2nm$ at $1.55\mu m$), a device length of $4.5nm$, and low-loss waveguide bands of $2mm$ radius of curvature.

Theoretical Analysis of the transmission function of a single stage Mach-Zehnder filter: proof of (3.10)

As shown in Figure 3-1, a Mach-Zehnder filter is consist of two directional couplers and a Mach-Zehnder. Where the transmission function of a directional coupler (in Figure 3-3(a)) is

$$\begin{pmatrix} a_{11} \\ a_{21} \end{pmatrix} = T_c \begin{pmatrix} a_{10} \\ a_{20} \end{pmatrix}$$

$$\tag{3.14}$$

T_c is a matrix of a directional coupler,

$$T_c = \begin{pmatrix} \cos \kappa L, & -j \sin \kappa L \\ -j \sin \kappa L, & \cos \kappa L \end{pmatrix}$$

$$\tag{3.15}$$

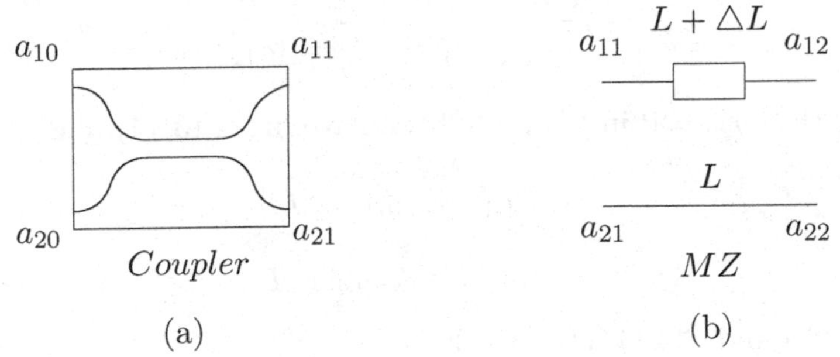

Figure 3-3   Coupler (a) and Mach-Zehnder (b)

The transmission function of a Mach-Zehnder (in Figure 3-3(b)) is

$$\begin{pmatrix} a_{12} \\ a_{22} \end{pmatrix} = T_{MZ} \begin{pmatrix} a_{11} \\ a_{21} \end{pmatrix}$$

(3.16)

The matrix of a Mach-Zehnder T_{MZ} is

$$T_{MZ} = \begin{pmatrix} e^{-j\frac{\triangle\phi}{2}}, & 0 \\ 0, & e^{-j\frac{\triangle\phi}{2}} \end{pmatrix}$$

(3.17)

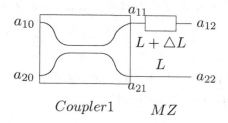

Figure 3-4   Single state Mach-Zehnder filter

Now, from Figure 3-4, the transmission function of a single state Mach-Zehnder filter is

$$\begin{pmatrix} a_{12} \\ a_{22} \end{pmatrix} = T_c \cdot T_{MZ} \cdot \begin{pmatrix} a_{10} \\ a_{20} \end{pmatrix}$$

(3.18)

And, from (3.16) and (3.17),

$$a_{12} = a_{11}e^{-j\frac{\triangle\phi}{2}} + a_{21}e^{-j\frac{\triangle\phi}{2}} = (a_{11} + a_{21})e^{-j\frac{\triangle\phi}{2}}$$

(3.19)

Considering that $a_{20} = 0$ in Figure 3-3, we have, from (3.14) and (3.15),

$$a_{11} = a_{10}\cos\kappa L$$

(3.20)

$$a_{21} = -ja_{10}\sin\kappa L$$

(3.21)

Inserting (3.20) and (3.21) into (3.19), we have

$$a_{12} = a_{10}(\cos\kappa L - j\sin\kappa L)(\cos\frac{\triangle\phi}{2} - j\sin\frac{\triangle\phi}{2})$$

or

$$a_{12} = (A \cos \frac{\triangle\phi}{2} - B \sin \frac{\triangle\phi}{2}) - j(A \cos \frac{\triangle\phi}{2} + B \sin \frac{\triangle\phi}{2}) \qquad (3.22)$$

with

$$A = a_{10} \cos \kappa L \qquad (3.23)$$

$$B = a_{10} \sin \kappa L \qquad (3.24)$$

Thus, the power at the output a_{12} is

$$
\begin{aligned}
P_1 &= a_{12} \cdot a_{12}^* \\
&= [(A \cos \frac{\triangle\phi}{2} - B \sin \frac{\triangle\phi}{2}) - j(A \cos \frac{\triangle\phi}{2} + B \sin \frac{\triangle\phi}{2})] \\
&\times [(A \cos \frac{\triangle\phi}{2} - B \sin \frac{\triangle\phi}{2}) + j(A \cos \frac{\triangle\phi}{2} + B \sin \frac{\triangle\phi}{2})]
\end{aligned}
$$

or

$$P_1 = (A^2 + B^2) \cos^2 \frac{\triangle\phi}{2} + (A^2 + B^2) \sin^2 \frac{\triangle\phi}{2} \qquad (3.25)$$

Now, if $\frac{\triangle\phi}{2} = \pi$, (3.25) becomes

$$P_1 = (A^2 + B^2) \cos^2 \frac{\triangle\phi}{2} \qquad (3.26)$$

Where $(A^2 + B^2) = a_{10}^2 = P_0$. Thus, the transmission of a single state of Mach-Zehnder filter is

$$T_1 = P_1/P_0 = \cos^2 \frac{\triangle\phi}{2} \qquad (3.27)$$

That is (3.10). Because of the power conservation, the two output power P_1 and P_2 are complementary, i.e. $P_1 + P_2 = P_0$, Thus,

$$T_2 = P_2/P_0 = 1 - T_1 = \sin^2 \frac{\triangle\phi}{2} \qquad (3.28)$$

From (3.27) and (3.28), we have

$$
\begin{aligned}
T_1 &= 1, \quad T_2 = 0 \quad if \quad \triangle\phi = (2m - 1)\pi \\
T_2 &= 1, \quad T_1 = 0 \quad if \quad \triangle\phi = 2m\pi
\end{aligned}
$$

$$(3.29)$$

Which gives the explanation of Figure 3-2, where $m = 1, 2, 3, \cdots$.

Cascaded Coupler Mach-Zehnder Filter

Here we would like to introduce the research of the Cascaded Coupler Mach-Zehnder Filter by M. Kuznetsov [23].

A high finesse filter is required for selecting every nth channel ($n = 3, 4, 5, \cdots$) out of the WDM signal. This can be accomplished with a cascaded coupler Mach-Zehnder (CMZ) filer, as illustrated in Figure 3-5.

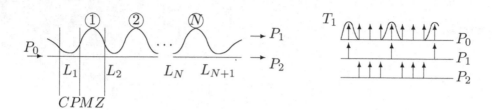

Figure 3-5   Cascaded Mach-Zehnder filter

Here:

a. The individual Mach-Zehnder state are identical (the same optical path length difference $\triangle\phi$ between two branches and the same coupling coefficient of the coupler).

b. However, the length (coupling strengths) of the directional couplers between stages need not be identical.

c. The cascaded filter operates on a same wavelengths such that the Mach-Zehnder optical path length difference $\triangle\phi$ is a multiple of 2π, the device acts as one long coupler with the total length of

$$L_{tot} = \sum L_i \tag{3.30}$$

Here, L_i are the lengths of the individual couplers ($i = 1, 2, \cdots, N + 1$) for the N-stage filter), and we have assumed that the coupling coefficient κ are the same in the different section. Thus, the total length L_{tot} should equal to one coupling length ($\kappa L_{tot} = \pi/2$) for a full power transfer to part P_1. Off resonance, power transfer to part P_1 drops.

d. Because of power conservation, the two outputs P_1 and P_2 are again complementary.

Such a direct analog of the grating-assisted co-directional coupler filter [19], [20] or the meander coupler [21], [22], in those case, the optical path length difference between the two interferometer arms is achieved by the waveguide propagation constant difference, whereas in our case, it is given by the geometrical path length difference.

Theoretical Analysis

One can describe the CMZ (coupler-Mach-Zehnder) filter operation using transmission matrices for Mach-Zehnder and coupler sections.

As shown in Figure 3-3,

a. For a directional coupler section the *ith* coupler is characterized by its transmission matrix $T_c(L_i)$ as

$$\begin{pmatrix} a_1 \\ a_2 \end{pmatrix}_{out} = T_c(L_i) \begin{pmatrix} a_1 \\ a_2 \end{pmatrix}_{in} = \begin{bmatrix} \cos(\kappa L_i), & -j\sin(\kappa L_i) \\ -j\sin(\kappa L_i), & \cos(\kappa L_i) \end{bmatrix} \begin{pmatrix} a_1 \\ a_2 \end{pmatrix}_{in} \quad (3.31)$$

Where a_1 and a_2 are the normalized fields in the top waveguide and the bottom waveguide, respectively.

b. The transmission matrix for the Mach-Zehnder is

$$T_{MZ} = \begin{bmatrix} \exp(j\triangle\phi/2), & 0 \\ 0 & , \exp(-j\triangle\phi/2) \end{bmatrix} \quad (3.32)$$

Where the phase delay $\triangle\phi$ is the same for all of the Mach-Zehnders. Thus, the transmission of the N-stage filter is

$$\begin{pmatrix} a_1 \\ a_2 \end{pmatrix}_{out} = T_c(L_{N+1}) \cdots T_{MZ} T_c(L_2) T_{MZ} T_c(L_1) \begin{pmatrix} a_1 \\ a_2 \end{pmatrix}_{in} \quad (3.33)$$

The power transmission of the N-stage filter is

$$T_1 = \frac{|a_1|_{out}^2}{|a_2|_{in}^2} \quad (3.34)$$

3.1.3   Cascaded CMZ filter with weighted coupler distribution

Weighted Coupler Distribution

For the N-stage CMZ filter, the total length L_{tot} can be distributed in different ways over the (N+1) individual couplers. This coupler length distribution controls the shape of the filter transmission characteristic [24], such as the filter passband width and the side-lobe level. A variety of continuous distribution has been considered in the context of grating-assisted filter [20], [22], [24]. In contrast, for CMZ filter, the distribution is discrete and the transformation that connects the weight distribution to the filter shape is different from that for the grating filters. Since an optimal discrete distribution is unknown, we have investigated several simple weight distributions to characterize the CMZ filter.

$$w_i = 1, \quad (i = 1, 2, \cdots, N+1), \qquad Uniform \quad (3.35)$$

$$w_i = \cos[\pi a(i - (N+2)/2)/2], \qquad Cosin \quad (3.36)$$

$$w_i = B(N, i-1) = \frac{N!}{(i-1)!(N-i+1)!}, \qquad Binomial \quad (3.37)$$

The individual coupler lengths are then

$$L_i = \frac{w_i L_{tot}}{\sum w_i} \qquad (3.38)$$

More generally, the coupler strengths of the individual sections, given by the products $\kappa_i L_i$, can follow the weight distribution by adjusting either the length L_i or the coupling coefficient κ_i of the individual sections.

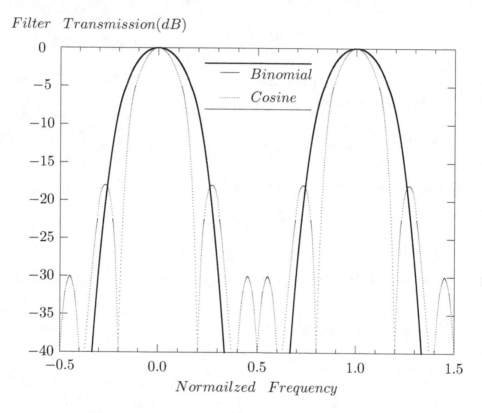

Figure 3-6   N=5 stage CMZ filter transmission(1) [25]

Figure 3-6 shows the N=5 state filter transmission as a function of normalized frequency for binomial and cosine weight distributions, respectively, and Figure 3-7 for binomial and uniform weight distributions, respectively. The transmission is periodic and the frequency normalization factor is the period $\triangle f_p$. Obviously,

a. The uniform distribution gives unacceptably high side-lobe level of -8dB.

b. The cosine distribution lowers them to -18dB.

c. While, for the binomial distribution, the side-lobes are very low at -47dB. The lower side-lobes are achieved at the expense of the wider transmission peak [24]. Side-lobe levels below -30dB would be required for WDM application.

As we increase the number of the stages, the width of the main peak decreases and, correspondingly, the filter finesse increases. For the bonomial distribution, the side-lobe level remain below -45dB and the transmission function T_1 appears to be closed to

$$T_1 \approx \cos^{2N}(\frac{\Delta\phi}{2}) \tag{3.39}$$

for the N-stage filter. This is just the transmission function of the single-stage raised to the N_{th} power.

Filter Transmission(dB)

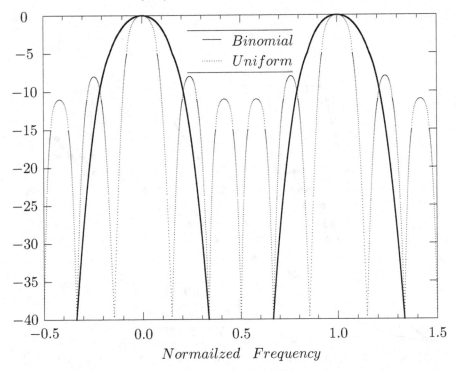

Normailzed Frequency

Figure 3-7   N=5 stage CMZ filter transmission(2) [26]

Approximating the cosine in (3.39) by a parabola, we obtain a relation, accurate for large N, between the filter finesse F and the number of stage N for the binomial distribution

$$N = B_b F^2 \tag{3.40}$$

where

$$B_b = \frac{4}{\pi^2} \ln(1/T_f) \tag{3.41}$$

Here we define the finesse F as the ratio of the transmission period Δf_p to the transmission peak width T_f, where the peal width is measured at the T_f transmission level below the transmission maximum.

$$F = \frac{\Delta f_p}{T_f} \tag{3.42}$$

The conventional finesse is a particular case for $T_f = 0.5 \to -3dB$. In Figure 3-8, we plot with open symbols the number of stages N required to achieve a given filter finesse F for the binomial coupling weight distribution; the finesse is measured for both the $T_f = -3dB$ and the $T_f = -20dB$ level. The points in the plot were obtained from the calculated transmission functions (3.39), while the lines through the open symbols are fits to (3.40) with $B_b = 0.25$ and 1.52 for $T_f = -3dB$ and $-20dB$, respectively. Equation (3.41) gives, correspondingly, the value $B_b = 0.28$ and 1.87, which is in good agreement with the fit value above.

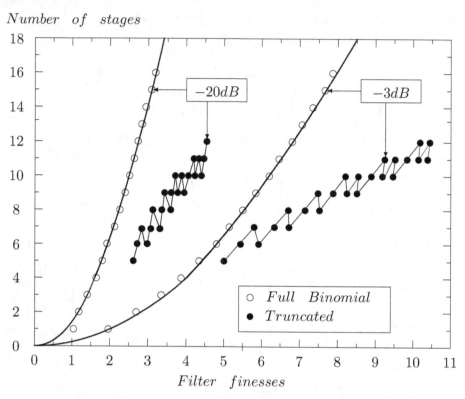

Figure 3-8   The required number of CMZ filter vs finesse
($-3dB$ and $-20dB$ levels) [27]

Truncated Binomial Weight Distribution

For the full binomial distribution, the number of the filter stages grows very rapidly (quadratically) as a function of the desired finesse as given in Figure 3-8 and equation (3.40). For a practical, physically compact implementation of the filter, a smaller number of stages is required. One way to reduce the number of stages is found by observing the binomial distribution. For instance, for $N = 10$ the binomial weight distribution is : {1 10 45 120 210 252 210 120 45 10 1}.

The first one and the last one, or say two weights are very relatively small; these couplers contribute negligible power exchange between two waveguides and thus can be eliminated. Therefore, we introduce the truncated binomial distribution for the $N = (M - 2r)$ stage filter as

$$
\begin{aligned}
w_i &= B_t(M, r, i) \\
&\equiv B(M, i - 1 + r), \quad i = 1, 2, \cdots, M - 2r + 1.
\end{aligned}
\tag{3.43}
$$

Which is just the order M binomial distribution with the first weight and last weight elements, r, dropped.

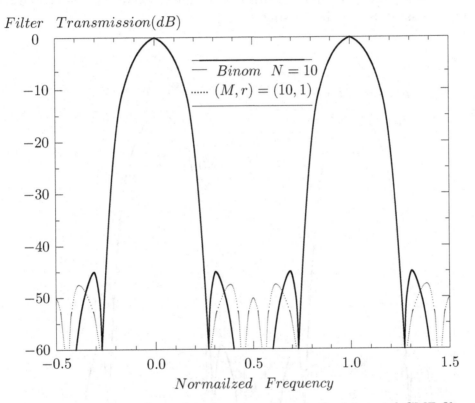

Figure 3-9   Effect of stage truncation on the binomial CMZ filter
transmission (1) [28]

Figure 3-9 shows the transmission of an $N = 10$ stage filter with full binomial weights, together with transmission of $(M, r) = (10, 1)$ eight-stage with truncated binomial distributions. Figure 3-10 shows the transmission of an $N = 10$ stage filter with full binomial weights, together with transmission of $(M, r) = (10, 2)$ six-stage filter with truncated binomial distributions. Truncating stages leaves the filter width essentially unchanged, while the Side-lobe level rises, reaching a remarkable uniform $-30dB$ leave for $(M, r) = (10, 2)$ distribution. Further truncation raises the Side-lobe level above $-30dB$. How many stages can be "truncated" depends on the Side-lobe level allowed for a particular application. We have tried a number of other weight distributions known from the antenna array theory, such as Chebyshev distribution; however, none give results as good as the truncated binomial distribution.

In Figure 3-8, the number of stages required to achieve a given filter finesse for the truncated binomial coupling weight distribution has been put on with filled symbols. For each data point, the binomial distribution was maximally truncated, keeping the sidebole level below $-30dB$. The first point corresponds to $(M, r) = (7, 1)$ five-stage distribution, and the last one corresponds to $(M, r) = (21, 10)$ eleven-stage distribution; note that the same number of stages $(M - 2r)$ can be obtained in several different ways. The author M.Kuznetsov has observed that the $-30dB$ sidebole level is achieved when the largest-to-smallest coupler weight ratio is of the order $6 - 7$, regardless of the number of stages.

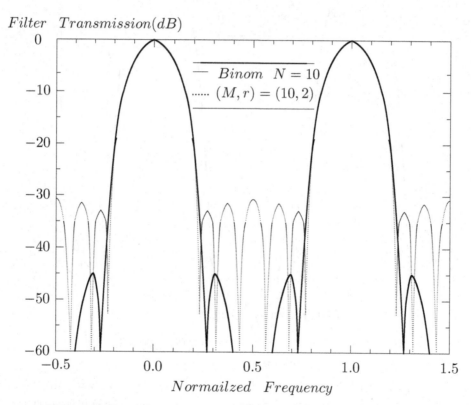

Figure 3-10   Effect of stage truncation on the binomial CMZ filter

transmission (2) [29]

To achieve lower sideboles, this ratio has to be higher. For the truncated binomial distribution and a fixed sidebole, the required number of stages appears to grow only linearly with finesse, as compared to the quadratic dependence for the full binomial distribution. This reduction in the number of stages, by as much as a factor of 3 for a filter with finesse of 12 is very important for reducing the physical size of the CMZ filter. in the integrated optical implementation.

The required number of stages

A WDM optical communication system requires channel adding/dropping filters that can select one out of every N_{ch} channels. For the truncated binomial distribution, which appears to be optimal, we now determine the analytical relation between the number N of CMZ filter stages and the desired filter finesse F or the channel number N_{ch}.

If the filter is to reject undesired channels with a cross-talk level of T_f, say, $-30dB$, then the number of selectable channels N_{ch} is given by twice the filter finesse F as defined at the T_f level

$$N_{ch} = 2 \cdot F \tag{3.44}$$

Of course, the filter Side-lobes also have to be kept below the T_f level.

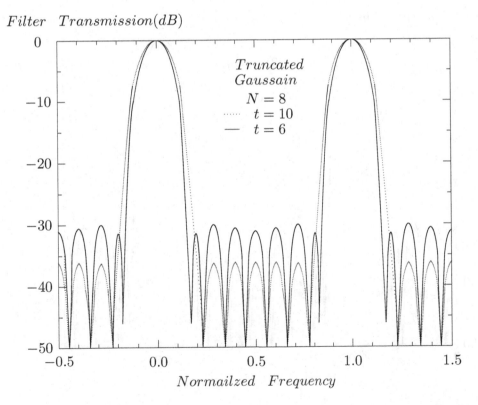

Figure 3-11   Transmission of the fixed-length $N = 8$ stage CMZ filter with truncated Gaussian weight distribution (1) [30]

It is well known that for the large N, the *binomial distribution* (3.37) can be approximated by the *Gaussian distribution*. Thus, we can approximate the truncated binomial by the truncated Gaussian distribution, both for convenient filter design and for analytically determining the number of stages dependence on the filter finesse. The *binomial distribution* (3.37) with a mean $\mu = N/2$ and variance $\nu = N/4$ can be approximated by a *Gaussian distribution*

$$G(\mu, \nu) = \exp\left[-\frac{(i - 1 - \mu)^2}{2\nu} \right] \tag{3.45}$$

with $i = 1, 2, \cdots, (N + 1)$, The truncated binomial distribution (3.43) can be approximated by the truncated Gaussian distribution

$$G_t(\mu, \nu_t) = \exp\left[-\frac{(i - 1 - \mu)^2}{2\nu_t} \right] \tag{3.46}$$

Where $i = 1, 2, \cdots, (N + 1)$, $\mu = N/2$, and $\nu_t = N^2/[8\ln(t)]$. Here, the center-to-wing ratio t of the distribution should be of the order $6 - 7$ in order to achieve the $-30dB$ filter Side-lobe level, as we have observed empirically.

Figure 3-11 illustrates the transmission of the $N = 8$ stage filter with truncated Gaussian weight distribution with $t = 14$ and 6.

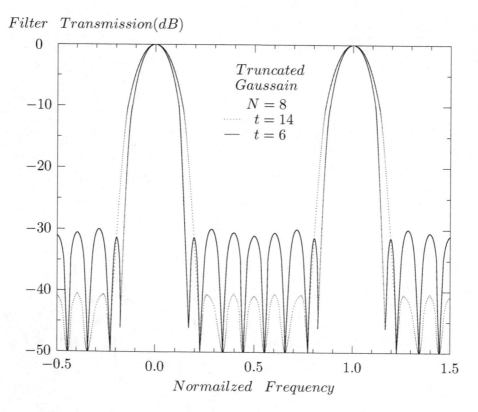

Figure 3-12   Transmission of the fixed-length $N = 8$ stage CMZ filter with truncated Gaussian weight distribution (2)[31]

Figure 3-12 illustrates the transmission of the $N = 8$ stage filter with truncated Gaussian weight distribution with $t = 10$ and 6. Obviously, for this fixed-length filter, as the center-to-wing ratio t of the distribution decrease, the filter transmission narrows down at the expense of the rising Side-lobe levels.

Now we will determine the analytical relation between the number of the stages N and the finesse F for the truncated binomial and Gaussian distributions. Approximating the full binomial distribution (3.37) by Gaussian (3.45), we truncate the wing ratio become t [(3.43) and (3.46)]. Starting with a full binomial with M stages, after truncation, we take the stages approximately as

$$N = \sqrt{2M \ln(t)}. \tag{3.47}$$

Assume that filter shape remains unchanged (3.39) and only the Side-lobe levels rise, the finesse is still given by $M = B_b F^2$ [see (3.40)]. Thus, for the truncated binomail and Gaussian distributions, the number of stages N as a function of the finesse F_f is given by

$$N = B_g F \tag{3.48}$$

Where

$$B_g = \sqrt{2B_b \ln(t)} = \frac{2}{\pi}\sqrt{1 \ln(1/F_f) \ln(t)}. \tag{3.49}$$

For the truncated distributions, the required number of CMZ filter stage grows only linearly with finesse F and the selectable number of channels N_{ch} [see (3.44)]. Compare this to the much worse quadratic growth (3.40) for the full binomial distribution.

These dependence are illustrated in Figure 3-13, where the bottom and top scales give the finesse F and the number of selectable channels N_{ch}, respectively, while the left scale is the required number of stages N. The finesse reference level is chosen to be $-30dB$. In Figure 3-13:

a. The solid circles corresponds to the full binomial distribution , and the solid line through them is a fit to the quadratic dependence $N = B_b F^2$ (3.40) . The fit gives $B_b = 2.1$, as compared to the analytical estimate of $B_b = 2.8$ from (3.41).

b. The solid circles with a solid line through it corresponds to the truncated binomial distribution, using maximal truncation with Side-lobe below $-30dB$.

c. The open circles corresponds to the truncated Gaussian distribution with the the truncated level of $t = 6$, on which, the Side-lobe level was approximately $-30dB$.

Obviously, the Gaussian truncated filters follow closely the behavior of the truncated binomial filters. As expected from (3.48), the required number of stages grows linearly with finesse.

d. The solid line through the open circles in Figure 3-13 is trunked Gaussian distributions, which is a line fitting to the points. The fit gives $B_g = 2.5$, as compared to $B_g = 3.2$ from the analytical estimate in (3.49). The numerical estimates from (3.49) and (3.41) work better for the moderate finesse level, say, $T_f \approx (3 - 10)dB$, and are worse for the low levels of $T_f \approx -30dB$, where the filter shape is more sensitive to the level of truncation, the exact weight distribution, and the deviates further from the approximation in (3.39).

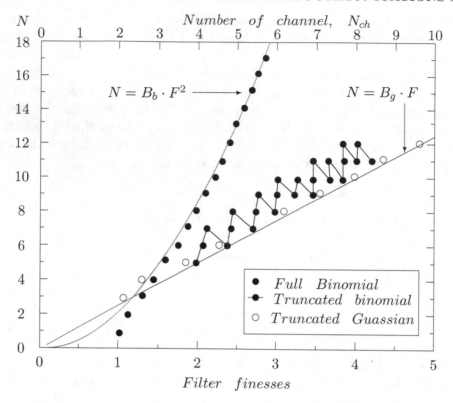

Figure 3-13   The required number of CMZ filter vs finesse [32]

Summery

We have described cascaded Coupler Mach-Zehnder (CMZ) channel adding/dropping filter for use in WDM system. Such filter allow selection of every n_{th} channel $(n = 1, 2, 3, 4, \cdots)$ from the WDM signal with very low $(-30 dB)$ cross-talk levels. Channel dropping function of the CMZ filters is periodic, which can be advantageous for certain applications. Using weighted coupler length with truncated binomial distribution, which appears to be optimal, and low Side-lobe characteristics.

The truncated Gaussian distribution can be used conveniently to approximate the truncated binomial distribution in designing such multistage filters. Importantly, we have shown that the number of CMZ filter stage grows linear with the required finesse and thus the number of selectable channels. This is important for keeping down the physical size of integrated optical CMZ. Physical device size will limit the ultimate number of stages and the achievable finesse of these filters. Integrated optical technology on silicon [15] or InP [16] can be used to implement the CMZ channel dropping filters and, perhaps, integrate them with other optoelectronic devices.

3.2 Optical Fiber Bragg Grating

3.2.1 Introduce to fiber Bragg grating (FBG)

Optical fiber Bragg grating (FBG)

As shown in Figure 3-14, an optical fiber Bragg grating (FBG) means that the **grating** is written inside the core of the single mode fiber, which can reflect one wavelength λ_B from several wavelengths input (λ_1, λ_2, λ_3, λ_4) and transmits all others. Where, λ_B can be any one of the input wavelengths λ_1, λ_2, λ_3, λ_4. The circle is an optical circulator, and the light go through the circulator by the right-hand direction.

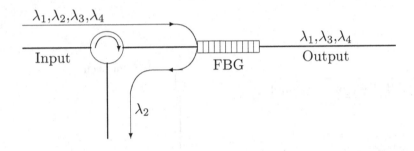

Figure 3-14 Optical fiber Bragg grating (FBG)

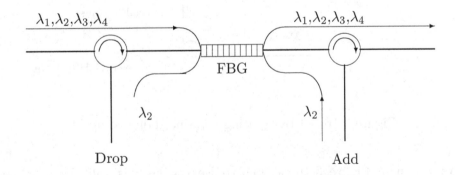

Figure 3-15 Optical Add/Drop

In this case, the optical fiber Bragg grating (FBG) can be used as the optical ADD/DROP as shown in Figure 3-15. Where:

a. The wavelength λ_2 is reflected by the FBG at the input end of the FBG, and then it go through the circulator to be dropped to the local networks. The others λ_1, λ_3, λ_4 go through the FBG to the output.

b. Meanwhile, the wavelength λ_2 come from the local networks is reflected by the FBG at the output end of the FBG, then it go through the circulator to combine with the wavelengths λ_1,λ_3,λ_4 to go to the output.

Note, that if the wavelength λ_2 can be reflected by the FBG at the input end of the FBG, definitely, it can be reflected by the FBG at the output end of the FBG since FBG is a linear component.

Principle of Fiber Bragg Grating

- A FBG is produced on the fly.

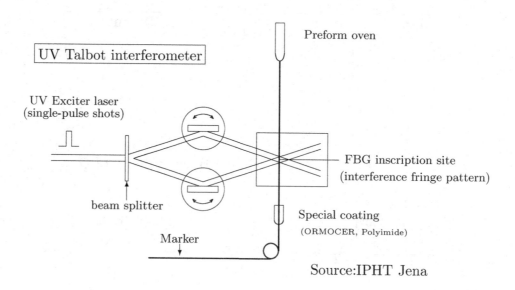

Figure 3-16   FBG gratings produced on the fly [33]

As shown in Figure 3-16, a single-pulse UV laser is split by a splitter, then combined together with an angle to cross a single mode fiber, scribes several fringes on the single mode fiber (see Figure 3-17(a)), when the fiber is in pulling process from the preform oven. The energy of the exposure (on the fiber) is in the range of several hundred J/cm^2 causing a 0.001% ... 0.1% change of the refractive index for the core of the single mode fiber as shown in Figure 3-17(b). In which, a Bragg grating is produced on the single mode fiber.

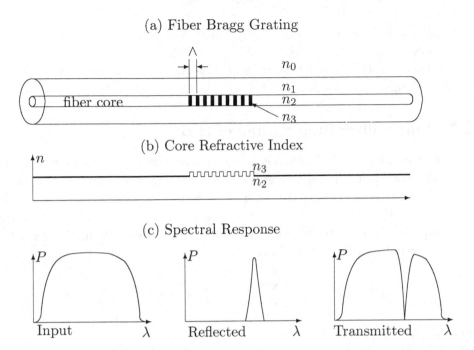

(a) Fiber Bragg Grating

(b) Core Refractive Index

(c) Spectral Response

Figure 3-17   Fiber Brag grating structure and the spectral response

- The spectral response of the FBG:

In this case, when the light with several wavelengths is input onto the left end of a FBG, the FBG reflected part of the light wavelengths and transmits all others because of the interference in the FBG. The interference means that the reflective wave from each fringe combined together with constructive interference for part of the wavelengths, but destructive interference for all others. Where, in Figure 3-17(c), the response of the reflected wavelengths is shown in the middle and the right one shows the response of all others.

The constructive interference condition is

$$\lambda_B = 2n_e \wedge \tag{3.50}$$

Where $\wedge$ is the the grating period (see Figure 3-17(a)) and λ_B is the wavelength reflected by the FBG in the free space. n_e is the effective refractive index of the grating. In this case,

$$\wedge = \frac{\lambda_B}{2n_e} \tag{3.51}$$

That means the period of the grating is equal to the half wavelength (λ_B) of the wave in the grating, and then the reflective waves from all of gratings are in constructive interference for the wavelength λ_B and in destructive for all others.

- The characteristic of a fiber Bragg grating (FBG)

Recall the characteristic of the FBG in a WDM transmission network: when several wavelengths λ_1, λ_1, λ_2, λ_3, λ_4, are incident onto the FBG, the FBG will reflect a wavelength λ_B and transmits all others. Where, λ_B can be any one of the input wavelengths λ_1, λ_2, λ_3, λ_4.

3.2.2 Single fiber Brag grating of (FBG)

As shown in Figure 3-17, when a light is incident into the FBG, the reflective light of the FBG travels forth and back in a grating spacing. And the forward wave $A^+(z)$ and the back wave $A^-(z)$ are coupling to each other due to the perturbation of the effective refractive index n_{eff}. Therefore, the electric fields in FBG are represented by the superposition of the normal mode (LP modes in single mode fiber) traveling in both forward and back directions,

$$E_y(x, y, z, t) = \frac{1}{2} \sum_m A_m(z) \varepsilon_y^m(x, y) e^{i(\omega t - \beta_m z))} + c.c. \qquad (3.52)$$

Where $c.c.$ is the radiation modes.

* $\varepsilon_y^m(x, y)$ might describe the LP modes, hence do no change energy.

* $A_m(z)$ is consist of the forward wave $A_m^+(z) exp(i\omega t - \beta_m z)$ and the back wave $A_m^-(z) \exp(i\omega t + \beta_m z)$.

Now, we try to find the coupling equations between the forward wave $A_m^+(z) \exp(i\omega t - \beta_m z)$ and the back wave $A_m^-(z) \exp(i\omega t + \beta_m z)$ by using the Maxwell equations. To this end, we'd better using the dielectric constant $\varepsilon'(\vec{r})$ to replace the effective refractive index n_{eff} in FBG. Namely in FBG,

$$\varepsilon'(\vec{r}) = \varepsilon(\vec{r}) + \Delta\varepsilon(\vec{r}) \qquad (3.53)$$

Where $\Delta\varepsilon(\vec{r})$ is the perturbation of $\varepsilon(\vec{r})$.

The wave equation in normal single mode fiber is

$$\nabla^2 E_y - \mu\varepsilon(\vec{r})\frac{dE_y}{dt^2} = 0 \qquad (3.54)$$

A perturbation polarization caused by $\Delta\varepsilon(\vec{r})$ of FBG is

$$\left[\vec{P}_{pert}(\vec{r}, t)\right]_y = \Delta\varepsilon(\vec{r})E_y(\vec{r}, t) = \Delta n^2(\vec{r})E_y(\vec{r}, t) \qquad (3.55)$$

Thus, the wave equation in FBG is

$$\nabla^2 E_y - \mu\varepsilon(\vec{r})\frac{dE_y}{dt^2} = \mu\frac{\partial^2}{dt^2}\left[\vec{P}_{pert}(\vec{r}, t)\right]_y \qquad (3.56)$$

The solution of (3.56) is the superposition of normal modes of single mode fiber (LP modes),

$$E_y(x, y, z, t) = \frac{1}{2} \sum_m A_m(z) \varepsilon_y^{(m)}(x, y) e^{i(\omega t - \beta_m z)} + c.c. \tag{3.57}$$

Where:

 * The normal modes $\varepsilon_y^{(m)}(x, y) \exp i(\omega t - \beta_m z)$ satisfies the wave equation (3.54) in normal single mode fiber,

$$\left(\frac{\partial^2}{\partial x^2} + \frac{\partial^2}{\partial y^2} - \beta_m^2 \right) \varepsilon_y^{(m)}(x, y) + \omega^2 \mu \varepsilon(\vec{r}) \varepsilon_y^{(m)}(x, y) = 0 \tag{3.58}$$

* Meanwhile, the normal modes $\varepsilon_y^{(m)}(x, y)$ satisfies the normalization condition in the single mode fiber

$$\int_\infty^\infty \int_\infty^\infty \varepsilon_y^{(l)}(x.y) \varepsilon_y^{(m)}(x, y) dx dy = \frac{2\omega\mu}{\beta_m} \delta_{l,m} \tag{3.59}$$

Insert (3.57) into (3.55), we have

$$\left[\vec{P}_{pert}(\vec{r}, t) \right]_y = \frac{\Delta n^2(\vec{r})}{2} \sum_m A_m(z) \varepsilon_y^{(m)}(x, y) e^{i(\omega t - \beta_m z)} + c.c. \tag{3.60}$$

Using (3.60) in (3.58), considering that $\frac{\partial}{\partial z} = -i\beta_m$, $\frac{\partial}{\partial t} = -i\omega$, we have [1]

$$\sum_m \left[\frac{A_m}{2} \left(\frac{\partial \varepsilon_y^{(m)}}{\partial x^2} + \frac{\partial \varepsilon_y^{(m)}}{\partial y^2} - \beta_m^2 \varepsilon_y^{(m)} + \omega^2 \mu \varepsilon(\vec{r}) \varepsilon_y^{(m)} \right) e^{-i\beta_m z} \right.$$

$$+ \frac{1}{2} \left(-2i\beta_m \frac{dA_m}{dz} + \frac{d^2 A_m}{dz^2} \right) \varepsilon_y^{(m)} e^{-i\beta_m z} \Big] e^{i\omega t}$$

$$+ \quad c.c = \mu \frac{\partial^2}{dt^2} \left[\vec{P}_{pert}(\vec{r}, t) \right]_y \tag{3.63}$$

Where:

 * from (3.58), $\left(\frac{\partial \varepsilon_y^{(m)}}{\partial x^2} + \frac{\partial \varepsilon_y^{(m)}}{\partial y^2} - \beta_m^2 \varepsilon_y^{(m)} + \omega^2 \mu \varepsilon(\vec{r}) \varepsilon_y^{(m)} \right) = 0$.
 * Considering that $A_m(z)$ is a slow varying amplitude, we have

$$\left| \frac{d^2 A_m}{dz^2} \right| << \left| \frac{dA_m}{dz} \right| \tag{3.64}$$

---

[1]Where

$$\frac{d}{dz} \left[A_m(z) e^{-i\beta_m z} \right] = -i\beta_m A_m(z) e^{-i\beta_m z} + \frac{dA_m}{dz} e^{-i\beta_m z} \tag{3.61}$$

$$\frac{d^2}{dz^2} \left[A_m(z) e^{-i\beta_m z} \right] = (-i\beta_m)^2 A_m(z) e^{-i\beta_m z} - 2i\beta_m \frac{dA_m}{dz} e^{-i\beta_m z} + \frac{d^2 A_m}{dz^2} e^{-i\beta_m z} \tag{3.62}$$

Thus, (3.63) becomes

$$\sum_m -i\beta_m \frac{dA_m}{dz} \varepsilon_y^{(m)} e^{i(\omega t - \beta_m z)} + c.c = \mu \frac{\partial^2}{\partial t^2} \left[\vec{P}_{pert}(\vec{r}, t) \right]_y \qquad (3.65)$$

Now, multiplying (3.65) by $\varepsilon_y^{(s)}(x, y)$, integral it all over the transverse plane, we have

$$\frac{dA_s^{(-)}}{dz} e^{i(\omega t + \beta_s z)} - \frac{dA_s^{(+)}}{dz} e^{i(\omega t - \beta_s z)} - c.c = \frac{i}{2\omega} \frac{\partial^2}{\partial t^2} \int_{-\infty}^{\infty} \int_{-\infty}^{\infty} \left[\vec{P}_{pert}(\vec{r}, t) \right]_y \varepsilon_y^{(s)}(x, y) dx dy \qquad (3.66)$$

Where:

* The normal mode condition (3.59) has been considered, and then only $m = s$ terms exit.

* $A_s^{(-)} e^{i(\omega t + \beta_s z)}$ is the back wave.

* $A_s^{(+)} e^{i(\omega t - \beta_s z)}$ is the forward wave.

And then, (3.66) is the coupling equation of the forward wave $A_s^{(+)}$ and the back wave $A_s^{(-)}$ caused by the perturbation of $\varepsilon(\vec{r})$, $(\Delta \varepsilon(\vec{r})$, or $\Delta n^2(\vec{r}))$.

Simplification of the coupling equation

After considering (3.60), the coupling equation (3.66) can be rewritten as

$$\frac{dA_s^{(-)}}{dz} e^{i(\omega t + \beta_s z)} - \frac{dA_s^{(+)}}{dz} e^{i(\omega t - \beta_s z)} - c.c = \frac{-i\varepsilon_0}{4\omega} \frac{\partial^2}{\partial t^2} \int_{-\infty}^{\infty} \int_{-\infty}^{\infty} \Delta n^2(z) \varepsilon_y^{(m)}(x, y) \varepsilon_y^{(s)}(x, y) dx dy \qquad (3.67)$$

Where in FBG, $\Delta n^2(z)$ is a rectangular function along z direction and its period is $\wedge$. Thus, [2]

$$\Delta n^2(z) = \Delta n^2 \sum_{-\infty}^{\infty} a_n \cos \frac{n\pi}{\frac{\wedge}{2}} z \qquad (3.69)$$

or

$$\Delta n^2(z) = \Delta n^2 \sum_{-\infty}^{\infty} a_q e^{i \frac{2\pi q}{\wedge} z} \qquad (3.70)$$

Insert (3.70) into (3.67), we find that only one term ($q = l, m = s$) in right side of the equality, $A_s^{(+)} \exp \left[i \left(\frac{2l\pi}{\wedge} - \beta_s \right) z \right]$, match one term in left side of the equality,

---

[2]The definition of Fourier series: If $f(x)$ is a period function with period $2l$ and the function is integrable in $[-l, l]$, then

$$f(x) = \frac{a_0}{2} + \sum_{n=1}^{\infty} (a_n \cos \frac{n\pi}{l} x + b_n \sin \frac{n\pi}{l} x) \qquad (3.68)$$

is called the Fourier series of $f(x)$.

$\frac{dA_s^{(-)}}{ds}e^{i\beta_s z}$, provided that

$$\frac{2l\pi}{\wedge} - \beta_s \approx \beta_s \tag{3.71}$$

This result can be written as, from (3.67),

$$\frac{dA_s^{(-)}}{dz} = -\kappa A_s^{(+)} e^{-i2(\Delta\beta)z} \tag{3.72}$$

Similarly, we have [3]

$$\frac{dA_s^{(+)}}{dz} = \kappa^* A_s^{(-)} e^{i2(\Delta\beta)z} \tag{3.74}$$

Where [34]

$$\kappa = \frac{i\omega\varepsilon_0 a_l}{4} \int_{-\infty}^{\infty} \int_{-\infty}^{\infty} \Delta n^2 [\varepsilon_y^{(s)}(x,y)]^2 dx dy \tag{3.75}$$

with

$$a_l = \begin{cases} \dfrac{-i}{l\pi} & : \quad l = odd \\ 0 & : \quad l = even \\ \dfrac{l}{2} & : \quad l = 0 \end{cases} \tag{3.76}$$

For $l = odd$

$$\kappa = \frac{-\omega\varepsilon_0}{4\pi l} \int_{-\infty}^{\infty} \int_{-\infty}^{\infty} \Delta n^2 [\varepsilon_y^{(s)}(x,y)]^2 dx dy \tag{3.77}$$

And, from (3.71)(3.73), the coupling equations is based on the condition that $\beta_s - \frac{l\pi}{\wedge} \approx 0$, which is near the phase-matching condition. A parameter $\Delta\beta$ is used to describe this situation as follows,

$$\Delta\beta = \beta_s - \frac{l\pi}{\wedge} = \beta_s - \beta_0, \quad \beta_0 = \frac{l\pi}{\wedge} \tag{3.78}$$

For phase-matching

$$\Delta\beta = 0 \tag{3.79}$$

we have, from (3.78),

$$\beta_s = \frac{l\pi}{\wedge} = \frac{2\pi}{\lambda_B} \tag{3.80}$$

---

[3]Insert (3.70) into (3.67), we find that only one term ($q = -l, m = s$) in right side of (3.67), $A_s^{(-)} \exp\left[i\left(\beta_s - \frac{2l\pi}{\wedge}\right)z\right]$, matches one term in left side of (3.67), $\frac{dA_s^{(-)}}{dz} \exp\left[-i\beta_s z\right]$, provided that

$$\beta_s - \frac{2l\pi}{\wedge} \approx -\beta_s \tag{3.73}$$

Where β_s is the phase constant of the normal mode in single mode fiber, and λ_B is the normal mode wavelength in single mode fiber. And then, the phase-matching condition is, from (3.80),

$$\wedge = l\frac{\lambda_B}{2} \tag{3.81}$$

Which means that the phase-matching can be reached if the grating period $\wedge$ equals the integer of $\frac{\lambda_B}{2}$. In this case, all reflected wave from each grating are in phase-matching and will be sum up to maximum.

Solutions of the coupling equations

Rewrite the coupling equations from (3.72) and (3.74) as follows,

$$\frac{dA_s^{(-)}}{dz} = -\kappa A_s^{(+)}e^{-i2\Delta\beta z} \tag{3.82}$$

$$\frac{dA_s^{(+)}}{dz} = \kappa^* A_s^{(-)}e^{+i2\Delta\beta z} \tag{3.83}$$

Set the back wave $A_s^{(-)}$ and the forward wave $A_s^{(+)}$ as follows,

$$A_s^{(-)}(z) = S(z)e^{-i\Delta\beta z} \tag{3.84}$$

$$A_s^{(+)}(z) = R(z)e^{+i\Delta\beta z} \tag{3.85}$$

Taking a derivative of (3.84) and (3.85) with respect to z, respectively, we have [4]

$$R'(z) + i\Delta\beta R(z) = \kappa^* S(z) \tag{3.86}$$

$$S'(z) - i\Delta\beta S(z) = \kappa R(z) \tag{3.87}$$

Where, from (3.75),

$$\kappa = \frac{i\omega\varepsilon_0 a_l}{4}\int_{-\infty}^{\infty}\int_{-\infty}^{\infty}\Delta n^2[\varepsilon_y^{(s)}(x,y)]^2 dxdy \tag{3.88}$$

---

[4] proof of (3.86) and (3.87):
First, from (3.84),

$$A_s'^{(-)}(z) = S'(z)e^{-i\Delta\beta z} - i\Delta\beta S(z)e^{-i\Delta\beta z} = \kappa R(z)e^{i\Delta\beta z}e^{-i2\Delta\beta z}$$

or

$$S'(z) - i\Delta\beta S(z) = \kappa R(z)$$

Now, from (3.85),

$$A_s'^{(+)}(z) = R'(z)e^{+i\Delta\beta z} + i\Delta\beta R(z)e^{+i\Delta\beta z} = \kappa^* S(z)e^{-i\Delta\beta z}e^{i2\Delta\beta z}$$

or

$$R'(z) + i\Delta\beta R(z) = \kappa^* S(z)$$

Now we set up a new κ_{new} as

$$\kappa_{new} = -i\kappa = \frac{\omega\varepsilon_0 a_l}{4} \int_{-\infty}^{\infty} \int_{-\infty}^{\infty} \Delta n^2 [\varepsilon_y^{(s)}(x,y)]^2 dxdy \qquad (3.89)$$

the coupling equations (3.86) and (3.87) may be rewritten as

$$R'(z) + i\Delta\beta R(z) = -i\kappa_{new}S(z) \qquad (3.90)$$

$$S'(z) - i\Delta\beta S(z) = i\kappa_{new}R(z) \qquad (3.91)$$

Take a derivative of (3.90) with respect to z, we have

$$R''(z) + i\Delta\beta R'(z) = -i\kappa_{new}S'(z) \qquad (3.92)$$

Considering (3.90), (3.91), we have

$$R''(z) + (\Delta\beta^2 - \kappa_{new}^2)R(z) = 0 \qquad (3.93)$$

Set

$$\gamma^2 = \kappa_{new}^2 - \Delta\beta^2 \qquad (3.94)$$

we have

$$R''(z) - \gamma^2 R(z) = 0 \qquad (3.95)$$

The solution of (3.95) is

$$R(z) = A\cosh\gamma z + B\sinh\gamma z \qquad (3.96)$$

Similarly, we have [5]

$$S''(z) - \gamma^2 S(z) = 0 \qquad (3.97)$$

And the solution of (3.97) is

$$S(z) = C\cosh\gamma z + D\sinh\gamma z \qquad (3.98)$$

Boundary conditions

---

[5]

Take a derivative of (3.91) with respect to z, we have

$$S''(z) - i\Delta\beta S'(z) = i\kappa_{new}R'(z)$$

Insert (3.90) and (3.91) into (3.97), we have

$$S''(z) + (\Delta\beta^2 - \kappa_{new}^2)S(z) = 0$$

or

$$S''(z) - \gamma^2 S(z) = 0$$

The boundary conditions at $z = 0$, is, from (3.96) (3.98),

$$S(0) = C, \quad R(0) = A \tag{3.99}$$

Where $z = 0$ is the start point of the FBG. Thus, we have, from (3.96) (3.98)

$$R(z) = R(0) \cosh \gamma z + B \sinh \gamma z \tag{3.100}$$

$$S(z) = S(0) \cosh \gamma z + D \sinh \gamma z \tag{3.101}$$

Insert (3.100) (3.101) into (3.90), we have

$$\gamma[R(0) \sinh \gamma z + B \cosh \gamma z] + i\Delta\beta[R(0) \cosh \gamma z + B \sinh \gamma z]$$

$$= -i\kappa_{new}[S(0) \cosh \gamma z + D \sinh \gamma z] \tag{3.102}$$

(3.102) at $z = 0$ gives

$$\gamma B + i\Delta\beta R(0) = -i\kappa_{new}S(0) \tag{3.103}$$

Thus

$$B = -\frac{i\kappa_{new}}{\gamma}S(0) - \frac{i\Delta\beta}{\gamma}R(0) \tag{3.104}$$

Insertion of (3.100) (3.101) into (3.91) produces

$$\gamma[S(0) \sinh \gamma z + D \cosh \gamma z] - i\Delta\beta[S(0) \cosh \gamma z + D \sinh \gamma z]$$

$$= -\kappa_{new}[R(0) \cosh \gamma z + B \sinh \gamma z] \tag{3.105}$$

(3.105) at $z = 0$ gives us

$$\gamma D - i\Delta\beta S(0) = i\kappa_{new}R(0)$$

Thus

$$D = \frac{i\kappa_{new}}{\gamma}R(0) + \frac{i\Delta\beta}{\gamma}S(0) \tag{3.106}$$

Insert (3.104) (3.106) into (3.100) (3.101), we have

$$R(z) = R(0) \cosh \gamma z + \left[-\frac{i\kappa_{new}}{\gamma}S(0) - i\frac{i\Delta\beta}{\gamma}R(0) \right] \sinh \gamma z$$

$$S(z) = S(0) \cosh \gamma z + \left[\frac{i\kappa_{new}}{\gamma}R(0) + \frac{i\Delta\beta}{\gamma}S(0) \right] \sinh \gamma z$$

or

$$R(z) = \left[\cosh \gamma z - \frac{i\Delta\beta}{\gamma} \sinh \gamma z \right] R(0) - \frac{i\kappa_{new}}{\gamma} \sinh \gamma z S(0) \tag{3.107}$$

$$S(z) = \left[\cosh \gamma z + \frac{i\Delta\beta}{\gamma} \sinh \gamma z \right] S(0) + \frac{i\kappa_{new}}{\gamma} \sinh \gamma z R(0) \tag{3.108}$$

Find the reflectivity of the FBG at $z = 0$

For the sake of simplification, from now on we change κ_{new} into κ. Where, $S(z)$ is the back wave in BFG, $R(z)$ is the forward wave in FBG. Thus, the reflectivity of FBG is

$$r_{pert} = \frac{S(0)}{R(0)} \qquad (3.109)$$

Using the boundary condition at the end of FBG $z = L_g$, we have, from (3.107) (3.108),

$$R(L) = \left[\cosh \gamma L_g - \frac{i\Delta\beta}{\gamma} \sinh \gamma L_g \right] R(0) - \frac{i\kappa}{\gamma} \sinh \gamma L_g S(0) \qquad (3.110)$$

$$S(L) = \left[\frac{i\kappa}{\gamma} \sinh \gamma L_g \right] R(0) + \left[\cosh \gamma L_g + \frac{i\Delta\beta}{\gamma} \sinh \gamma L_g \right] S(0) \qquad (3.111)$$

Now, the real situation is that the incident wave is the forward wave $R(z)$, the back wave caused by FBG is $S(z)$. And no incident wave from the right side of FBG. In this case, $S(L_g) = 0$. And the reflectivity of the FBG (3.110) is, from (3.111)

$$r_{pert} = \frac{S(0)}{R(0)} = \frac{-\dfrac{i\kappa}{\gamma} \sinh \gamma L_g}{\cosh \gamma L_g + \dfrac{i\Delta\beta}{\gamma} \sinh \gamma L_g} \approx \frac{i\kappa L_g}{1 + i\Delta\beta L_g} \qquad (3.112)$$

In which, two conditions have been considered:

a. $\kappa \to \Delta\beta$, and then $\gamma \to 0$.

b. $\gamma L_g \to 0$, and then $\sinh \gamma L_g \to \gamma L_g$ Thus, the power reflectivity is, from (3.112),

$$R_{pert} = |r_{pert}|^2 = \frac{\left(-\dfrac{i\kappa}{\gamma} \right)\left(-\dfrac{i\kappa}{\gamma} \right) \sinh^2 \gamma L_g}{\cosh^2 \gamma L_g + \left(\dfrac{\Delta\beta}{\gamma} \right)^2 \sinh^2 \gamma L_g}, \qquad (3.113)$$

More in detail:

* For $\kappa < \Delta\beta, \gamma L_g \to 0$,

$$R_{pert} = |r_{pert}|^2 = \frac{(\kappa L_g)^2 \sin^2 \sqrt{(\Delta\beta L_g) - (\kappa L_g)^2}}{(\Delta\beta)^2 - (\kappa L_g)^2}, \quad For \quad \kappa < \Delta\beta, \gamma L_g \to 0 \qquad (3.114)$$

or [6]

$$P_{pert} = |r_{pert}|^2 \approx \frac{(\kappa L_g)^2 \sin^2 \sqrt{(\Delta\beta L_g)^2 - (\kappa L_g)^2}}{(\Delta\beta L_g)^2 - (\kappa L_g)^2}, \quad For \quad \kappa < \Delta\beta, \gamma L_g \to 0$$

(3.115)

* For $\kappa > \Delta\beta$, $\gamma L_g \to 0$, we have $\gamma = \sqrt{\kappa^2 - (\Delta\beta)^2}$, and then, from (3.113),

$$P_{pert} = |r_{pert}|^2 = \frac{\kappa^2 \sinh^2 \gamma L_g}{\gamma^2 \cosh^2 \gamma L_g + (\Delta\beta)^2 \sinh^2 \gamma L_g}$$

(3.116)

In which, the denominator is

$$
\begin{aligned}
\gamma^2 \cosh^2 \gamma L_g + (\Delta\beta)^2 \sinh^2 \gamma L_g &= (\kappa^2 - \Delta\beta^2) \cosh^2 \gamma L_g + (\Delta\beta)^2) \sinh^2 \gamma L_g \\
&= \kappa^2 \cosh^2 \gamma L_g - (\Delta\beta)^2 (\cosh^2 \gamma L_g - \sinh^2 \gamma L_g) \\
&= \kappa^2 \cosh^2 \gamma L_g - (\Delta\beta)^2 \\
&= \kappa^2 \cosh^2 \gamma L_g, \quad Since \quad \kappa > \Delta\beta
\end{aligned}
$$

Thus,

$$P_{pert} = |r_{pert}|^2 \approx \tanh^2 \gamma L_g = \tanh^2 \sqrt{(\kappa L_g)^2 - (\Delta\beta L_g)^2}, \quad For \quad \kappa > \Delta\beta, \gamma L_g \to 0$$

(3.117)

In summery, from (3.115), (3.117), the power reflectivity of FBG is

$$P_{pert} = |r_{pert}|^2 \approx \begin{cases} \dfrac{(\kappa L_g)^2 \sin^2 \sqrt{(\Delta\beta L_g)^2 - (\kappa L_g)^2}}{(\Delta\beta L_g)^2 - (\kappa L_g)^2} & : \quad For \quad \Delta\beta > \kappa, \gamma L_g \to 0 \\[4mm] \quad\quad\quad\quad : \\[2mm] \quad\quad\quad\quad : \\[2mm] \tanh^2 \sqrt{(\kappa L_g)^2 - (\Delta\beta L_g)^2} & : \quad For \quad \Delta\beta < \kappa, \gamma L_g \to 0 \end{cases}$$

(3.118)

---

[6]Proof of (3.115):

$$P_{pert} \approx \frac{\left(\frac{\kappa}{\gamma}\right)^2 \sinh^2 \gamma L_g}{1 + (\Delta\beta)^2} \approx \left(\frac{\kappa}{\gamma}\right)^2 \frac{\sinh^2 \sqrt{(\kappa L_g)^2 - (\Delta\beta L_g)^2}}{1}, \quad Since \quad 1 >> (\Delta\beta L)^2$$

$$= \frac{\kappa^2 \left\{ sinh[i\sqrt{(\Delta\beta L_g)^2 - (\kappa L_g)^2}] \right\}^2}{\kappa^2 - \Delta\beta^2}$$

$$= \frac{(\kappa L_g)^2 \left\{ i \sin \sqrt{(\Delta\beta L_g)^2 - (\kappa L_g)^2} \right\}}{(\kappa L_g)^2 - (\Delta\beta L_g)^2}$$

since $\sinh i\theta = i \sin\theta$.
Thus,

$$R_{pert} = |r_{pert}|^2 \approx \frac{(\kappa L_g)^2 \left\{ \sin \sqrt{(\Delta\beta L_g)^2 - (\kappa L_g)^2} \right\}}{(\Delta\beta L_g)^2 - (\kappa L_g)^2}$$

and the reflectivity of FBG is

$$r_{pert} \approx \begin{cases} \dfrac{\kappa L_g \sin \sqrt{(\Delta\beta L_g)^2 - (\kappa L_g)^2}}{\sqrt{(\Delta\beta L_g)^2 - (\kappa L_g)^2}} & : \quad For \quad \Delta\beta > \kappa, \gamma L_g \to 0 \\[1em] & : \\ & : \\ \tanh \sqrt{(\kappa L_g)^2 - (\Delta\beta L_g)^2} & : \quad For \quad \Delta\beta < \kappa, \gamma L_g \to 0 \end{cases} \qquad (3.119)$$

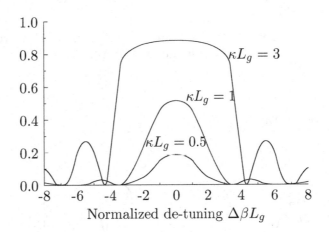

Figure 3-18   Reflectivity r_{per} vs normalized de-tuning $\Delta\beta L_g$

The calculation results from (3.119) is shown in Figure 3-18. In which,

† $|r_{per}|$ is increasing with the increasing of κL_g.
† $|r_{per}|$ is decreasing with the increasing of $\Delta\beta$.
† The zero point of $|r_{per}|$ is at, from (3.119),

$$\gamma L_g = N\pi \qquad (3.120)$$

Consider that $\gamma = \sqrt{\kappa^2 - \Delta\beta^2}$, we have, from (3.120), the zero points of r_{pert} is at

$$\Delta\beta L_g = \sqrt{(\kappa L_g)^2 + (N\pi)^2}, \quad N = 1, 2, 3, \cdot \qquad (3.121)$$

as shown in Figure 3-18.

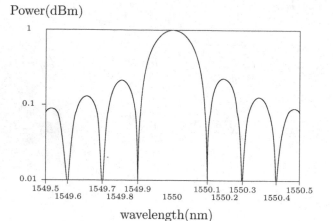

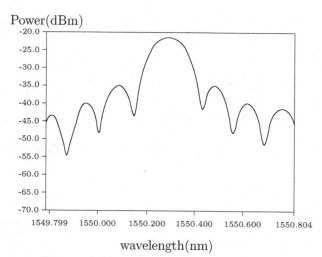

Figure 3-19   A 6-mm Bragg grating (15%) reflection
calculated from (3.122) [35]

Figure 3-20   A 6-mm Bragg grating (15%) reflection
measured with interrogator sm125 from Micron Optics [36]

In fact, from (3.119), the function of the reflectivity r_{pert} is a function of $\sin x/x$. A paper "Strain Measurement with Fiber Bragg Grating Sensors" by Manfred Kreuzer, in HBM, Darmstadt, Germany also show us that the function of the reflectivity r_{pert} is a function of $\sin x/x$ as follows

$$r_{pert} = \frac{\lambda_0}{\pi \cdot N \cdot \Delta\lambda_i} \cdot \sin\left(\frac{\pi \cdot N \cdot \Delta\lambda_i}{\lambda_0}\right) \qquad (3.122)$$

Where

$$N = \frac{2 \cdot n \cdot Gl}{\lambda_0} \qquad (3.123)$$

$Gl = L_g$ is the length of the grating, n is the refraction index of the fiber, λ_0 is the peak wavelength of the FBG in free space. The wavelength λ_i of the x-axis is $\lambda_i = \lambda_0 + \Delta\lambda_i$. $\Delta\lambda_i$ runs over a range of 1000pm in Figure 3-19 and Figure 3-20. Where Figure 3-19 is the calculation results from (3.122), and Figure 3-20 shows a real measured spectrum of a 6mm FBG. The author said: "Both of them show almost identical characteristics. Only the deep poles in the calculated plot are different. The reason for this is that the interrogator obviously can not follow the very fast signal slops of the poles".

The reflection bandwidth of the FBG

From Figure 3-20, the peak point of reflectivity r_{pert} in FBG is at the central frequency ω_0 (or the central wavelength λ_B), which corresponds to $\Delta\beta = 0$. Now the reflectivity bandwidth corresponds to the points that r_{pert} drops down to 0.707 (or 3dB) of the peak point. These points can be obtained by, from (3.116),

$$\cosh^2\gamma L_g + \left(\frac{\Delta\beta}{\gamma}\right)^2 \sinh\gamma L_g = 2 \qquad (3.124)$$

since the peak point corresponds to $\Delta\beta = 0$, and then $\cosh^2\gamma L + \left(\frac{\Delta\beta}{\gamma}\right)^2 \sinh\gamma L = 1$ at peak point. From (3.124), we may obtain the bandwidth of the reflectivity r_{pert} of the FBG as follows [7]

$$\Delta\lambda = 2\Delta\lambda_r = \frac{2\lambda_B^2\kappa}{\pi n_{eff}} \qquad (3.125)$$

Where n_{eff} is the effective refractive index of the normal mode LP in single mode fiber. And the bandwidth $\Delta\lambda$ is proportional to κ. Where from (3.77),

$$\kappa = \frac{-\omega\varepsilon_0}{4\pi l} \int_{-\infty}^{\infty} \int_{-\infty}^{\infty} \Delta n^2 [\varepsilon_y^{(s)}(x,y)]^2 dxdy = \Delta n^2\eta \qquad (3.126)$$

$\Delta n = n_3 - n_2$ is the refractive index different in FBG, as shown in Figure 3-17, and

$$\eta = \frac{-\omega\varepsilon_0}{4\pi l} \int_{-\infty}^{\infty} \int_{-\infty}^{\infty} [\varepsilon_y^{(s)}(x,y)]^2 dxdy \qquad (3.127)$$

Conclusion

---

[7]The reflection bandwidth of the FBG obtained by Manfred Kreuzer, in HBM, Darmstadt, Germany is

$$\Delta\lambda = FWHM = \frac{1.8955 \cdot \lambda_0^2}{\pi \cdot n \cdot Gl}$$

Where $\lambda_0 = \lambda_B$ is the central wavelength of the FBG spectral, $Gl = L_g$ is the grating length, $n = n_{eff}$ is the refractive index of the single mode fiber.

The analysis of the FBG come to a conclusion as follows:

As shown in Figure 3-14, the function of the FBG is to reflect the wavelength $\lambda_B(=\lambda_2)$ from multiple wavelengths $(\lambda_1,\lambda_2,\lambda_3,\lambda_4)$ input. In other words, the function of the FBG is a **back wave filter**, which reflect the central wavelength and let others go through.

The central wavelength of the FBG, λ_B is obtained from the phase-matching condition, namely, from (3.81),

$$\lambda_B = 2\wedge \tag{3.128}$$

and the reflectivity spectral is shown in Figure 3-20. The bandwidth of the reflectivity r_{pert} is

$$\Delta\lambda = \frac{2\lambda_B^2 \Delta n^2 \eta}{\pi n_{eff}} \tag{3.129}$$

The pole points of the reflectivity spectral in Figure 3-20 is at, from (3.121),

$$\Delta\beta L_g = \sqrt{(\Delta n^2 \eta L)^2 + (N\pi)^2}, \quad N = 1,2,3,\cdots \tag{3.130}$$

Where $\kappa = \Delta n^2 \eta$ has been considered.

3.2.3 Cascaded FBG filter with index change

The optical add/drop multiplexer (OADM) with FBG

Based on the conclusion, the Fiber Bragg Grating (FBG) can be used as the optical De-multiplexer, optical multiplexer and optical Add/Drop in WDM optical fiber communications system.

A de-multiplexer can be achieved by cascading multiple drop sections of the optical add/drop multiplexer (OADM) (Figure 3-15) as shown in Figure 3-21.

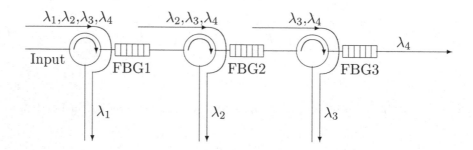

Figure 3-21   Optical De-multiplexer by using FBG and circulators

In which, the input is four channels $(\lambda_1 + \lambda_2 + \lambda_3 + \lambda_4)$ combined.

The FBG1 is used to reflect wavelength λ_1 to drop down and let the others $(\lambda_2, \lambda_3, \lambda_4)$ to go through.

The FBG2 is used to reflect the wavelength λ_2 to drop down, and let the others to go through.

The FBG3 is used to reflect λ_3 to drop down, and let λ_4 to go through.

And then the one input $(\lambda_1, \lambda_2, \lambda_3, \lambda_4)$ has been De-multiplexed into four outputs.

Which can be down by the phase-matching design as follows:

$$\wedge_1 = 2\lambda_{B1},$$

$$\wedge_2 = 2\lambda_{B2},$$

$$\wedge_3 = 2\lambda_{B3},$$

$$\wedge_4 = 2\lambda_{B4},$$

Where, λ_{B1}, λ_{B2}, λ_{B3} are the central wavelength of reflectivity for FBG1, FBG2, FBG3, respectively. And $\lambda_{B1} = \lambda_1$, $\lambda_{B2} = \lambda_2$, $\lambda_{B3} = \lambda_3$.

The grating length should be long enough, for instance $L = 6mm$, so as to obtain a large ratio between the central peak and the side peak as shown in Figure 3-20.

Now, how to make a multiplexer by using the FBG and the circulators?

Figure 3-22 is one option for the multiplexer by using FBG. In which, the inputs is four channels λ_1, λ_2, λ_3, λ_4 independently.

The FBG1 is used to reflect wavelength λ_1 to go to output and let the other (λ_4) to go through.

The FBG2 is used to reflect the wavelength λ_2 to go to output, and let the others (λ_1, λ_4) to go through.

The FBG3 is used to reflect λ_3 to go to output, and let others $(\lambda_1, \lambda_2, \lambda_4)$ to go through.

And then the four independent inputs λ_1, λ_2, λ_3, λ_4 has been multiplexed to one output $(\lambda_1 + \lambda_2 + \lambda_3 + \lambda_4)$.

Which can be down by the phase-matching design of FBG:

$$\wedge_n = 2\lambda_n, \quad n = 1, 2, 3 \tag{3.131}$$

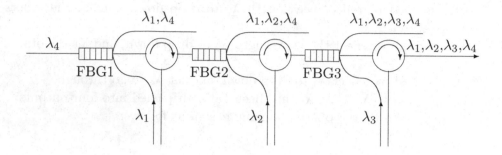

Figure 3-22   Optical multiplexer by using FBG and circulators

The De-multiplexer and multiplexer by using FBG can be used to a long haul DWDM optical fiber telecommunications system as illustrated in Figure 3-23.

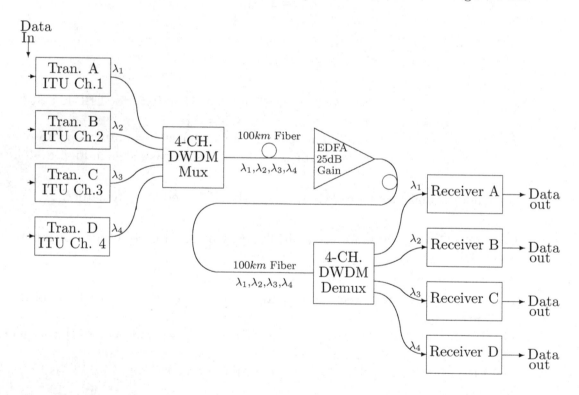

Figure 3-23   DWDM optical fiber telecommunications system

Where,

    a. A multiplexer (as shown in Figure 3-22) is used to multiplexer 4 independent channels, λ_1, λ_2, λ_3, and λ_4 into a combined channel $\lambda_1 + \lambda_2 + \lambda_3 + \lambda_4$.

b. The long-haul optical fiber transmissions the combined channel $\lambda_1 + \lambda_2 + \lambda_3 + \Lambda_4$ to the input of a de-multiplexer.

c. The de-multiplexer (as shown in Figure 3-21) separate the combined channel $\lambda_1 + \lambda_2 + \lambda_3 + \lambda_4$ into 4 independent channels λ_1, λ_2, λ_3, and λ_4.

Thus, the 4 channels λ_1, λ_2, λ_3, and λ_4 transmissions independently from one city to another city without considering the protocol and rate different issue.

EDFA is an optical Er-Doped-Fiber-Amplifier. Which is an optical amplifier with using Er-Doped single mode fiber, which will be discussed in section 3.5.2. The input and output of EDFA are both single mode fibers. Thus, it is convenient to be connected to the optical multiplexer and optical de-multiplexer.

For a site needed to adding/dropping some channels, we may use the Optical Add/Drop with using FBG as shown in Figure 3-24.

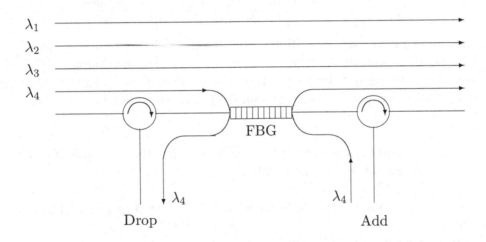

Figure 3-24   Optical Add/Drop with using FBG

Where,

a. A channel λ_4 is dropped down by the reflection of a FBG.

b. A channel λ_4 is added up by the reflection of the same FBG.

c. The channels λ_1, λ_2, λ_3 go through the FBG independently.

DWDM

DWDM is the Density-Wavelength-Division-Multiple system. In which, the FBG can be used for the DWDM multiplexer and the DWDM De-multiplexer. However, the FBG is not the normal FBG, but a special FBG:

In normal FBG, the index change $\Delta n = n2 - n3$ is uniform along all grating length L and the period of the FBG $\wedge$ is a constant. The spectral of these FBGs is shown in Figure 3-19 and Figure 3-20. Which can not meet the requirement in the DWDM system.

In special FBG, the structure of the FBG can vary via *the index change* $\Delta n(z) = n2(z) - n3$ and *the period change* $\wedge(z)$ along the grating length. These FBGs can be designed to meet two specifications for the DWDM system:

a. A narrow flat response at central frequency.

b. The Side-lobe low down to $-30dB$ compared with the peak point at central frequency.

Where, the peal width is measured at the T_f transmission level below the transmission maximum, in conventional, this T_f transmission level is $3dB$. However, the ratio between the peak amplitude at the central wavelength and the peak amplitude at the side band T_f should be at least $30dB$ for the DWDM system.

Here, T_f is defined as

$$T_f = \frac{Peak \quad amplitude \quad at \quad central \quad wavelength}{Peak \quad amplitude \quad at \quad the \quad side \quad band} \qquad (3.132)$$

A DWDM optical communication system requires channel adding/dropping filters that can select one out of every N_{ch} channels. And then the channel adding/dropping filter is to reject undesired channels with a cross-talk level of T_f, say, $30dB$ at least. This is beyond the capability of the uniform FBG as shown in Figure 3-17.

To meet the requirement of the filter, we have to use the special FBG to change the refractive index $\Delta n = n_2 - n_1$ in FBG.

There are six common structures for FBG [37] as shown in Figure 3-25 and Figure 3-26:

a. Uniform positive-only index change.

b. Gaussian index change.

c. Raiaed-cosine index change.

d. Chirped index change.

f. Discrete phase shift.

g. Superstructure index change.

In fact, there are basically three quantities that control the properties of the FBG:

a. The period of FBG $\wedge$.

b. The grating length L_G, given as

$$L_G = N\wedge \qquad (3.133)$$

c. The grating strength Δn.

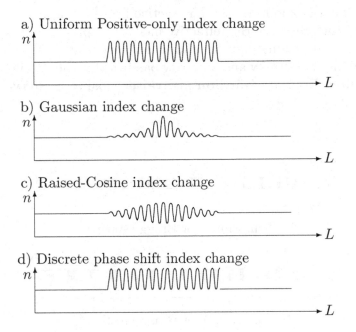

a) Uniform Positive-only index change

b) Gaussian index change

c) Raised-Cosine index change

d) Discrete phase shift index change

Figure 3-25   Refractive index profile in the core

However, there are four properties need to be controlled:

a. Central wavelength λ_B.

b. Reflectivity or the peak value at the central wavelength.

c. Bandwidth $\Delta\lambda$.

d. Side-lobe strength.

The central wavelength λ_B is mainly controlled by the period of the FBG $\wedge$ as, from (3.128),

$$\lambda_B = 2\wedge \tag{3.134}$$

This is the phase-matching condition of FBG, and then it is the constructive interference condition in FBG.

As shown in Figure 3-18, the peak value at central wavelength can be controlled by the grating length L_G and the grating strength Δn. This means that the grating length L_G and the grating strength Δn can be used to set the peak reflectivity at the central wavelength.

The bandwidth of the reflectivity r_{pert}, from (3.130),

$$\Delta\lambda = \frac{2\lambda_B^2 \Delta n^2 \eta}{\pi n_{eff}} \tag{3.135}$$

can be controlled by the grating strength Δn. This means that the grating strength Δn can be used to set the bandwidth.

Now we need a fourth quantity that can be varied to help with the side-lobe suppression. That is apodized of the refractive index change. The term "apodization" refers to the grating of the refractive index to approach zero at the end of the grating. Apodized grating offer significant improvement in side-lobe suppression while maintaining reflectivity and a narrow bandwidth. The two functions typically used to apodized FBG are **Gansion** and **raised-cosine** as shown in Figure 3-25 and Figure 3-26.

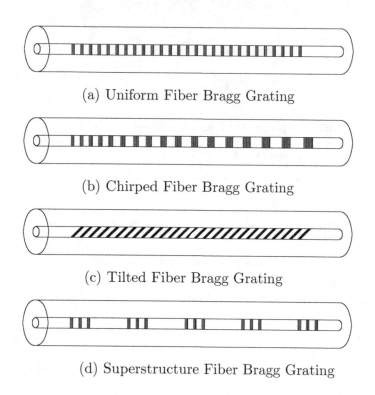

(a) Uniform Fiber Bragg Grating

(b) Chirped Fiber Bragg Grating

(c) Tilted Fiber Bragg Grating

(d) Superstructure Fiber Bragg Grating

Figure 3-26   Structure of the refractive index changes in a uniform FBG

The refractive index profile of the grating may be modified to add other feather as shown in Figure 3-25 and Figure 3-26, such as

- **Chirped fiber Bragg grating**:

In which, the refractive index profile of the grating is a linear variation in the grating period, called a chirp. Recall the phase-matching condition (3.134), the central wavelength λ_B changing with the grating period $\wedge$ will broadening the reflected spectrum. Which means that each period of the grating will form a constructive interference at different wavelength or each period of the grating will form a different central wavelength. The sum of the spectrums will be broadened.

- **Tiled fiber Bragg grating:**

In a tiled FBG (TFBG), the variation of the refractivive index is at an angle to the optical axis. The angle of tile in a TFBG has an effect on the reflected wavelength and the bandwidth.

- **Long-period gratings:**

Typically the grating period $\wedge$ is the same size as the central wavelength λ_B as given in (3.134). For a grating that reflects at $1500nm$, the grating period is $500nm$, using refractive index of 1.5. Longer periods can be used to achieve much broader response than the possible with a standard FBG. These grating are called the long-period fiber grating. They typically have grating period on the order of 100 micrometers, to a millimeter, and are therefore much easier to manufacture.

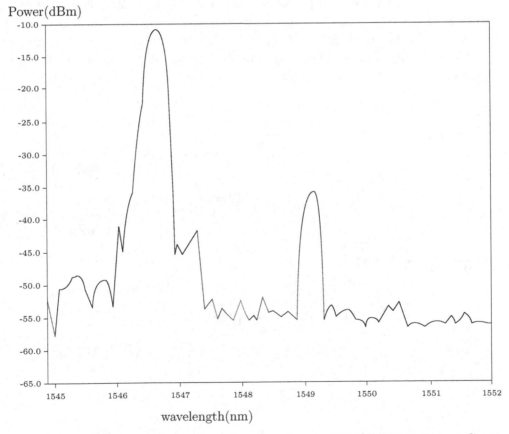

Figure 3-27   Reflection peak of a refractive change (0.01%) Fiber Bragg Grating [38]

Now, we would like to show you an experimental response of the FBG by Manfred Kreuzer, HBM, Department, Germany as shown in Figure 3-27. [38] In which, the Side-lobe low down to $-25dB$ compared with the peak point at central wavelength, and this side-lobe level is reached by the grating length L_G and the high grating change $(n_2 - n_1)/n_1 = 1\%$ only. If we use the special optical fiber Bragg grating such as Gaussian or Raised-Cosine index change (see Figure 3-25) in it, the side-lobe low down to $-30dB$ is not so difficult to be obtained.

Different coating of diffractive structure are used for fiber Bragg gratings in order to reduce the mechanical impact on the Bragg wavelength shift for $1.1 - 1.5$ times as compared to an un-coated waveguide.

3.3 Optical (de)multiplexer

3.3.1 Optical (de)multiplexer using Mash-Zhnder filter

The Mash-Zhnder filter may also be cascaded to construct a multilevel filter used for an optical de-multiplexer as shown in Figure 3-27. For example, eight wavelength, λ_1 to λ_8, are separated by one filter into two groups, λ_1, λ_3, λ_5, λ_7 and λ_2 λ_4, λ_6 λ_8. Each group is separated by two filters into four subgroups (λ_1, λ_5), (λ_3, λ_7), (λ_2, λ_6), and (λ_4, λ_8), and so on, until all the wavelengths have been separated.

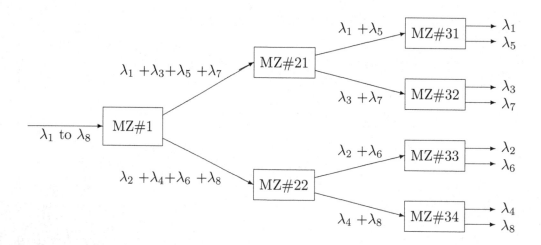

Figure 3-28   Optical De-multiplexer (Mach-Zehnder filter chain) [39]

3.3.2 Array Waveguide Grating (AWG)

Introduction

Array Waveguide Grating (AWG) multiplexers/de-multiplexers are planar device which are based on an array of waveguides with both imaging and dispersive poperties. They image the field in an input waveguide onto an array of output waveguides in such a way that the different wavelegnth signal present in the input waveguide are imaged onto different output waveguide.

AWGs were first reported by Smit [40](1988) and later by Takahashi [41] (1990) and Dragone [42], [43](1991). They are known under different names: Phased Array, Arrayed Waveguide Gratings (AWGs), and Waveguide Grating Routers (WGRs). The acronym AWG, introduced by Takahashi [41] is the most frequently used name today and will also be used in the text. Together with the Thin-Film Filters and Fiber Bragg Grating, AWGs are the most important filter type applied in WDM networks, and with the advance of Photonic Integrated Circuits technology they are expected to become the most important one.

The most important technology used for realization of AWGs today are silica-on-silicon technology and Indiumphosphide (In-P)-based semiconductor technology. In addition, reach on silicon-based polymer technology, and on lithium niobate have been reported as well [43], [44], [45].

Silica-on-silicon (SoS) AWGs have been introduced to the market in 1994 and currently hold the largest share of the AWG market. Their modal field matches well with that if a fiber, and therefore it is relatively easy to couple them to fibers. They combine low propagation loss ($< 0.05dB/cm$) with a high fiber-coupling efficiency (losses in the order of $0.1dB$). A disadvantage is that they are relatively large due to their fiber matched waveguide properties, which prohibit the use of short bands. This is presently being improved by using higher index contrast and spot-size converters to keep fiber coupling losses low.

AWGs can be utilized to accomplish complex functions in fiber optic WDM networks. They are also increasingly used in other areas such as signal processing, measurement, and characterization or sensing as well. Integration of AWGs enables compact, high functionality devices.

Semiconductor-based devices have the potential to integrate a wide variety of functions on single chip; they are suitable for integration of passive device such as AWGs with active ones such as electro-optical switches, modulators, and optical amplifiers, and also non-linear devices such as wavelength converters. The dominant technology for operation in the telecom window is based on InP. InP-based AWGs can be very compact due to the large index-contract of InP-based waveguides.

Principle

Figure 3-29 and Figure 3-30 shows the schematic layout of an AWG de-multiplexer. When a beam propagating through the transmitter waveguide enters, the first Free

Propagation Region (FPR) is no longer laterally confined but becomes divergent. On arriving at the input aperture the beam is coupled into the waveguide array and propagates through the individual waveguides towards the output aperture.

The length of the array waveguides is chosen such that the optical path length difference between adjacent waveguides equals an integer multiple of the central wavelength λ_c of the de-multiplexer. For λ_c, the field in the individual waveguides arrive at the output aperture with equal phase (mod. 2π), and thus the field distribution at the input aperture is reproduced at the output aperture.

The divergent beam at the input aperture is thus transformed into a convergent one with equal amplitude and phase distribution. And the input field at object plane gives rise to a corresponding image at the center of the image plane. The spatial separation of different wavelengths is obtained by linearly increasing the lengths of the array waveguides, which introduces a wavelength-dependant tilt of the outgoing beam associated with a shift of the focal point along the image plane. If receiver waveguide are placed at proper positions along the image plane, different wavelengths are led to different output ports.

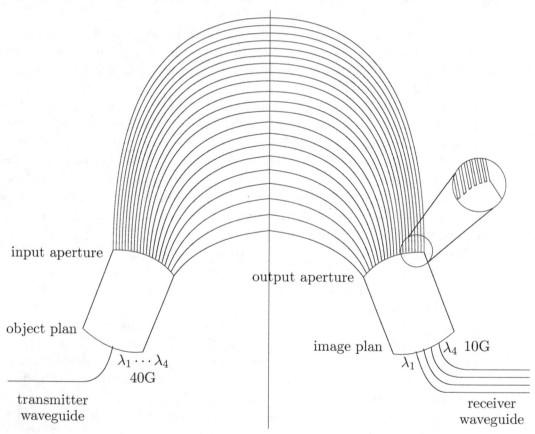

Figure 3-29   AWG (Arrayed waveguide gratings) de-multiplexer [46]

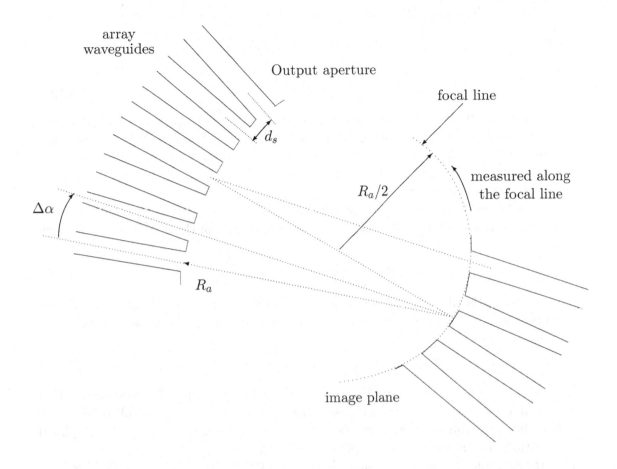

Figure 3-30   Beam focusing geometry in the free propagation region [47]

Focusing, Spatial Dispersion, and Free Spectral Range

Focusing of the fields propagating in an AWG is obtained if the length different ΔL between adjacent waveguides is equal to an integer number m of wavelengths inside the AWG:

$$\Delta L = m \cdot \frac{\lambda_c}{n_{eff}} \tag{3.136}$$

The integer m is called the order of the array, λ_c is the central wavelength (in vacuo) of the AWG, n_{eff} is the effective refractive index of the guide mode, and λ_c/n_{eff} corresponds to the wavelength inside the array waveguide. *Under these circumstances the array behave like a lens with image and object planes at distance R_a of the array aperture* as shown in Figure 3-30. On the other hand, the focal line (which defines the image plane) follows a circle with radius $R_a/2$, and transmitter and receiver waveguides should be located on this line. This geometry is equivalent to a Rowland-type mounting.

The length increment ΔL of the array gives rise to a phase difference according to

$$\Delta\phi = \beta\Delta L \tag{3.137}$$

Where

$$\beta = 2\pi\nu n_{eff}/c \tag{3.138}$$

is the propagation constant in the waveguides, $\nu = c/\lambda$ is the frequency of the propagating wave, and c is the speed of the light in vacuo. The wavelength-dependent phase difference $\Delta\phi$ introduce the wavelength-dependent tilt of the outgoing wave front associated with a wavelength-dependent shift of the corresponding image (as already mentioned above).

The lateral displacement d_s of the focal spot along the image plane per unit frequency change $d\nu$ is called the spatial dispersion D_{sp} of the AWG, which is given by

$$D_{sp} = \frac{ds}{d\nu} = \frac{1}{\nu_c} \cdot \frac{n_g}{n_{FPR}} \cdot \frac{\Delta L}{\Delta\alpha} \tag{3.139}$$

Where n_{FPR} is the (slab) mode index in the Free Propagation Region, $\Delta\alpha$ is the divergence angle between the array waveguides, and n_g is the group index of the waveguide mode:

$$n_g = n_{eff} + \nu\frac{dn_{eff}}{d\nu} \tag{3.140}$$

According to (3.136) and (3.139) the (spatial) dispersion is fully determined by the order m and the divergence angle $\Delta\alpha$, and as a consequence filling-in of the space between the array waveguides near the apertures due to finite lithographical resolution does not affect the dispersion properties of the AWG.

If the input wavelength change is such that the phase difference $\Delta\phi$ between adjacent waveguides (3.137) has increased by 2π, the transfer will be the same as before, i.e. the response of the AWG is periodic. The period in the frequency domain is called the Free Spectral Range (FSR) as shown in Figure 3-31 and Figure 3-32 shows order m at the central wavelength. $\Delta\beta\Delta L = 2\pi$ in combination with (3.136) leads to the Free Spectral Range

$$FSR = \frac{\nu_c}{m}\left\{\frac{n_{eff}}{n_g}\right\} \tag{3.141}$$

In order to avoid crosstalk problems with adjacent orders, the free spectral range should be larger than the whole frequency spanned by all channels, and thus for a de-multiplexer with 8 channels and $200GHz$ channel spacing, for example, the FSR should be at least $1600GHz$, where channels are centred around $1500nm$, (3.141) yields that this requires an array with an order of about 120 (with $n_{eff}/n_g \approx 0.975$ for the SoS- and $n_{eff}/n_g \approx 0.9$ for the InP-material system).

If the device is to be used in combination with Erbium-Doped Fiber Amplifier

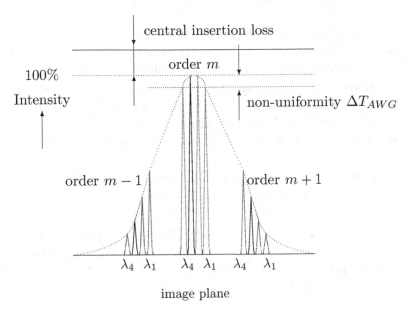

Figure 3-31   The field in image plane for different wavelength [48]

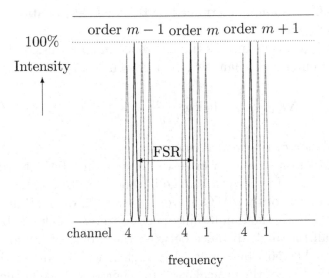

Figure 3-32   The corresponding frequency curve for different channels [49]

(EDFAs), the FSR should be chosen such that adjacent orders do not coincide with the peak of the EDFA gain spectrum in order to avoid accumulation of amplified Spontaneous Emission (ASE).

Insertion loss and non-uniformity

Fields propagating through an AWG are attenuated due to various loss mechanisms. The most important contribution to this loss is coming from the junctions between the free propagation regions and the waveguide array. For low-loss, the fan-in and fan-out sections should be a sooth transition from the guided propagation in the array to the free-space propagation in the FPRs (Free Propagation Regions) and vice versa. This will occur if the divergence angle $\Delta\alpha$ between the array waveguides is sufficiently small and the vertex between the waveguides is sufficiently sharp. Where:

a. Due to the finite resolution of the lithographical process, blunting of the vertex will occur. Junction losses for practical devices are between 1 and $2dB$ per junction (i.e. between $2-4dB$ for the total device).

b. Propagation loss in the AWG and coupling losses due to a mismatch between the imaged field and the receiver waveguide mode are usually much smaller.

If the transmission for the central channel of the AWG is T_c, then the attenuation of the central channel is given by

$$A_0 = -10\log T_c \tag{3.142}$$

From Figure 3-31 it is seen that the outer channels have more loss than the central one. This reduction is caused by the fact that the far field of the individual array waveguides drops in direction different from the main optical axis. The envelope, as indicated in the Figure 3-32, is mainly determined by the far field radiation pattern of the individual array waveguides.

The non-uniformity ΔT_{AWG} is defined as the difference in transmission between the central channel(s) and outermost channels (i.e. between No.1 and No.N):

$$\Delta T_{AWG} = -10\log \frac{T_{1,N}}{T_c} \tag{3.143}$$

in which T_c is the transmission of the central channel(s).

The power, which is lost from the main lobe, will appear in adjacent orders as shown in in Figure 3-31. If the free spectral range is chosen N times the channel spacing, then the outer channels of the AWG will experience almost $3dB$ more loss than the central channels, i.e. the non-uniformity is close to $3dB$. This is because at a deflection angle at half the angle distance between the orders of the array the power in the image is reduced by 50% because at that angle it will be equally divided over the two orders at both sides of the optical axis. In a periodical AWG the outer channels will be close to these $3dB$ points.

From above it is clear that the non-uniformity of an AWG can be reduced by increasing the FSR, however, at the expense of a larger device size.

Bandwidth

As explained in the introduction, AWGs are lens-like imaging devices; they form an image of the field in the object plane at the image plane. Because of the linear length increment of the array waveguides the lens exhibit dispersion: if the wavelength changes, the image moves along the image plane without changing shape, in principle. Most of the AWG properties can be understood by considering the coupling behavior of the focal field in the image plane to the receiver waveguide(s). This coupling is described by the overlap integral of the normalized receiver waveguide mode $U_r(s)$ and the normalized focal field $U_f(s)$ in image plane, as illustrated in Figure 3-33:

$$\eta(\Delta s) = \left| \int U_f(s - \Delta s) U_r(s) ds \right|^2 \qquad (3.144)$$

in which Δs is the displacement of the focal field relative to the receiver wavegui8de center. If $U_r(s)$ and the image field $U_f(s)$ have the same shape, which will be the case if identical waveguide are used for the transmitter and the receiver waveguides, then the coupling efficiency η can be close to 100% on proper design of the AWG. The power transfer function $T_i(\nu)$ for the *ith* receiver waveguide is found by substituting

$$\Delta s = D_{sp} \cdot (\nu - \nu_i) \qquad (3.145)$$

in (3.144), which yields

$$T_i(\nu) = T_c \cdot \eta\{D_{sp} \cdot (\nu - \nu_i)\} \qquad (3.146)$$

where ν_i is the frequency corresponding to the ith channel.

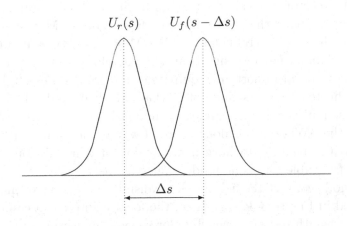

Figure 3-33   The receiver waveguide mode profile $U_r(s)$ and the focal field $U_f(s)$

It should be noted:

a. The power transmission of the (virtual) central channel T_c is normally smaller than 1 due to the transmission losses in the AWG

b. For an even number of channels there will be no receiver waveguide at the center of the image plane.

Figure 3-33 and Figure 3-30 show an example of a de-multiplexer response curve for different channels, in which the most important characteristic of the device are indicated.

To analyze the transmission loss at the input and output aperture of the array, which form the main contribution to the total insertion loss for the central channels of the device, the most usual approach is to overlap (3.144) the far field of the transmitter with the sum field of the individual array waveguides, with excitation coefficients proportional to the incident field strength. This analysis ignores the transmitter losses at input and output aperture of the arrays.

Passband shape

An important feature of the AWG-field characteristics is the passband shape. Without special measures, the channel response has a more or less Gaussian shape. This is because the mode profiles of the transmitter and receiver waveguides of the AWG can be described to a good approximation by a Gaussian function, so that their overlap (3.144) is also Gaussian. The $1dB$ bandwidth is usually $25 - 30\%$ of the channel spacing.

The Gaussian shape of the channel response impose tight restrictions on the wavelength tolerance of the emitters (e.g. laser diodes) and requires accurate temperature control for both the AWGs and the laser diodes. Moreover, when signals are transmitted through several fields in a WDM network, the cumulative passband width for each channel narrows significantly. Therefore, flat and broadened channel transmissions are an important requirement for AWG de-multiplexer. Different approaches to flatten the passbands of AWGs have been published. The simplest method is to use multimode waveguides at the receiver side of an AWG. If the focal spot moves at the AWG output along a broad waveguide, almost 100% of the light is coupled into the receiver to have a flat region of transmission. However, this approach is unfavorable for single mode fiber systems.

Other methods convert the field at the transmitter or receiver into a double image as indicated in Figure 3-34 (in which, the focal field U_f is a camel-shaped focal field). The wavelength response, which follows from the overlap of the field with the normal mode of a receiver/transmitter waveguide, will get a flat region as shown in Figure 3-34 (where the dashed curve indicates the non-flattened response obtained by applying a non-modified focal field, i.e. $U_f = U_r$). The double image is created by use of a short MMI coupler [63], a Y-junction or a non-adiabatic parabolic horn [64], where the best results are achieved with the parabolic horn.

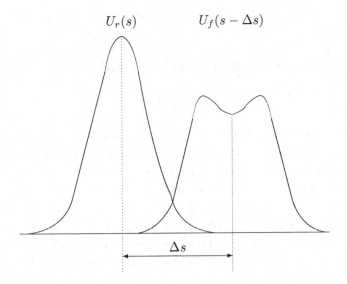

Figure 3-34 Wavelength response flattening: Shape of the focal field U_f required for obtaining a flat region in the overlap with the receiver mode U_r.

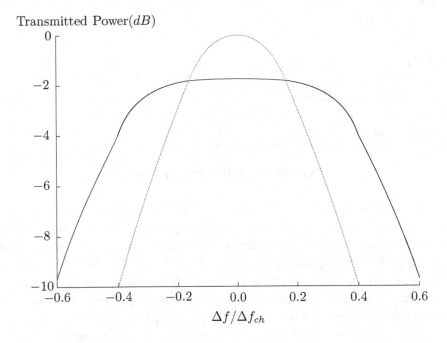

Figure 3-35 Wavelength response obtained by applying a camel-shape focal field

Crosstalk

One of the most important characteristics of the device is the inter-channel crosstalk. Which is the contribution of unwanted signals, e.g. in the case of adjacent channels, the contribution of the (unwanted) signal at frequency ν_{i+1} to the (detected) channel f_i. The theoretical adjacent-channel crosstalk A_x follows from the overlap of the focal field with the unwanted mode (3.144) as

$$A_x = \eta(d) \tag{3.147}$$

with d being the distance between the receiver waveguide and the adjacent waveguide. From this formula it is seen that arbitrary large crosstalk attenuation is possible by positioning the receiver waveguides sufficiently far apart. *Usually a gap of $1-2$ times the waveguide width is sufficient for more than $40dB$ adjacent interchannel crosstalk attenuation.* However, in practice other mechanisms appear to limit the crosstalk attenuation. The most important ones are errors in the phase transfer of the array waveguides. They are due to non-uniformities in layer thickness, waveguide width, and refractivity index and cause a rather noisy "crosstalk floor", which is better than $35dB$ for good devices. Semiconductor-based devices exhibit slightly inferior crosstalk figures compared to silica-based ones.

Crosstalk figures provided for experimental and commercial devices usually refer to single-channel crosstalk level, i.e. the crosstalk resulting from a single channel. In an operational environment crosstalk contributions from all active channels will impair the crosstalk level compared to the single-channel crosstalk attenuation.

Typical values of AWG's parameters are list in Table 6-6:

Table 6-6

Parameters	AWG
Insertion Loss in dB	3.6 - 6
Cross-talk attenuation in dB	25 - 45
Channel spacing in GHz	12.5 - 200

Wavelength Routing Properties

An interesting device is obtained if the AWG is designed with N input and N output waveguides and a free spectral region equalling N times the channel spacing. With such an arrangement the device behaves cyclical: (a signal disappearing from output N will appear at output 1 if the frequency is increased), namely, a cyclical wavelength router [42], [43](1991). It provides an important additional functionality compared to multiplexers and demultiplexers and plays a key role in more complex devices such as add-drop multiplexers and wavelength switches. Figure 3-35 illustrates its functionality. Each of the N input ports can carry N different frequencies. The N frequencies carried by input channel 1 (signals $a_1 - a_4$ in Figure 3-36) are distributed among output channels 1 to N in such a way that output channel 1 carries frequency N (signal a_4) and channel N carries frequency

1 (signal a_1). The N frequencies carried by input 2 (signal $b_1 - b_4$) are distributed in the same way, but cyclically shifted by 1 channel in such a way that frequencies $1 - 3$ are coupled to ports $3 - 1$ and frequency 4 to port 4. In this way each output channel receives N different frequencies, one from each input channel.

wavelength router

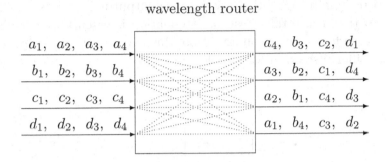

Figure 3-36   Wavelength router

To realize such an inter-connectivity scheme in a strictly non-blocking way, using a single frequency, a large number of switches would be required. With a wavelength router, this functionality can be achieved using only a signal component. Wavelength routers are key components also in multi-wavelength add-drop multiplexers and crossconnects (seen below).

Configuration-dependent crosstalk in Add-drop multiplexer

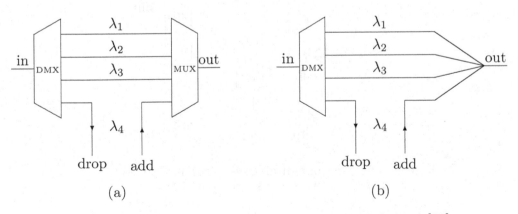

Figure 3-37   Add-drop multiplexers with fixed add-drop channel [50]

One key application of AWGs is adding or dropping wavelengths from a multiplexer. AWGs in different configurations can be used for this propose, and the corresponding interchannel crosstalk is a parameter of particular importance. One standard configuration consist of two symmetrically connected $1 \times N$ AWGs with identical wavelength response (Figure 3-37(a)). This architecture exhibits an almost perfect isolation between the add port and the drop port, and spurious intensity of the dropped signal propagating towards the output is suppressed again by the multiplexer, so that the total crosstalk attenuation is essentially doubled. The latter is not the case for cheaper configuration illustrated in Figure 3-37(b), which uses a power combiner instead of a multiplexing AWG.

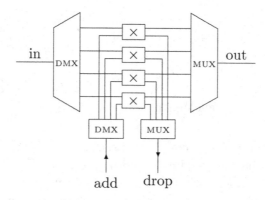

Figure 3-38   Reconfigurable add-drop multiplexer
based on one 1×2 switch per channel

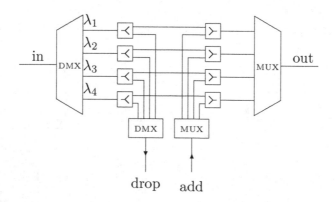

Figure 3-39   Reconfigurable add-drop multiplexer
based on two 1×2 switch per channel

An add-drop multiplexer as shown in Figure 3-39(a) can be made configurable, i.e. the added and dropped waveguides can be selected with an external control signal by combining the (de)multiplexers with switches as shown in Figure 3-40.

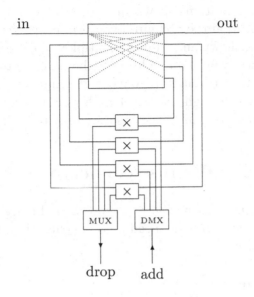

Figure 3-40   Ad-drop multiplexer configurations
based on a single wavelength router in a loop-back configuration

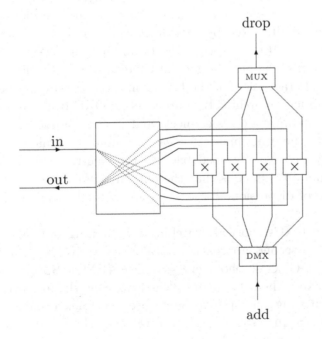

Figure 3-41   Ad-drop multiplexer configurations
based on a single wavelength router in a fold-back configuration

In the configuration of Figure 3-41 the add-signal may end up in the drop-port at an undesirably high level through-crosstalk in the switch. In the configuration of Figure 3-41, which requires two 1×2 switches per channel instead of one 2×2, the isolation is almost perfect.

A disadvantage of the configurations shown in Figure 3-38 and Figure 3-39 is that for low insertion loss operation they require two (de)multiplexers.

Figure 3-40 and Figure 3-41 shown an add-drop multiplexer realized with a single $(N + 1) \times (N + 1)$ AWG, respectively. From add port, they are fed to four different switches with de-multiplexer, then through the wavelength router to go to out port. Meanwhile, from input port, go through the wavelength router, they are fed to the multiplexer to be dropped down. In which, they require only one (de)multiplexer.

3.4 Optical ADD/DROP cross-connector

An opticalcross-connect (OXC) is a device as shown in Figure 3-42 used by telecommunications carriers to switch high-speed optical signals in a fiber network, such as an optical mesh network.

There are several ways to realize an OXC:

a. Opaque OXCs (electronic switching): One can implement an OXC in the electronic domain, namely, all the input optical signals are converted into electric signals after they are demultiplexed by multiplexers. The electronic signals are then switched by an electronic switch module. Finally, the switched electronic are converted back into optical signals by using them to modulate lasers and then the resulting optical are multiplexed by optical multiplexers onto outlet optical fibers. This is known as an "OEO" (Optical-Electrical-Optical) design. Cross-connects based on an OEO switching process generally have a key limitation: the bandwidth of the modulation characteristic of the laser limits the maximum bandwidth of the signal by $10Gbps$. Such an architecture prevents an OXC from performing with the same speed as an all-optical cross-connect, and is not transparent to the network protocols used . On the other hand, it is easy to monitor signal quality in an OEO device, since everything is converted back to the electronic format at the switch node. An additional advantage is that the optical signals are regenerated, so they leave the node free of dispersion and attenuation. An electronic OXC is also called an opaque OXC.

b. Transparent OXCs (optical switching): Switching optical signal in an all-optical device is the second approach to realize an OXC. Such a switch is often called a transparent OXC or photo cross-connect (PXC). Specifically, optical signals are demultiplexed, then the demutiplexed wavelengths are switched by switch modules. After switching, the optical signals are multiplexed onto output fibers by optical multiplexes. Such a switch architecture keeps the features of data rate and protocol transparency. However, because the signals are kept in optical format, the transparent OXC architecture does not allow easy optical signal quality monitoring.

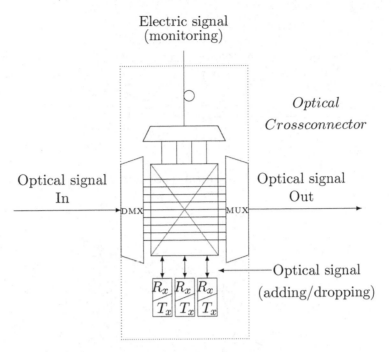

Figure 3-42   Optical Add/Drop cross-connector [51]

c. Translucent OXCs (optical and electronic switching): As a compromise between opaque and transparent OXCs, there is a type of OXC called a tranlucent OXC. In such a switch architecture, there is a switch stage which consist of an optical switch module and an electronic switch module. Optical signal passing through the switch stage can be switched either by the optical switch module or electronic switch module. In most case, the optical switch module is preferred for the purpose of transparency. When the optical switch module's switching interface are all busy or an optical switch signal needs signal regeneration through an OEO conversion process, the electronic module is used. Translucent OXC nodes provide a compromise of full optical signal transparency and comprehensive optical signal monitoring. It also provides the possibility of signal regeneration at each node.

For the principle explanation, Figure 3-43 shows a simplest example for the optical Add/Drop cross-connector. This is an all-optical cross-connector, in which:

a. The WDM optical signal $(\lambda_1, \lambda_2, \lambda_3, \lambda_4, \lambda_5)$ input onto the cross-connector.

b. When the switch holds, the optical signal go through the solid line to the output,

c. Meanwhile, some wavelengths $(\lambda_3, \lambda_4, \lambda_5)$ are dropping down to the local network, some wavelengths (λ_1, λ_5) are adding to the output along the solid lines.

d. When the switch is to be switching down, all optical signal go through the cross-connector along the dash lines.

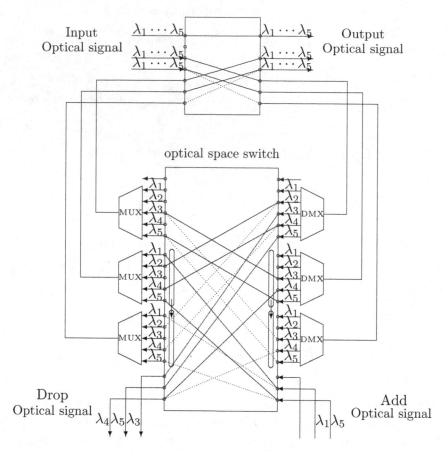

Figure 3-43   Optical Add/Drop cross-connector

3.5 Optical amplifier

The all-optical WDM networks mainly based on two technologies:

a. All-optical add/drop cross-connector, which includes optical multiplexer, optical de-multiplexer, and optical add/drop, such as optical Mach-Zehnder (de)multiplexer, optical Bragg grating (de)multiplexer technologies.

b. Optical fiber amplifiers based on the nonlinearity and the interaction between light wave with the material in single mode fiber.

3.5.1 Nonlinearity in single mode fiber

The light wave traveling in a long haul optical fiber communication system is multi-channels signal such as wavelength division multiplexing (WDM) signal, in which the power of light is limited by the nonlinearity in single mode fiber since the nonlinearity in single mode fiber will cause the cross-talk between the central channel and the adjacent channels. However, the nonlinearity in single mode fiber can be used to design the single mode fiber amplifiers for the all-optical transmission system. To discuss the nonlinearity in single mode fiber, we have to mention the nonlinear wave equation.

Nonlinear wave equation

Linear case

Till now the single mode fiber is considered to be a linear dispersion medium, in this case, the wave equation is,

$$\nabla^2 \mathbf{E} + n^2 k_0{}^2 \mathbf{E} = 0 \tag{3.148}$$

where the refractive index in single mode fiber is

$$n = n_1 \tag{3.149}$$

This is valid only if the power of light is less than a threshold point.

Nonlinear case

However, if the power of light is over a threshold point , the single mode fiber has to be considered as a nonlinear dispersion medium. In this case, the refractive index of single mode fiber become

$$n = n_1 + \triangle n = n_1 + \left(\frac{n_2}{2n_1}\right)^2 |\mathbf{E}|^2 + \cdots \tag{3.150}$$

Substitution of (4.77) into (4.76) give us the nonlinear wave equation as follows:

$$\nabla^2 \mathbf{E} + \left(n_1 + [\frac{n_2}{2n_1}]^2 |\mathbf{E}|^2 + \cdots\right)^2 k_0{}^2 \mathbf{E} = 0 \tag{3.151}$$

The solution of the light wave in single mode fiber has to satisfy the nonlinear wave equation (4.78). And different nonlinearity phenomenon has to satisfy different nonlinear equation. [52]

At beginning, we'd better to point out some nonlinear phenomena in single mode fiber as follows:

For example, when light wave travels in a single mode fiber:
a. Stimulated Brillouin scattering (**SBS**) has no measurable effect to a signal of $3mW$. However, a significant effect can be measured if the power of the signal is increased to $6mW$. Where the signal to be considered is one channel signal. That means that the power of one channel signal need to be kept below $3mW$ if SBS is to

be avoided. However, Stimulated Brillouin scattering (**SBS**) can be used to make **Brillouin scattering amplifier** in a single mode fiber when a pump light inject into the single mode fiber and is above a "certain threshold value"..

b. Stimulated Raman Scattering (**SRS**) is not an issue in single-channel systems. However, it can be a significant problem in wavelength division multiplexing (WDM) systems when the signal power is above a "threshold value". However, Stimulated Raman Scattering (**SRS**) can be used to make **Raman scattering amplifier** in single mode fiber when an incident pump light inject into the single mode fiber and the pump power is above a "certain threshold value".

c. A doped single mode fiber, such as Erbium-doped single mode fiber, can be used to make **Erbium-doped fiber amplifier**. The idea of the Erbium-doped fiber amplifier is from the semiconductor laser since a semiconductor laser is a doped-silica multi-layers waveguide.

3.5.2   Erbium-Doped Fiber Amplifier

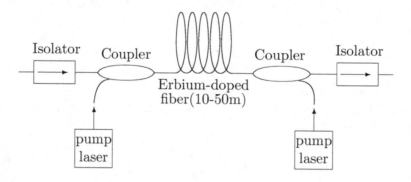

Figure 3-44   Erbium-Doped Fiber Amplifier

Erbium is a rare-earth element that, when excited, emit light around $1,500nm$, which is the low-loss wavelength for optical fibers used in DWDM. Figure 3-44 shows a simplified diagram of EDFA (Erbium-Doped Fiber Amplifier). A weak signal enters the erbium-doped fiber, into which light at $980nm$ or $1480nm$ is injected using a pump laser. this injected light stimulates the erbium atoms to release their stored energy as additional $1500nm$ light. As this process continues down the fiber, the signal grows stronger. The spontaneous emission in the EDFA also add noise to the signal; this determines the noise figure of an EDFA.

EDFA are typically capable of gains of $30dB$ or more and output power of $+17dBm$ or more. The target parameters when selecting an EDFA, however, are low noise and flat gain. While the signal gain provided with EDFA technology is inherently wavelength-dependent, it can be corrected with gain flattening filter. Such filters are often built into modern EDFAs.

Photons radiated follow the frequency, phase, and direction of the incident optical signal, which effectively leads to an amplification of the input optical signal. This is the mechanism of the semiconductor laser amplifier. Now we will introduce the doped-fiber amplifier. The amplification of light wave is achieved by the stimulated scattering (which cause the photons radiated), which is the same for both of semiconductor laser and the doped-fiber amplifier. However, the pump process is different for both of them.

Doped fiber amplifiers (DFAs) are optical amplifiers that use a doped single mode fiber as a gain medium to amplify an optical signal. In principle, the doped fiber amplifiers (DFAs) are based on the semiconductor laser. The signal to be amplified and a pump laser are multiplexed into the doped fiber, and then the signal is amplified through interaction with the doping ions as shown in Figure 3-44.

The most common example is the Erbium Doped Fiber Amplifier (EDFA), where the core of a silica fiber is doped with trivalent erbium ions and can be efficiently pumped with a laser at a wavelength of 980nm or 1,480nm, and exhibits gain in the 1,5550nm region.

Amplification of light wave in Erbium-Doped Fiber is achieved by the stimulated emission from dopant irons in the doped fiber. To understand this point, we'd better to recall some based knowledge of the amplification mechanism in semiconductor laser as follows:

For semiconductor laser

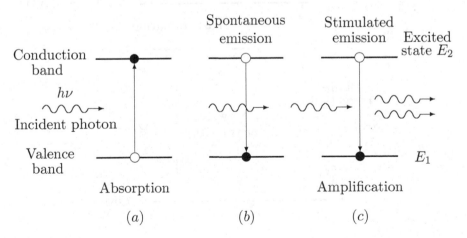

Figure 3-45   The interaction of photon and electron in semiconductor

Basically, the amplification mechanism in semiconductor laser is involved the interaction of photon and electron in semiconductor as shown in Figure 3-45. In which, three processes involved:

a. Absorption: as shown in Figure 3-45(a), an electron transits from the ground level (E_1) into the exited level (E_2), which is induced by an incident of a photon based on $E_2 - E_1 = h\nu$, as the result the photon is absorbed. This is called "absorption"

b. Spontaneous emission: as shown in Figure 3-45(b), an electron at exited level returns from the excited level to the ground level (E_1) emitting a photon with $h\nu = E_2 - E_1$. Which is called "spontaneous emission", which will cause the noise.

c. Stimulated emission: as shown in Figure 3-45(c), an external photon induces an electron returning from the excited level (E_2) into the ground level (E_1) and emitting a second photon according to $h\nu_{21} = E_2 - E_1$. Which is called "stimulated emission".

For a stimulated emission as shown in Figure 3-45(c), an external photon incident causes two photons output, which means that the light has been amplified. Which means that the stimulated emission occurs if the number of electron N_2 in excited state E_2 is larger than the number of electron N_1 in ground state E_1. In this case, the amplification of light wave in semiconductor laser is achieved . And the condition to obtain the light amplification in semiconductor laser is

$$N_2 > N_1 \tag{3.152}$$

this condition is called the "population inverse". Usually, the most population should stay in ground state, and now it is required that most population should stay in excited state.

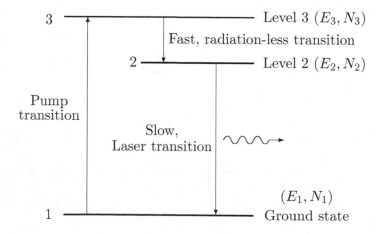

Figure 3-46   Three-levels energy diagram

The stimulated emission is achieved through several steps as shown in Figure 3-46.

a. First step, most electrons should be excited from the ground level E_1 into the exited level E_3 by a "pumping process", which can be achieved by an external bias current applied to the semiconductor diode.

b. Second step take place after about $.1\mu s$, when electrons slide down from the excited level E_3 to the meta-stable energy level E_2.

c. Third step: The electrons can live up to $11ms$ at meta-stable level E_2 before the radiate the energy (or stimulated emission) and fall back to the ground level E_1.

For Erbium Doped Fiber Amplifier (EDFA)

For Erbium Doped Fiber Amplifier (EDFA), the situation is the same as the semiconductor laser amplifier. Namely, the stimulated emission in Erbium Doped Fiber is achieved through several steps:

a. First step, most electrons should be excited from the ground level $(4_{T_{15/2}})$ into the exited level $(4_{I_{11/2}})$ by a "pumping process", which can be achieved by the external laser pumping. However, in semiconductor laser, the similar process is achieved by an external bias current applied to the semiconductor diode.

b. Second step take place after about $.1\mu s$, when electrons slide down from the level $(4_{I_{11/2}})$ to the metastable energy level $(4_{I_{13/2}})$.

c. Third step: The electrons can live up to $11ms$ at meta-stable level before the radiate the energy (or stimulated emission) and fall back to the ground level.

The population densities equation in EDFA

Now we will analyze the stimulated emission from the meta-stable level $(4_{I_{13/2}})$ into the ground level $(4_{I_{15/2}})$ in Erbium Doped Fiber Amplifier (EDFA).

The start point is that the population densities N_2 and N_1 at state $(4_{I_{13/2}})$ and $(4_{I_{15/2}})$ should satisfy the following equations, respectively:

$$\frac{\partial N_2}{\partial t} = (\sigma_p^a N_1 - \sigma_p^e N_2)\psi_p + (\sigma_s^a N_1 - \sigma_p^e N_2)\psi_s - N_2/\tau_1 \qquad (3.153)$$

$$\frac{\partial N_1}{\partial t} = (\sigma_p^e N_2 - \sigma_p^a N_1)\psi_p + (\sigma_s^e N_2 - \sigma_p^a N_1)\psi_s + N_2/\tau_2 \qquad (3.154)$$

Where:

"e" and "a" stand for emission and absorption, respectively.

"s" and "p" stand for signal and pump, respectively.

"α" and "ψ" stand for cross-section in Erbium Doped Fiber and photon flux, respectively.

"τ_1" is spontaneous lift-time of the electron at the excited state ($\approx 10ms$).

The photo flux ψ is defined as the ratio of photons number to cross-section area of the fiber, $\sigma_{mo,s}$ for signal mode, $\sigma_{mo,p}$ for pump mode, respectively. Which are

$$\psi_s = \frac{P_s}{h\nu_s\sigma_{mo,s}} \tag{3.155}$$

$$\psi_p = \frac{P_p}{h\nu_p\sigma_{mo,p}} \tag{3.156}$$

Where:

"h" is the Planck constant.

"s" and "p" stand for signal and pump, respectively.

"ν_s" and "ν_p" stand for optical frequency of the signal and pump, respectively.

The power of the signal and pump vary along the doped fiber length since there are three processes involved:

Stimulated emission,

Absorption,

Spontaneous emission.

Assuming that spontaneous emission can be neglected, the following equations can be written for the power of the signal and the pump

$$\frac{\partial P_s}{\partial z} = \Gamma_s(\sigma_s^e N_2 - \sigma_s^a N_1)P_s - \alpha P_s \tag{3.157}$$

$$\pm\frac{\partial P_p}{\partial z} = \Gamma_p(\sigma_p^e N_2 - \sigma_p^a N_1)P_s - \alpha' P_p \tag{3.158}$$

Where:

α and α' are the fiber loss at signal and pump wavelength, respectively.

Γ_s and Γ_p are the confinement factors at signal and pump wavelength, respectively.

The sign in front of (3.158) take into account the pumping direction:

It is plus for forward direction pumping,

It is minus for backward direction pumping.

Equation (3.153)(3.154)(3.157) and (3.158) can be solved analytically by assuming that fiber losses can be neglected since it is measured in tens of meters. By substitution back to (3.153), a steady state solution can be written as

$$N_2(z) = -\frac{\tau_1}{\sigma_{mo,d}\cdot h\nu_s}\frac{\partial P_s}{\partial z} - \frac{\delta\cdot\tau_1}{\sigma_{mo,d}\cdot h\nu_p}\frac{\partial P_P}{\partial z} \tag{3.159}$$

Where $\sigma_{mo,d} = \sigma_{mo,s} = \sigma_{mo,p}$ is the cross-section area of the doped portion of the fiber core. The solution can be utilitized to fine the solutions of (3.153) and (3.154).

Equation of (3.154) describes the generic case for signal value that are high enough to force amplifier into saturation. However, for small signal value, parameters ψ_s from (3.153) and (3.154) can be neglected and the total gain of an EDFA having the length L_a can be expressed as

$$G = \exp\left[\Gamma_s \int_0^{L_a} (\sigma_s^e N_2 - \sigma_s^a N_1 - \alpha)\right]dz \tag{3.160}$$

There is an optimum length of doped fiber, that depends on the pump power level. Which should be taken account when designing EDFA for different applications.

Optical gain of EDFA

The optical gain of EDFA is related to the stimulated light scattering and will occur after an inverse population is achieved through the pumping process.

The amplification coefficient that characterize stimulated emission process is given as

$$g(\nu) = \frac{g_0}{1 + P_{in}/P_{sat} + [2\pi T_2(\nu - \nu_0)]^2} \qquad (3.161)$$

Where:

ν is the frequency of the incident optical signal,

ν_0 is the atomic transition frequency related to the two level energy diagram,

g_0 is the value of the amplification peak,

P_{sat} is the saturation power, and

T_2 is the dipole relaxation time that takes sub-picosecond values.

The total gain of EDFA depends on the concentration of Erbium ions in the glass, the length and cross-section of the doped fiber active area, as well as the pump power. The exact analysis of the stimulated emission in EDFA can be done by using the same approach as in semiconductor laser in Chapter 5.

3.5.3 Raman amplifier

This subsection is going to discuss Stimulated Raman Scattering in a long-haul optical fiber communication system studied by author and his students [53][54][55][56].

As shown in Figure 3-47(a), the Stimulated Raman Scattering (**SRS**) is caused by the interaction between the light and the glass molecular variation in the fiber under an injection of a large pumping power. Scattering light appears in both of the forward and backward directions. In a single-channel system, the "Raman Threshold" (the power level at which Raman Scattering begins to take place) is very high.

As shown in Figure 3-47(b), based on the formula $h\nu_{12} = E_2 - E_1$, we see that the frequency of the light ν_{12} or ω is a measurement of the energy difference between two energy levels $E_2 - E_1$. Now Figure 3-47(a),(b) shows us three process of the Raman scattering:

a. Transfer of energy from some pump photons (ω_p) to vibrating silica molecules.

b. The interaction between the pump light mode (ω_p) and the glass molecular variation in the fiber will create new photons with frequency ($\omega_p - \omega_s$) both in forward direction and the backward direction, which are called the forward Stocks wave and the backward Stocks wave, respectively. Where ω_s is the frequency of the signal.

c. The nonlinearity of the silica fiber will cause the amplitude coupling between the pump wave ω_p and the Stocks wave $(\omega_p - \omega_s)$ along the whole length of the silica fiber, which will generate the new photons with the signal frequency ω_s, i.e. $\omega_s = \omega_p - (\omega_p - \omega_s)$. These new photons with ω_s will make amplification of signal light mode (ω_s) along the whole silica fiber. Again, photons radiated follow the frequency, phase, and direction of the optical signal mode, which effectively leads to an amplification of the optical signal along the silica fiber.

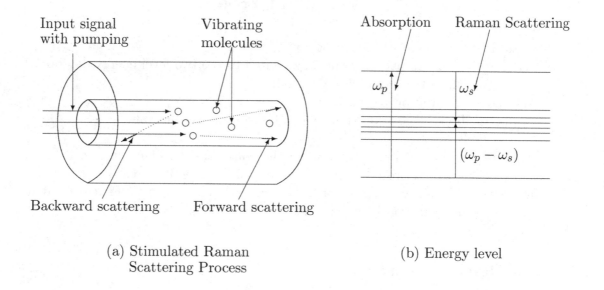

(a) Stimulated Raman
Scattering Process

(b) Energy level

Figure 3-47   Simulated Raman scattering process [57]

The problem and the benefit of SRS

(1) Stimulated Raman Scattering is not an issue for single-channel transmission systems.

(2) However, Stimulated Raman Scattering can be a significant problem in WDM systems. When multiple channels are present, power is transferred from shorter wavelength to longer ones because of the stimulated Raman scattering.

(3) Fortunately, Stimulated Raman Scattering can be a useful effect for building up an optical amplifier based on Raman Scattering Scattering. This is called "**Raman Effect Amplifier**". In Raman Effect Amplifier, power has been transferred from high power shorter wavelength (pump wavelength) to low power longer wavelength (signal wavelength).

The Important characteristics of SRS

The Important characteristics of the Stimulated Raman Scattering are:

(1) SRS can take affect over about $40THz$ (a very wide range) below the higher frequency (short wavelength) involved. That means it can extend over a range of wavelengths of about $300nm$ longer than the shortest wavelength involved. The effect is maximized when the two frequencies are $13.2THz$ apart.

(2) SRS increases exponentially with increased power. When the power is very high, it is possible that all of the signal power is transferred to the Stocks Wave.

(3) In a 10-channel WDM system with $1nm$ channel spacing, the power level must be less than $3mW$ (per channel) if SRS is to be avoided.

Nonlinear wave equations

As we mentioned above, the amplification of light in silica fiber caused by the Stimulated Raman Scattering is an amplitude coupling of pump wave mode ω_p with the Stocks wave mode $(\omega_p - \omega_s)$, or simple say is cased by the nonlinearity in silica fiber.

In the nonlinear dispersion medium such as an optical single mode fiber, the Maxwell equation is able to be simplified to

$$\nabla^2 \mathbf{E} - \left(\frac{1}{c^2}\right)\frac{\partial^2}{\partial t^2}(\mathbf{E} + 4\pi\mathbf{P}) = \left(\frac{4\pi}{c^2}\right)\frac{\partial^2}{\partial t^2}\mathbf{P}^{(NL)} \tag{3.162}$$

Where $\mathbf{P}$ and $\mathbf{P}^{(NL)}$ denotes the linear polarization vector and the nonlinear polarization vector, respectively.

For the sake of simplification, it is reasonable to assume that the input signal in a single-mode fiber has an uniform distribution of linear polarization states over the transverse cross section of the single mode fiber. And the direction of the polarization is x, the propagation direction is z. Then

$$P = \chi^{(1)}E(r, \theta, z; t) \tag{3.163}$$

$$P^{(NL)} = \chi^{(3)}E^3(r, \theta, z; t) \tag{3.164}$$

Note, the single-mode fiber is made of the crystal and the crystal is an antisymmetrical material. Therefore, the second order of electrical susceptibility $\chi^{(2)} = 0$.

Substituting (4.80) and (4.81) into (4.79), we have

$$\frac{d^2}{dz^2}E(r, \theta, z; t) - \left(\frac{n_L^2}{c}\right)^2\frac{d^2}{dt^2}E(r, \theta, z; t) = \frac{16\pi\chi^{(3)}}{c^2}\frac{d^2}{dt^2}[E^3(r, \theta, z; t)] \tag{3.165}$$

Where $\dfrac{d}{dx} = \dfrac{d}{dy} = 0$ since it is assumed that the input signal in single-mode fiber has an uniform distribution of linear polarization states over the transverse cross section of the single mode fiber.

It is easy to understand that the linear polarization vector $\mathbf{P}$ will cause the wave dispersion, and the nonlinear polarization vector $\mathbf{P}^{(\mathbf{NL})}$ will cause the new frequency components. And then the field in the single-mode fiber is able to be expressed as

$$
\begin{aligned}
|E(r,\theta,z;t)| &= \sum_m \psi_m(r,\theta)\,|A_m(z,t)|\,\cos(k_m z - \omega_m t + \phi_m) \\
&+ \sum_{m'} \psi_{m'}(r,\theta)\,|A_{m'}(z,t)|\,\cos(k_{m'} z - \omega_{m'} t + \phi_{m'})
\end{aligned}
$$

$$(3.166)$$

Where m denotes the wave propagating in the forward direction with a frequency ω_m, and m' denotes the wave propagating in the backward direction with another frequency $\omega_{m'}$. Both of them belong to the single transverse waves in single-mode fiber .

Assume that the power of the pump wave (p) injected into the single-mode fiber is above *the threshold power*, and then the pump wave will stimulate the first order of stocks wave (s) and the first order of anti-stocks wave (a), however, it is not large enough to stimulate the higher order of stocks waves. In this case, $m = p, s, a; \quad m' = s, a$.

Next, we may consider that the field amplitude

$$
F_m(z,t) = N_m A_m(z,t) \tag{3.167}
$$

so that the power of the field will be

$$
P_m(z,t) = |F_m(z,t)|^2
$$

And then the normalization condition gives

$$
N_m^2 = \frac{n_{mL}\,c}{8\pi} \int_0^\infty \int_0^{2\pi} \psi_m^2(r,\theta)\,r\,dr\,d\theta \tag{3.168}
$$

Where n_{mL} is the linear refraction index at the frequency ω_m.

Substituting (4.84) into (4.83), we have

$$
\begin{aligned}
E(r,\theta,z;t) &= \frac{1}{2}\sum_m \psi_m(r,\theta)\left[\frac{F_m(z,t)}{N_m}e^{i(k_m z - \omega_m t)}\right. \\
&\quad \left.+\frac{F_m^*(z,t)}{N_m}e^{-i(k_m z - \omega_m t)}\right] \\
&+\frac{1}{2}\sum_{m'}\psi_{m'}(r,\theta)\left\{\frac{F_{m'}(z,t)}{N_{m'}}e^{i[k_{m'}(L-z)-\omega_{m'}t]}\right. \\
&\quad \left.+\frac{F_{m'}^*(z,t)}{N_{m'}}e^{-i[k_{m'}(L-z)-\omega_{m'}t]}\right\}
\end{aligned}
$$

$$(3.169)$$

Where L is the length of the single-mode fiber from the inject point of the pump wave to the end point of the single-mode fiber.

For the sake of simplification, we may consider the forward waves alone without considering the backward waves for the time been, i.e. let $m' = 0$ in (4.86). And then substituting (4.86) into (4.82), collect the same frequency terms in both sides, we may obtain the coupling wave equation for the Stocks wave as following:

$$\frac{ik_s\psi_s}{N_s}\left(\frac{\partial F_s}{\partial z} + \frac{n_{sL}}{c}\frac{\partial F_s}{\partial t}\right) =$$

$$-\frac{16\pi\omega_s^2}{c}\Big[\chi_{spps}^{(3)}\frac{6\psi_p^2\psi_s}{8N_pN_s}|F_p|^2F_s$$

$$+\chi_{spps}^{(3)}\frac{3\psi_p^2\psi_a}{8N_p^2N_a}|F_a|^2F_a^*e^{i(2k_p-k_a-kc)z}$$

$$+\chi_{ssss}^{(3)}\frac{3\psi_s^3}{8N_s^3}|F_s|^2F_s$$

$$+\chi_{saas}^{(3)}\frac{6\psi_s\psi_a^2}{8N_sN_a^2}|F_a|^2F_s\Big] \tag{3.170}$$

Where

$$\chi_{spps}^{(3)} = \chi^{(3)}(-\omega_s, -\omega_p, \omega_p, -\omega_s) \tag{3.171}$$

$$\chi_{ssss}^{(3)} = \chi^{(3)}(-\omega_s, -\omega_s, \omega_s, -\omega_s) \tag{3.172}$$

$$\chi_{saas}^{(3)} = \chi^{(3)}(-\omega_s, -\omega_a, \omega_a, -\omega_s) \tag{3.173}$$

Of course, we understand that the forward waves contains not only the Stocks wave F_s but also the pumping wave F_p and the anti-Stocks wave F_a. And the coupling wave equations for the pumping wave F_p or the anti-Stocks wave F_a is similar to (4.87). Therefore, we will concentrate to deal with the equation (4.87) first.

Now, to eliminate the transverse distribution function ψ_a and ψ_p etc. from (4.87), it is able to multiply (4.87) by ψ_m and then take an integral over the transverse cross section of single mode fiber, and then make a definition

$$< ijkl >= \frac{\int_0^{2\pi}\int_0^{\infty}\psi_i\psi_j\psi_k\psi_l\,rdrd\theta}{N_iN_jN_kN_l} = A_e^{-1} \tag{3.174}$$

we have

$$\frac{\partial F_s}{\partial z} + \frac{n_{sL}}{c}\frac{\partial F_s}{\partial t}$$

$$= i\frac{3\omega_s}{4}\Big[2\chi_{spps}^{(3)} < spps > |F_p|^2F_s$$

$$+\chi_{spps}^{(3)} < spap > |F_p|^2 F_a^* e^{i(\Delta kz + 2\phi_p)}$$

$$+\chi_{ssss}^{(3)} < ssss > |F_s|^2 F_s$$

$$+2\chi_{saas}^{(3)} < saas > |F_a|^2 F_s] \tag{3.175}$$

Similarly, we have

$$\frac{\partial F_p}{\partial z} + \frac{n_{pL}}{c}\frac{\partial F_p}{\partial t}$$

$$= i\frac{3\omega_p}{4}\left[\chi_{pppp}^{(3)} < pppp > |F_p|^2 F_p\right.$$

$$+2\chi_{pssp}^{(3)} < pssp > |F_s|^2 F_p$$

$$+2\chi_{paap}^{(3)} < paap > |F_a|^2 F_p\right] \tag{3.176}$$

and

$$\frac{\partial F_a}{\partial z} + \frac{n_{aL}}{c}\frac{\partial F_a}{\partial t}$$

$$= i\frac{3\omega_a}{4}\left[2\chi_{appa}^{(3)} < appa > |F_p|^2 F_a\right.$$

$$+\chi_{apsp}^{(3)} < apsp > |F_p|^2 F_s^* e^{i(\Delta kz + 2\phi_p)}$$

$$+\chi_{aaaa}^{(3)} < aaaa > |F_a|^2 F_a$$

$$+2\chi_{assa}^{(3)} < assa > |F_s|^2 F_a\right] \tag{3.177}$$

Where

$$\Delta k = 2k_p - k_s - k_a$$

$$2\omega_p = \omega_s + \omega_a$$

And in $\chi^{(3)}(-\omega_i, -\omega_j, \omega_k, -\omega_l),\quad (i,j,k,l,=p,s,a)$, we have

$$-\omega_i = -\omega_j + \omega_k - \omega_l$$

$$\mathbf{Im}\left[\chi^{(3)}(-\omega_i, -\omega_j, \omega_k, -\omega_l)\right] = -\mathbf{Im}\left[\chi^{(3)}(\omega_i, \omega_j, -\omega_k, \omega_l)\right]$$

It is worth to point out, from (4.91) to (4.93), that the amplitude coupling between the pumping wave F_p, Stocks wave F_s is caused by the imaginary part of $\chi^{(3)}$, which results in the Stimulated Raman Scattering, and the phase coupling between the pumping wave F_p, Stocks wave F_s. And the anti-Stocks wave F_a is caused by the real part of $\chi^{(3)}$, which results in the self-phase modulation phenomenon.

Stimulated Raman Scattering (SRS)

From wave point of view, when the pumping power is above a threshold value in a single-mode fiber, it will stimulate the forward Stock wave and the anti-Stocks wave. This phenomenon is said to be *the Stimulated Raman Scattering*.

Step 1: Nonlinear coupling wave equation:

The nonlinear coupling wave equation described the Stimulated Raman Scattering in a single-mode fiber is able to be obtained from (4.93) and (4.92) as follows:

$$\frac{\partial F_p}{\partial z} + \frac{n_{pL}}{c}\frac{\partial F_p}{\partial t} = -\frac{3\omega_p}{2}|\mathbf{Im}[\chi^{(3)}_{pssp}]| <pssp> |F_s|^2 F_p \tag{3.178}$$

$$\frac{\partial F_s}{\partial z} + \frac{n_{sL}}{c}\frac{\partial F_s}{\partial t} = \frac{3\omega_s}{2}|\mathbf{Im}[\chi^{(3)}_{spps}]| <spps> |F_p|^2 F_s \tag{3.179}$$

Where F_p and F_s represent the amplitude factor of the forward pumping wave and the forward Stocks wave, respectively.

Multiplying (4.95) and (4.96) by F_p^* and F_s^*, respectively, considering that $F_p \cdot F_p^* = P_P$, $F_s \cdot F_s^* = P_s$, we have the coupling wave equations of the Stimulated Raman Scattering as follows:

$$\frac{\partial P_p}{\partial z} + \frac{1}{v_p}\frac{\partial P_p}{\partial t} = -\frac{g_p}{KA_e}P_s P_p - \alpha_p P_p \tag{3.180}$$

$$\frac{\partial P_s}{\partial z} + \frac{1}{v_s}\frac{\partial P_s}{\partial t} = \frac{g_s}{KA_e}P_p P_s - \alpha_s P_s \tag{3.181}$$

Where

$$g_p = 3\omega_p|\mathbf{Im}[\chi^{(3)}(-\omega_p, -\omega_s, \omega_s, -\omega_p)]|$$

$$g_s = 3\omega_s|\mathbf{Im}[\chi^{(3)}(-\omega_s, -\omega_p, \omega_p, -\omega_s)]|$$

are the gain coefficients of the pumping wave and the Stocks wave, respectively. And

$$v_p = c/n_{pL}, \quad v_s = c/n_{sL}$$

are the propagation velocity of the pumping wave and the Stocks wave, respectively. K is the polarization factor. For a single-mode fiber with a holding polarization,

$K = 1$. For a single-mode fiber with non-holding polarization, $K = 2$, which is the worst case. Therefore, $1 \leq K \leq 2$. The definition of A_e is in (4.91).

Step 2: Parameter transformation:

Now we will find the solution of (4.97) and (4.98). Normally, the single-mode fiber is a low dissipation fiber. In this case, it is reasonable to consider that the linear attenuation coefficient for both the pumping wave and the Stocks wave are approximately the same, therefore, it is reasonable to assume that $\alpha_p = \alpha_s = \alpha$. In this case, set a parameter transformation as follows [8]

$$z' = z, \quad t'_p = t - z/v_p, \quad t'_s = t - z/v_s \tag{3.182}$$

And then, in (4.97) and (4.98),

$$\frac{\partial}{\partial z} = \frac{\partial}{\partial z'}\frac{\partial z'}{\partial z} + \frac{\partial}{\partial t'}\frac{\partial t'}{\partial z} = \frac{\partial}{\partial z'} + \frac{\partial}{\partial t'}\left(-\frac{1}{v}\right)$$

$$\frac{1}{v}\frac{\partial}{\partial t} = \frac{1}{v}\left(\frac{\partial}{\partial z'}\frac{\partial z'}{\partial t} + \frac{\partial}{\partial t'}\frac{\partial t'}{\partial t}\right) = \frac{1}{v}\left(\frac{\partial}{\partial t'}\right)$$

Therefore,

$$\frac{\partial}{\partial z} + \frac{1}{v}\frac{\partial}{\partial t} = \frac{\partial}{\partial z'}$$

And then (4.97) and (4.98) become

$$\frac{\partial \tilde{P}_p}{\partial z'} = -\frac{g_p}{KA_e}\tilde{P}_p\tilde{P}_s - \alpha\tilde{P}_p \tag{3.183}$$

$$\frac{\partial \tilde{P}_s}{\partial z'} = \frac{g_s}{KA_e}\tilde{P}_p\tilde{P}_s - \alpha\tilde{P}_s \tag{3.184}$$

Where

$$\tilde{P}_p = P_p(z', t'_p), \quad \tilde{P}_s = P_s(z', t'_s)$$

And the boundary conditions are

$$\tilde{P}_p(z', t'_p)|_{z'=0} = P_p(0, t'_p)$$

$$\tilde{P}_s(z', t'_s)|_{z'=0} = P_s(0, t'_s)$$

Obviously, the parameter transformation (4.99) transforms the partial differential equations (4.97) and (4.98) into the ordinary differential equations (4.100) and (4.101).

---

[8] This is from the contour method (or the characteristic curve method).

Step 3: Find the solution:

To fine the solution of (4.100) and (4.101), suppose that

$$\tilde{P}_p(z', t_p') = C_p(z', t_p')e^{-\alpha z'} \tag{3.185}$$

$$\tilde{P}_s(z', t_s') = C_s(z', t_s')e^{-\alpha z'} \tag{3.186}$$

Substituting (4.102) and (4.103) into (4.100) and (4.101), we have

$$\frac{\partial C_p(z', t_p')}{\partial z'} = -\tilde{g}_p C_p(z', t_p') \, C_s(z', t_s') \, e^{-\alpha z'} \tag{3.187}$$

$$\frac{\partial C_s(z', t_s')}{\partial z'} = \tilde{g}_s C_p(z', t_p') \, C_s(z', t_s') \, e^{-\alpha z'} \tag{3.188}$$

Where $\tilde{g}_p = g_p/KA_e$, $\tilde{g}_s = g_s/KA_e$. And the boundary conditions are

$$C_p(z', t_p')|_{z'=0} = P_p(0, t_p')$$

$$\tag{3.189}$$

$$C_s(z', t_s')|_{z'=0} = P_s(0, t_s')$$

From (4.104) and (4.105), we have

$$\frac{\partial}{\partial z'}(\tilde{g}_s C_p + \tilde{g}_p C_s) = 0 \tag{3.190}$$

Therefore,

$$\tilde{g}_s C_p + \tilde{g}_p C_s = G(t') \tag{3.191}$$

Where $G(t')$ is an integral constant in the integral of (4.107) with respect to z', and then $G(t')$ is independent of z'. Therefore, we may consider that

$$G(t') = G(t')|_{z'=0} = [\tilde{g}_s C_p(z', t_p') + \tilde{g}_p C_s(z', t_s')]_{z'=0}$$

and then

$$G(t') = \tilde{g}_s P_p(0, t_p') + \tilde{g}_p P_s(0, t_s') \tag{3.192}$$

Substituting (4.109) into (4.108), we have

$$C_p = \frac{G(t')}{\tilde{g}_s} - \frac{\tilde{g}_p}{\tilde{g}_s} C_s$$

$$\tag{3.193}$$

$$C_s = \frac{G(t')}{\tilde{g}_p} - \frac{\tilde{g}_s}{\tilde{g}_p} C_p$$

Substituting (4.110) into (4.104) and (4.105), we have

$$\frac{\partial C_p}{\partial z'} = -G(t') \, C_p \, e^{-\alpha z'} + \tilde{g}_s \, C_p^2 \, e^{-\alpha z'}$$

$$\frac{\partial C_s}{\partial z'} = G(t')\, C_s\, e^{-\alpha z'} - \tilde{g}_p\, C_s^2\, e^{-\alpha z'}$$

Or

$$\frac{\partial}{\partial z'}\left(\frac{1}{C_p}\right) - G(t')\left(\frac{1}{C_p}\right)e^{-\alpha z'} = -\tilde{g}_s\, e^{-\alpha z'}$$

$$\frac{\partial}{\partial z'}\left(\frac{1}{C_s}\right) - G(t')\left(\frac{1}{C_s}\right)e^{-\alpha z'} = \tilde{g}_p\, e^{-\alpha z'} \tag{3.194}$$

These are an ordinary differential equations with the form of

$$Y' + P(x)Y = Q(x)$$

and the solution is

$$Y = \exp\left\{-\int P dx\right\}\left(\int Q \exp\left\{\int P dx\right\}dx + C\right)$$

Following this equation, considering the boundary conditions (4.106) and (4.109), we have

$$C_p(z', t'_p) = \frac{G(t')}{\tilde{g}_s}\,\frac{1}{1 + H(z', t')} \tag{3.195}$$

$$C_s(z', t'_s) = \frac{G(t')}{\tilde{g}_p}\,\frac{H(z', t')}{1 + H(z', t')} \tag{3.196}$$

Where

$$H(z', t') = \frac{\omega_p}{\omega_s}\,\frac{P_s(0, t'_s)}{P_p(0, t'_p)}\,\exp[G(t')L_e] \tag{3.197}$$

$$L_e = [1 - e^{-\alpha z'}]/\alpha$$

$$G(t') = \tilde{g}_s[P_p(0, t'_p) + (\omega_p/\omega_s)P_s(0, t'_s)]$$

$$t'_p = t - \frac{n_{pL}}{c}z, \qquad t'_s = t - \frac{n_{sL}}{c}z \tag{3.198}$$

Substituting (4.112) and (4.115) into (4.102) and (4.103), considering, from (4.99), that $z' = z$, we have

$$P_p\left(z,\, t - \frac{z}{v_p}\right) = P_0(0, t')\exp(-\alpha z)\frac{1}{1 + H} \tag{3.199}$$

$$P_s\left(z,\, t - \frac{z}{v_s}\right) = \left(\frac{\omega_s}{\omega_p}\right)P_0(0, t')\exp(-\alpha z)\frac{1}{1 + H} \tag{3.200}$$

Where

$$H = H(z', t') = \frac{\omega_p}{\omega_s} \frac{P_s(0, t'_s)}{P_p(0, t'_p)} \exp[G(t')L_e] \qquad (3.201)$$

$$P_0(0, t') = P_p\left(0, t - \frac{z}{v_p}\right) + \frac{\omega_p}{\omega_s} P_s\left(0, t - \frac{z}{v_s}\right) \qquad (3.202)$$

(4.116) and (4.117) are *the transient-state solution* of the coupling wave equations (4.97) and (4.98) under the condition of $\alpha_p = \alpha_s = \alpha$.

Now we will consider two situations:

(1) **Quasi-steady-state solution:**

For a pulse signal in a single-mode fiber, if the duration of the pulse signal is larger than the time delay [9] for several times, the dispersion in the single-mode fiber is able to be ignored, in this case, the transient-state solution is said to be *the quasi-steady-state solution*. And then,

$$v_p = v_s = v, \qquad \alpha_p = \alpha_s = \alpha$$

and the Stimulated Raman Scattering coupling wave equations (4.97) and (4.98) become

$$\frac{\partial P_p}{\partial z} + \frac{1}{v} \frac{\partial P_p}{\partial t} = -\frac{g_p}{KA_e} P_p P_s - \alpha P_p \qquad (3.203)$$

$$\frac{\partial P_s}{\partial z} + \frac{1}{v} \frac{\partial P_s}{\partial t} = \frac{g_s}{KA_e} P_s P_p - \alpha P_s \qquad (3.204)$$

The solutions of (4.119) and (4.120) can be obtained by substituting the following transformation

$$t' = t - \frac{z}{v_p} = t - \frac{z}{v_s} = t - \frac{z}{v} \qquad (3.205)$$

into (4.116) and (4.117) to get

$$P_p\left(z, t - \frac{z}{v}\right) = P_0\left(0, t - \frac{z}{v}\right) \exp(-\alpha z) \frac{1}{1 + H} \qquad (3.206)$$

$$P_s\left(z, t - \frac{z}{v}\right) = \frac{\omega_s}{\omega_p} P_0\left(0, t - \frac{z}{v}\right) \exp(-\alpha z) \frac{H}{1 + H} \qquad (3.207)$$

Where $P_0\left(0, t - \frac{z}{v}\right)$, H are able to be obtained by substituting (4.121) into (4.118) and (4.114).

---

[9]   The time delay is caused by the wave travelling the total length L of the single-mode fiber.

(2) **Steady-state solution:**

For a continuous wave in a single-mode fiber, the solution of the Stimulated Raman Scattering is said to be *the steady-state solution.* In this case, the coupling wave equations are

$$\frac{\partial P_p}{\partial z} = -\frac{g_p}{KA_e} P_p P_s - \alpha P_p \tag{3.208}$$

$$\frac{\partial P_s}{\partial z} = \frac{g_s}{KA_e} P_p P_s - \alpha P_s \tag{3.209}$$

Where the time factor $\exp(i\omega t)$ has been considered in (4.86). And the $\partial/\partial t$ term is absent since the wave amplitude is invariant along the single-mode fiber and then it is said to be *the steady-state solution.* The solution is able to be obtained by substituting

$$t = z/v \tag{3.210}$$

into (4.122) and (4.123) to get

$$P_p(z) = P_0(0) \exp(-\alpha z) \frac{1}{1+H} \tag{3.211}$$

$$P_s(z) = \frac{\omega_s}{\omega_p} P_0(0) \exp(-\alpha z) \frac{H}{1+H} \tag{3.212}$$

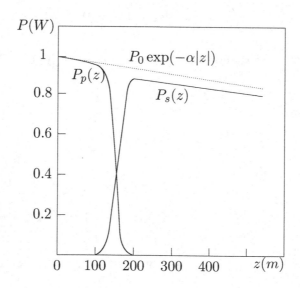

Figure 3-48   The steady-state solution of SRS [53]

Where H is able to be obtained by substituting (4.126) into (4.114). $P_0(0)$ is able to be obtained by substituting (4.126) into (4.118) to get

$$P_0(0) = P_p(0) + \frac{\omega_p}{\omega_s} P_s(0) \tag{3.213}$$

The steady-state solution (4.127) and (4.128) describes the transformation process between the pumping power $P_p(z)$ and the Stocks power $P_s(z)$ as shown in Figure 3-48. Obviously, the energy transfer from pump photons ($P_P(z)$) into the Stocks photons ($P_s(z)$) along the whole silica fiber, which results in the amplification of the signal photons ($P_0(z)$) along the whole silica fiber according to (4.129).

The structure of the Raman Scattering Amplifiers

The Raman Scattering Amplifiers are implemented as shown in Figure 3-49. Where:

    a. The pump source is consist of a combination of the semiconductor laser diodes.

    b. The forward Stocks wave is been used in Figure 3-49(a).

    c. The backward Stocks wave is been used in Figure 3-49(b).

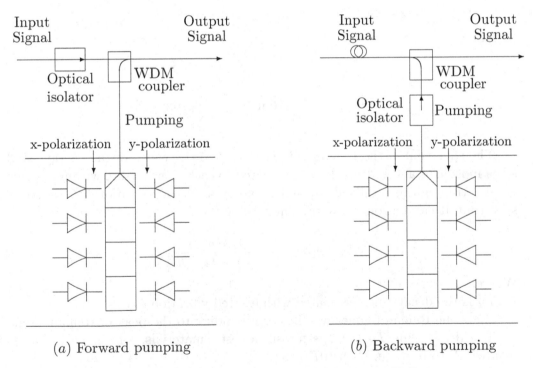

(a) Forward pumping   (b) Backward pumping

Figure 3-49   Raman amplifier structure [58]

The spectrum of the Raman gain

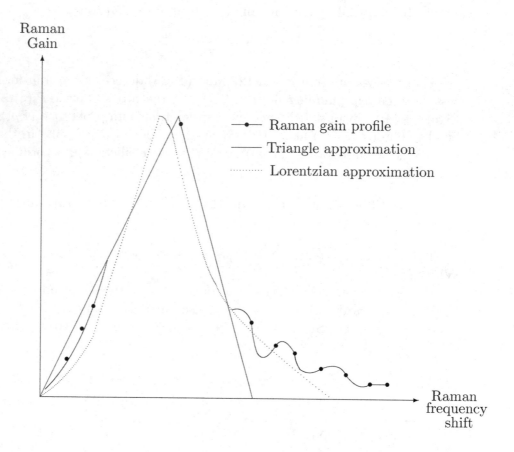

Figure 3-50   Raman gain profile [59]

The spectrum of the Raman gain is related to the width of the energy band of silica molecules and to time decay associated to each energy state within the energy band. Although it is difficult to find an analytical representation for Raman gain spectrum, it can be roughly approximately by the *Lorentzian Spectral profile* given as

$$g_g(\omega_R) = \frac{g_g(\Omega_R)}{1 + (\omega_R - \Omega_R)^2 T_R^2} \tag{3.214}$$

Where:

T_R is the decay time associated with excited vibration states.

Ω_R is the Raman frequency shift corresponding to the peak of Raman gain.

The decay time T_R is $0.1ps$ for silica-based materials, which makes the gain bandwidth to be wider than $10THz$.

The raman gain peak $g_R(\Omega_R) = g_{Rmax}$ is roughly between $10^{-12}m/W$ and $10^{-13}m/W$ for the wavelengths about $1,300nm$.

The approximation of the Ram gain with *Lorentizian curve* for silica fiber as well as a typical shape of the actual gain profile are shown in Figure 3-50. The actual gain profile extends over frequency range of about $40THz$ (which is approximately $300nm$), with a peak around $13.2THz$. There are also several smaller peaks located around frequency of $15, 18, 24, 32,$ and $37THz$.

The gain profile can also be approximately by a *triangle function* as

$$g_g(\omega_R) = \frac{g_g(\Omega_R)\omega_R}{\Omega_R} \tag{3.215}$$

This approximation is also shown in Figure 3-50.

Threshold value of pump power

It it important to estimate a threshold value of the pump power, above which the Raman Scattering takes a Stimulated character. The *threshold power* is usually defined as the incident power at which the half power is eventually converted to the Stocks signal.

The threshold can be estimated by solving (4.124) and (4.125). For that purpose, value g_g from (4.124), and (4.125) should be approximately by the peak value $g_R(\Omega_R)$. Accordingly, the amplification of the Stocks power along the distance L can be expressed as

$$P_s(L) = P_{s0} \exp\left[\frac{g_{gmax}P_{s0}L}{2A_{eff}}\right] \tag{3.216}$$

The value P_{s0} that corresponds to the Raman threshold P_{th}, as defined above, is

$$P_{th} = P_{s0} \approx \frac{16A_{eff}}{g_{Rmax}L_{eff}} \tag{3.217}$$

Where:

L_{eff} is the effective length.

The estimate Raman threshold is about $500mW$ for typical value of the fiber parameters (for $A_{eff} = 50\mu m^2$, $L_{eff} = 20km$, and $g_R(\Omega_R) = 7. \cdot 10^{-13}m/W$).

Summery

Stimulated Raman Scattering (SRS) can be effectively used for optical signal amplification since it can enhance the optical signal level by transferring the pump to the signal. And then Raman amplifier can improve the performance of optical transmission systems by providing an additional optical power margin.

However, SRS effect could be quite detrimental in the dense WDM transmission systems since the Raman gain spectrum is very broad, and energy transfer occurs from the low wavelength channels to the high-wavelength ones. In this situation, the optical fiber acts as the Raman amplifier since longer wavelengths are amplified by using power carried by lower wavelengths, which serve as the multiple Raman pumps.

The cost-effective design of Raman amplifier was enabled by the availability of reliable high-power pump lasers as shown in Figure 3-49.

3.5.4   Brillouin amplifier

Brillouin amplifiers are based on Brillouin Scattering, which is a physical process that occurs when an optical signal mode interaction with acoustical phonons, rather than with the glass molecules in Raman amplifier. During this process an incident optical signal reflects backwards from the grating formed by acoustic vibrations and down shifts in frequency (from ω_p to ω_s), as illustrated in Figure 3-51(a) and (b), respectively.

The acoustic vibrations originate from the thermal effect if the power of an incident optical signal is relatively small. In this case, the amount of backward scattered Brillouin signal is also small. If the power of the incident optical signal goes up, it increases the material density through the electrostrictive effect. The change in density enhances acoustic vibrations and forces the Brilliouin scattering to take a form of Stimulated Brillouin Scattering (SRS).

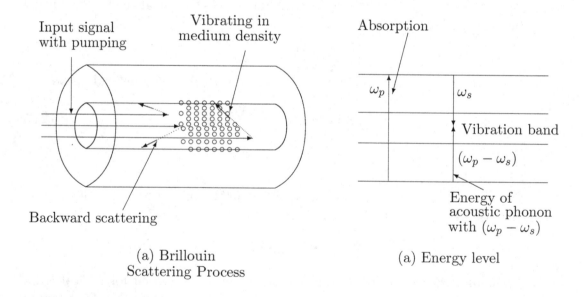

(a) Brillouin
Scattering Process

(a) Energy level

Figure 3-51   Brillouin scattering process [60]

The parametric interaction among Pump, Stocks signal and acoustical waves require both energy and momentum conservation. The energy is effectively conserved through the frequency downshift, while the momentum conservation occurs through the backward direction of the Stocks signal. The frequency downshift is expressed by the Brilliouin Shift Ω_B, which is given as

$$\Omega_B = 2n\omega_p V_A/c \qquad (3.218)$$

Where:

    n is the refractive index of the fiber material,

    V_A is the acoustic wave velocity,

    c is the light speed in vacuum, and

    ω_p is the optical pump frequency.

Equation (3.218) can also be written as

$$f_B = \frac{\Omega_B}{2\pi} = \frac{2nV_A}{\lambda_p} \qquad (3.219)$$

Where $\omega_p = 2\pi c/\lambda_g$ was used. By inserting typical values of parameters ($V_A = 5.96km/s$, $\lambda_p = 1150nm$, $n = 1.45$) in Equation (3.219), the frequency shift becomes $f_B = 11.5GHz$. This frequency shift is fiber material dependent and vary from $10.5GHz$ to almost $12GHz$ for different fiber material.

Nonlinear wave equations

As we mentioned above, the amplification of light in silica fiber caused by the Stimulated Brillouin Scattering is an amplitude coupling of pump wave mode with the Stocks wave mode , or simple say is caused by the nonlinearity in silica fiber.

In the nonlinear dispersion medium such as an optical single mode fiber, the Maxwell equation is able to be simplified to

$$\nabla^2 \mathbf{E} - \left(\frac{1}{c^2}\right)\frac{\partial^2}{\partial t^2}(\mathbf{E} + 4\pi\mathbf{P}) = \left(\frac{4\pi}{c^2}\right)\frac{\partial^2}{\partial t^2}\mathbf{P}^{(NL)} \qquad (3.220)$$

Where $\mathbf{P}$ and $\mathbf{P}^{(NL)}$ denotes the linear polarization vector and the nonlinear polarization vector, respectively.

For the sake of simplification, it is reasonable to assume that the input signal in a single-mode fiber has an uniform distribution of linear polarization states over the transverse cross section. And the direction of the polarization is x, the propagation direction is z. Then

$$P = \chi^{(1)}E(r, \theta, z; t) \qquad (3.221)$$

$$P^{(NL)} = \chi^{(3)}E^3(r, \theta, z; t) \qquad (3.222)$$

Note, the single-mode fiber is made of the crystal and the crystal is an antisymmetrical material. Therefore, the second order of electrical susceptibility $\chi^{(2)} = 0$.

Substituting (3.221) and (3.222) into (3.220), we have

$$\frac{d^2}{dz^2}E(r, \theta, z; t) - \left(\frac{n_L^2}{c}\right)^2\frac{d^2}{dt^2}E(r, \theta, z; t) = \frac{16\pi\chi^{(3)}}{c^2}\frac{d^2}{dt^2}[E^3(r, \theta, z; t)] \qquad (3.223)$$

Where $\dfrac{d}{dx} = \dfrac{d}{dy} = 0$ since it is assumed that the input signal in single-mode fiber has an uniform distribution of linear polarization states over the transverse cross section.

It is easy to understand that the linear polarization vector $\mathbf{P}$ will cause the wave dispersion, and the nonlinear polarization vector $\mathbf{P^{(NL)}}$ will cause the new frequency components. And then the field in the single-mode fiber is able to be expressed as

$$
\begin{aligned}
|E(r, \theta, z; t)| \;=\; & \sum_{m} \psi_m(r, \theta) \, |A_m(z, t)| \, \cos(k_m z - \omega_m t + \phi_m) \\
& + \sum_{m'} \psi_{m'}(r, \theta) \, |A_{m'}(z, t)| \, \cos(k_{m'} z - \omega_{m'} t + \phi_{m'})
\end{aligned}
$$

$$(3.224)$$

Where m denotes the wave propagating in the forward direction with a frequency ω_m, and m' denotes the wave propagating in the backward direction with another frequency $\omega_{m'}$. Both of them belong to the single transverse waves in single-mode fiber .

For the Stimulated Brillouin Scattering (SRS), assume that the power of the pump wave (p) injected into the single-mode fiber is above *certain threshold power*, and then the pump wave will stimulate the first order of anti-stocks wave (a), but no forward Stocks wave, however, it is not large enough to stimulate the higher order of stocks waves.

Next, we may consider that the field amplitude

$$
F_m(z, t) = N_m A_m(z, t) \tag{3.225}
$$

so that the power of the field will be

$$
P_m(z, t) = |F_m(z, t)|^2
$$

And then the normalization condition gives

$$
N_m^2 = \frac{n_{mL}\, c}{8\pi} \int_0^\infty \int_0^{2\pi} \psi_m^2(r, \theta) r \, dr \, d\theta \tag{3.226}
$$

Where n_{mL} is the linear refraction index at the frequency ω_m.

Substituting (3.225) into (3.223), we have

$$
\begin{aligned}
E(r, \theta, z; t) \;=\; & \frac{1}{2} \sum_m \psi_m(r, \theta) \left[\frac{F_m(z, t)}{N_m} e^{i(k_m z - \omega_m t)} \right. \\
& \left. + \frac{F_m^*(z, t)}{N_m} e^{-i(k_m z - \omega_m t)} \right]
\end{aligned}
$$

$$+\frac{1}{2}\sum_{m'}\psi_{m'}(r,\theta)\left\{\frac{F_{m'}(z,t)}{N_{m'}}e^{i[k_{m'}(L-z)-\omega_{m'}t]}\right.$$

$$\left.+\frac{F_{m'}^{*}(z,t)}{N_{m'}}e^{-i[k_{m'}(L-z)-\omega_{m'}t]}\right\} \tag{3.227}$$

Where L is the length of the single-mode fiber from the inject point of the pump wave to the end point of the single-mode fiber.

Stimulated Brillouin Scattering

Stimulated Brillouin Scattering is caused by the interaction between the optical fiber mode and the acoustic wave under an injection of a narrow pumping source. **Since the matching condition of the wave vector, the Stimulated Brillouin Scattering only cause the backward Stocks wave but no forward Stocks wave. And then the Brillouin amplification occurs only in the backward direction..**

Therefore, substituting $m=p$, (means pumping wave), $m'=s=B$, (means Stocks wave "$s=B$"), into (3.227) and then into (3.224), we have the coupling wave equations for the Stimulated Brillouin Scattering as follows:

$$\frac{\partial P_p}{\partial z}+\frac{1}{v_p}\frac{\partial P_p}{\partial t}=\frac{g_p}{KA_e}P_BP_p-\alpha_pP_p$$

$$\tag{3.228}$$

$$\frac{\partial P_B}{\partial z}-\frac{1}{v_s}\frac{\partial P_B}{\partial t}=-\frac{g_s}{KA_e}P_pP_B+\alpha_sP_B$$

Where

$$g_p=3\omega_p\left|Im[\chi^{(3)}(-\omega_p,-\omega_s,\omega_s,-\omega_p)]\right|$$

$$g_s=3\omega_s\left|Im[\chi^{(3)}(-\omega_s,-\omega_p,\omega_p,-\omega_s)]\right|$$

That means that the Stimulated Brillouin Scattering is also caused by the imaginary part of $\chi^{(3)}$.

The process to find the solution is almost the same as that in the Stimulated Raman Scattering . Here, we only take a very short words to explain this process. Now we will consider two situations:

(1) **Quasi-steady-state solution:**

For a pulse signal in a single-mode fiber, if the duration of the pulse signal is larger than the time delay [10] for several times, the dispersion in the single-mode

---

[10] The time delay is caused by the wave traveling the total length L of the single-mode fiber.

fiber is able to be ignored, in this case, the transient-state solution is said to be *the quasi-steady-state solution*. And then,

$$v_p = v_s = v, \qquad \alpha_p = \alpha_s = \alpha$$

and the Stimulated Brillouin Scattering coupling wave equations (3.228) becomes

$$\frac{\partial P_p}{\partial z} + \frac{1}{v} \frac{\partial P_p}{\partial t} = - \frac{g_p}{KA_e} P_p P_B - \alpha P_p \qquad (3.229)$$

$$\frac{\partial P_B}{\partial z} + \frac{1}{v} \frac{\partial P_B}{\partial t} = \frac{g_s}{KA_e} P_B P_p - \alpha P_B \qquad (3.230)$$

The solutions of (3.229) and (3.230) can be obtained by substituting the following transformation

$$t' = t - \frac{z}{v_p} = t - \frac{z}{v_s} = t - \frac{z}{v} \qquad (3.231)$$

into (3.229) and (3.230) to get

$$P_p\left(z, t - \frac{z}{v}\right) = P_0\left(0, t - \frac{z}{v}\right) \exp(-\alpha z) \frac{1}{1+H} \qquad (3.232)$$

$$P_B\left(z, t - \frac{z}{v}\right) = \frac{\omega_s}{\omega_p} P_0\left(0, t - \frac{z}{v}\right) \exp(-\alpha z) \frac{H}{1+H} \qquad (3.233)$$

Where $P_0\left(0, t - \frac{z}{v}\right)$, H are able to be obtained by substituting (3.231) into (4.118) and (4.114).

(2) **Steady-state solution:**

For a continuous wave in a single-mode fiber, the solution of the Stimulated Raman Scattering is said to be *the steady-state solution*. In this case, the coupling wave equations are

$$\frac{\partial P_p}{\partial z} = - \frac{g_p}{KA_e} P_p P_B - \alpha P_p \qquad (3.234)$$

$$\frac{\partial P_B}{\partial z} = \frac{g_s}{KA_e} P_p P_B - \alpha P_B \qquad (3.235)$$

Where the $\partial / \partial t$ term is absent since the wave amplitude is invariant along the single-mode fiber and then it is said to be *the steady-state solution*. The solution is able to be obtained by substituting

$$t = z/v \qquad (3.236)$$

into (3.232) and (3.233) to get

$$P_p(z) = P_0(0) \exp(-\alpha z) \frac{1}{1+H} \qquad (3.237)$$

$$P_B(z) = \frac{\omega_s}{\omega_p} P_0(0) \exp(-\alpha z) \frac{H}{1+H} \tag{3.238}$$

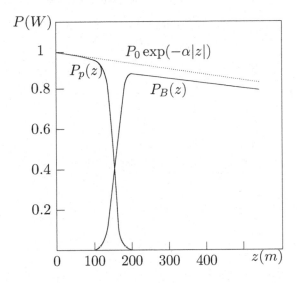

Figure 3-52   The steady-state solution of SBS [53]

Where H is able to be obtained by substituting (3.236) into (4.114). $P_0(0)$ is able to be obtained by substituting (3.236) into (4.118) to get

$$P_0(0) = P_p(0) + \frac{\omega_p}{\omega_s} P_B(0) \tag{3.239}$$

The steady-state solution (3.237) and (3.238) describes the transformation process between the pumping power $P_p(z)$ and the Stocks power $P_B(z)$ as shown in Figure 3-52. Obviously, the energy transfer from pump photons ($P_P(z)$) into the Stocks photons ($P_B(z)$) along the whole silica fiber, which results in the amplification of the signal photons ($P_0(z)$) along the whole silica fiber according to (4.129).

Spectrum of the Gain (3.240)

The scattering Stocks photons will have equals frequency but will be dispersed within a frequency band. The number of photons corresponding to any specified frequency within the band determines the value of the Brillouin gain with respect to that frequency. The spectrum of the Brillouin gain is related to the lifttime of acoustic photons that can be characterized by the time constant T_B. The gain spectrum can be approximated by Lorentzian spectral profile as

$$g_B(\omega_B) = \frac{g_B(\Omega_B)}{1 + (\omega_B - \Omega_B)^2 T_B^2} \tag{3.241}$$

and shown in Figure 3-53. Where Ω_B is the Brillouin frequency shift calculated by equation (3.218).

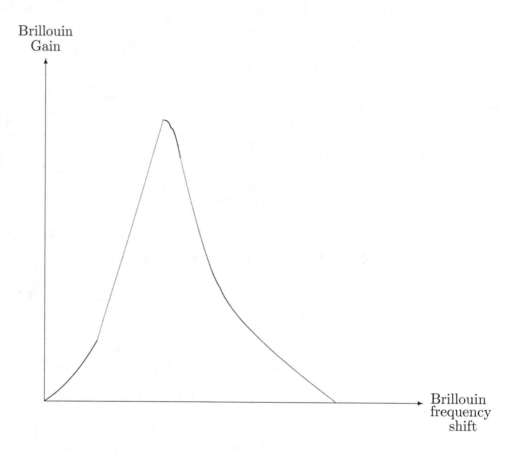

Figure 3-53   Brillouin gain profile [90], [62]

It is known that:

Not only just Ω_B, but also the width of the function given by (3.218), depend on the characteristics of the fiber material.

In addition, the Brillouin gain is dependent to the fiber waveguide characteristics.

The Stimulated Brillouin Scattering (SBS) gain bandwidth will be $17MHz$ at $\lambda = 1550nm$, while it can be almost $100MHz$ in doped silica fiber. The typical value of the SBS gain bandwidth is about $50MHz$.

The maximum value of the SBS gain is about $30dB$.

Chapter 4

Signals in Optical Fiber

4.1 Introduction

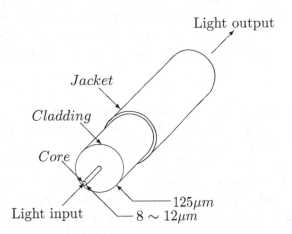

Figure 4-1   Single mode fiber

An optical fiber is a very thin silicon glass cylinder, consist of core and cladding. Cladding is surrounding the core and has a tubular jacket surrounding it as indicated in Figure 4-1.

When light enters one end of the fiber, it is confined within the fiber and traveling until it leaves the fiber at the another end of the fiber. During the traveling, the light suffers:

(1) Attenuation: The pulse will be weaker because all glass absorb light, making recovery of a bit stream impossible.

135

(2) Dispersion: The pulse of light will be spread out. A narrow pulse become wider and ultimately joins with the pulse behind, making recovery of a reliable bit stream impossible.

(3) Polarization unstable: Conventional communication optical fiber is cylindrically symmetry but contains some imperfection. The light traveling in this fiber will be changed in polarization, making the photo-detector can not receive a maximum signal by the end of the fiber.

(4) Nonlinear effect: The power been sent in a single mode fiber is limited by the nonlinearity in single mode fiber. This is about half a watt in stand single mode fiber.

(5) Noise: One of great benefits of optical fiber communications is that the fiber doesn't pick up the noise from outside circumstance, such as electromagnetic field noise from the power station and radio from the wireless station. However, there are various kinds of noise that can come from components within the system itself, such as mode partition noise in single mode fiber and modal noise, which is a phenomena in multi-mode fiber.

Now, we will discuss the transmission characteristic of optical fiber:
(1) Why the light can be confined within the core of fiber?
(2) The attenuation of optical fiber.
(3) The dispersion in optical fiber.
(4) The polarization un-stability.
(5) The nonlinearity effect in single mode fiber.
At beginning, we have to discuss how the light can be confined in the core of the fiber and the cladding look likes a mirror for the light inside the core.

4.1.1   Critical angle θ_c

The light traveling in an idea optical fiber can be explained by the critical angle θ_c as indicated in Figure 4-2. Where an idea optical fiber contains a core (n_1) and a cladding (n_2), and the end of the fiber is covered by air.

Now, when a light with a certain acceptance angle θ_a is entering a fiber, it must fall in order to be totally reflected inside the fiber. If light comes in along a ray path at an angle smaller than θ_a, then when the ray path hits the core-cladding boundary at point B, **it must be total reflected** and bound back to point C under the condition of

$$\theta_i \geq \theta_c \qquad\qquad (4.1)$$

If

$$\theta_i \leq \theta_c \qquad\qquad (4.2)$$

the light will go over the core-cladding boundary and get into the cladding and lost.

Where θ_c is the critical angle.

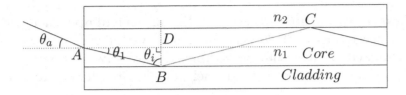

Figure 4-2   Light traveling in an optical fiber

Critical Angle θ_c is an incident angle such that the transmission angle $\theta_t = 90^0$ and total reflection occurs. Which occurs if the wave travels from a medium with **higher** index n_1 into a medium with **lower** index n_2. In this case, if the incident angle

$$\theta_i \geq \theta_c, \tag{4.3}$$

total reflection occurs and the transmission angle

$$\theta_t = 90^0. \tag{4.4}$$

Therefore, critical angle θ_c can be found by Snell's law of refraction (4.226) in Section 4.7.5.2 this Chapter,

$$n_1 \sin \theta_i = n_2 \sin \theta_t \tag{4.5}$$

with $\theta_t = 90^0$, namely,

$$\theta_c = \sin^{-1} \frac{n_2}{n_1} \tag{4.6}$$

4.1.2   Numerical aperture NA

As indicated in Figure 4-2, there is a certain acceptance angle θ_a within which light entering a fiber must fall in order to be totally reflected inside the fiber. If the light comes in with a angle $\theta > \theta_a$, then when the ray path hits the interface of core-cladding at the point B, we found, from Figure 4-2, that the incident $\theta_i < \theta_c$ and the light will be passed into the cladding and lost.

There are three indices (n_0, n_1, n_2) involved in determining the acceptance angle θ_a, those are the air (n_0) outside the fiber, core (n_1) and cladding (n_2), respectively.

By Snell' law of refraction (4.5), we have

$$n_0 \sin \theta_a = n_1 \sin \theta_1 \tag{4.7}$$

which, by considering the right triangle ABD, can be written as

$$n_0 \sin \theta_a = n_1 \cos \theta_c = n_1 \sqrt{1 - \sin^2 \theta_c} \tag{4.8}$$

Using (4.6), (4.8) becomes

$$n_0 \sin \theta_a = \sqrt{n_1{}^2 - n_2{}^2} \tag{4.9}$$

Where the quantity $n_0 \sin \theta_a$ is called NA, the **numerical aperture** of the fiber.

In fact, the acceptance angle θ_a is a vertex angle with a cone of acceptance and any lens system to couple light into the fiber must delivery its light within this cone. If the light is arriving at wider angles, namely, $\theta > \theta_a$, it will not propagate down the fiber.

Considering that $n_0 = 1$ and the difference in core and cladding indices $n_1 - n_2 << 1$, we may replace $\sin \theta_a$ by θ_a, and then, from (4.9),

$$NA = \theta_a = \sqrt{n_1{}^2 - n_2{}^2} \approx n_1 \sqrt{2\triangle} \tag{4.10}$$

Where $\triangle$ is the fractional index difference

$$\triangle = \frac{n_1 - n_2}{n_1} \tag{4.11}$$

4.2 Modes in optical fiber

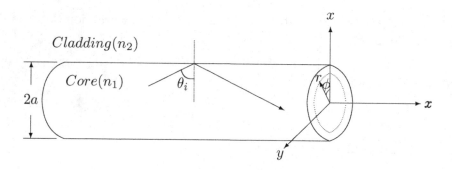

Figure 4-3   A view of an idea optical fiber

Figure 4-3 shows a view of an idea optical fiber. In which a light is incident into the core-cladding boundary with incident angle θ_i.

Considering that the wavelength of the light λ is much smaller than the fiber radius a, i.e. $\lambda << a$, the core-cladding interface can be considered plane. Therefore, we may use the conclusion of light obliquely incident onto a plane boundary in Section 4.7.5.2 this chapter as follows:

(1) When $\theta_i > \theta_c$, total reflection occurs in n_1-side of the boundary and then the n_2-side of the boundary presents an evanescent wave as shown in Figure 4-4.

(2) When $\theta_i < \theta_c$, the wave propagation in n_2-side of the boundary with a propagation constant k_r = real number in n_2-side of the boundary. We say this wave is lost in n_2-side of the boundary.

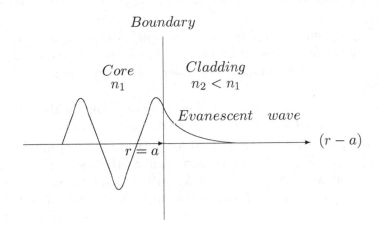

Figure 4-4   The boundary condition at the interface between two dielectrics.

Where, from (4.6) [1],

$$\theta_c = \sin^{-1}\frac{n_2}{n_1} \qquad (4.13)$$

Therefore, we have to control the incident angle $\theta_i > \theta_c$ so that the incident ray will be total reflection at interface of the core-cladding boundary inside the fiber core. In this case, the n_2-side of the boundary will present an evanescent wave, which is a wave with no power transmission into the n_2-side and the field strength in n_2-side decays exponentially with distance $(r - a)$ away from the boundary in n_2-side, which can be expressed as $e^{-\alpha(r-a)}$ and, from (4.236) in Section 4.7.5.2 this Chapter,

$$\alpha = \frac{2\pi}{\lambda}\sqrt{n_1{}^2\sin^2\theta_i - n_2{}^2}, \qquad (4.14)$$

and then the thickness of the cladding could be computed so that the cladding could be designed to be thick enough to keep the leakage of light to be negligible amount.

Figure 4-4 shown us a boundary condition at the core-cladding interface. This is a typical boundary condition at the interface between two dielectrics such as the optical fiber, namely, **the $E_\tau \neq 0$ at the interface between two dielectrics** instead of $E_\tau = 0$ at the metal surface. Therefore, the solution of the fields inside

---

[1]   Considering that $\varepsilon = \varepsilon_r\varepsilon_0$ and $n = \sqrt{\varepsilon_r}$ we have

$$\theta_c = \sin^{-1}\sqrt{\frac{\varepsilon_2}{\varepsilon_1}} = \sin^{-1}\frac{n_2}{n_1} \qquad (4.12)$$

and outside the optical fiber is much complicated than that in the metal waveguide.

Now, as shown in Figure 4-4, the total reflection at any interface of core-cladding boundary will result in a stand wave inside the core, which results in a numbers of field patterns inside the fiber core, each field pattern is so called a mode in optical fiber.

Mode propagation in fiber core corresponds to a standing waves across the diameter ($2a$) of the fiber core. Which results in multi-mode propagation inside the core. This fiber is called the multi-mode fiber, where the diameter of the multi-mode is ($2a = 50\mu m$).

However, for the single mode fiber, only single mode allow to be propagating in the optical fiber, where the diameter of the single mode is $2a = 10\mu m$.

Single mode fiber has been used for optical fiber communications system since the single mode fiber posses very important feathers such as very low attenuation ($0.2dB/km$) at $\lambda = 1.5\mu m$ and very low dispersion (to be zero at $1.3\mu m$). Therefore, we will pay more attention to the single mode fiber after then.

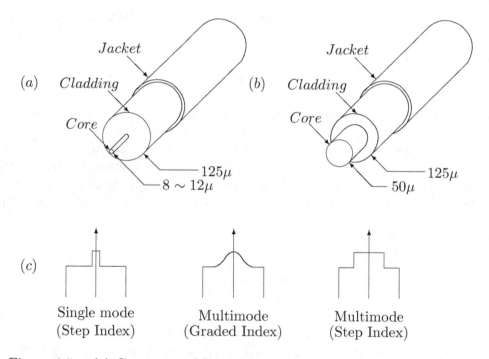

Figure 4-5   (a) Geometry of Single mode fiber,
                      (b) Geometry of Multimode fiber, (c)Index Profile of fiber

To analyze the modes in optical fiber, we have to go over the mathematic analysis of optical fiber. At beginning, we will confine ourself to the step index profile in single mode fiber as shown in Figure 4-5(a). Where

(1) The geometry of the single mode fiber is shown in Figure 4-5(a). In which, the core and cladding are made of silicon and the jacket is made of plastic or rubber to protect the core and cladding. The diameter of single mode fiber is $8 \sim 12\mu m$.

(2) The geometry of the multi-mode fiber is shown in Figure 4-5(b). The diameter of the multi-mode fiber is $50\mu m$.

(3) The step profile of the index versus radius of the multi-mode fiber is shown in Figure 4-5(c). In which the index of the core is n_1 and the index of the cladding is n_2.

4.2.1 Mode analysis of multi-mode optical fiber

Wave Equation

We may assume that the mediums, core and cladding, in multi-mode fiber in Figure 4-5(c) are lossless and a sinusoidal signal fields E, H is propagating in multi-mode fiber. Then the solution of the fields in multi-mode fiber can be obtained by the wave equation [2]

$$\nabla^2 \mathbf{E} + \left(\frac{n\omega}{c}\right)^2 \mathbf{E} = 0 \qquad (4.16)$$

and the boundary conditions in multi-mode fiber. Considering that

$$k_0 = \omega\sqrt{\mu_0\varepsilon_0} = \frac{\omega}{c} \qquad (4.17)$$

(4.16) can be rewritten as

$$\nabla^2 \mathbf{E} + n^2 k_0{}^2 \mathbf{E} = 0 \qquad (4.18)$$

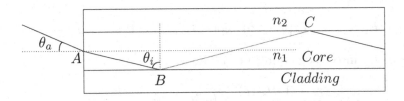

Figure 4-6   Wave propagation in an optical fiber

---

[2]   Wave equation, from (4.163) in Section 4.7.5.2 this Chapter,

$$\nabla^2 \mathbf{E} - \mu\epsilon\frac{\partial^2 \mathbf{E}}{\partial t^2} = 0 \qquad (4.15)$$

can be used for sinusoidal signal. In this case, $\frac{\partial}{\partial t} = j\omega$. Considering that $\varepsilon = \varepsilon_0\varepsilon_r = \varepsilon_0 n^2$ and light velocity $c = 1/\sqrt{\mu_0\varepsilon_0}$, (4.15) can be rewritten as (4.16).

In cylindrical coordinates system as shown in Figure 4-6, it becomes

$$
\begin{aligned}
\nabla^2 \mathbf{E} &= \mathbf{a}_r\left(\nabla^2 E_r - \frac{2}{r^2}\frac{\partial E_\phi}{\partial \phi} - \frac{E_r}{r^2}\right) + \mathbf{a}_\phi\left(\nabla^2 E_\phi - \frac{2}{r^2}\frac{\partial E_r}{\partial \phi} - \frac{E_\phi}{r^2}\right) + \mathbf{a}_z \nabla^2 E_z \\
&= -n^2 k_0{}^2(\mathbf{a}_r E_r + \mathbf{a}_\phi E_\phi + \mathbf{a}_z E_z)
\end{aligned} \tag{4.19}
$$

This equation can be separated into three equation as follows:

$$
\nabla^2 E_r - \frac{2}{r^2}\frac{\partial E_\phi}{\partial \phi} - \frac{E_r}{r^2} = -n^2 k_0{}^2 E_r \tag{4.20}
$$

$$
\nabla^2 E_\phi - \frac{2}{r^2}\frac{\partial E_r}{\partial \phi} - \frac{E_\phi}{r^2} = -n^2 k_0{}^2 E_\phi \tag{4.21}
$$

$$
\nabla^2 E_z = -n^2 k_0{}^2 E_z \tag{4.22}
$$

From which we understand that those are complicated equations since the components E_ϕ and E_r are coupling to each other in (4.20) and (4.21), only E_z component is in particulary simple form in (4.22), which can be expended to

$$
\frac{\partial^2 E_z}{\partial r^2} + \frac{1}{r}\frac{\partial E_z}{\partial r} + \frac{1}{r^2}\frac{\partial^2 E_z}{\partial \phi^2} + \frac{\partial^2 E_z}{\partial z^2} + n^2 k_0{}^2 E_z = 0 \tag{4.23}
$$

Given the solution of E_z and H_z, we can obtain the solution of other components by the Maxwell equations.

The solution of E_z for multi-mode fiber

Considering that

(1) E_z is propagation along z direction with the propagation constant β, so the z-dependance should be $\exp(-j\beta z)$,

(2) E_z is periodic in ϕ direction, so the ϕ-dependance should be $\exp(-j\phi)$, E_z can be written as

$$
E_z = E_z(r)\exp(-jl\phi)\exp(-\beta z), \quad l = 0, \pm 1, \pm 2, \cdots \tag{4.24}
$$

Substitution of E_z into wave equation (4.23), we have

$$
\frac{d^2 E_z(r)}{dr^2} + \frac{1}{r}\frac{dE_z(r)}{dr} + \left(n^2 k_0{}^2 - \beta^2 - \frac{l^2}{r^2}\right) = 0 \tag{4.25}
$$

This is a Bessel function, which is suitable not only in core ($n = n_1$),

$$
\frac{d^2 E_z(r)}{dr^2} + \frac{1}{r}\frac{dE_z(r)}{dr} + \left(n_1{}^2 k_0{}^2 - \beta^2 - \frac{l^2}{r^2}\right) = 0 \tag{4.26}
$$

but also in cladding ($n = n_2$),

$$
\frac{d^2 E_z(r)}{dr^2} + \frac{1}{r}\frac{dE_z(r)}{dr} + \left(n_2{}^2 k_0{}^2 - \beta^2 - \frac{l^2}{r^2}\right) = 0 \tag{4.27}
$$

Now, the wave should be propagating power in the core and be an evanescent wave in the cladding, which can be down by the definition of

$$U = (n_1{}^2 k^2 - \beta^2)^{1/2} a \qquad (4.28)$$

$$W = (\beta^2 - n_1{}^2 k^2)^{1/2} a \qquad (4.29)$$

in (4.26), (4.27) to get

$$\frac{d^2 E_z(r)}{dr^2} + \frac{1}{r}\frac{dE_z(r)}{dr} + \left[\left(\frac{U}{a}\right)^2 - \frac{l^2}{r^2}\right] E_z(r) = 0 \qquad (4.30)$$

and

$$\frac{d^2 E_z(r)}{dr^2} + \frac{1}{r}\frac{dE_z(r)}{dr} - \left[\left(\frac{W}{a}\right)^2 + \frac{l^2}{r^2}\right] E_z(r) = 0 \qquad (4.31)$$

Obviously, (4.30) is a Bessel function of the first kind and order l and (4.31) the modified Bessel function of the first kind and order l.

The solution of the Bessel function (4.30) is for core ($r < a$) and the solution of the modified Bessel function (4.31) is for cladding ($r > a$). Combining both of them gives a completed solution of E_z in multi-mode fiber as follows:

$$E_z \propto \begin{cases} J_0\left(U\dfrac{r}{a}\right)\cos l\phi & r \leq a \\[2mm] \dfrac{J_0(U)}{K_0(W)} K_0\left(W\dfrac{r}{a}\right)\cos l\phi & r \geq a \end{cases} \qquad (4.32)$$

Obviously, this solution matches the boundary condition at interface between core and cladding ($r = a$).

As shown in Figure 4-7, the Bessel function $J_l(x)$ for ($r \leq a$) look somewhat like decaying sinusoidal function, while the modified Bessel function $K_l(x)$ look like decaying exponent function. Which is exactly what we would expect the wave guiding condition in fiber, namely, it should be propagating power in the core and be an evanescent wave in the cladding. The ϕ-dependence of E_z is a periodic function $\cos l\phi$.

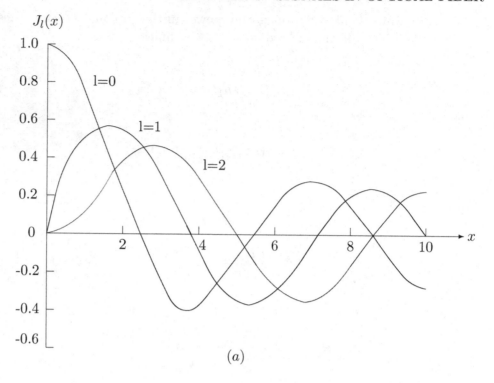

(a)

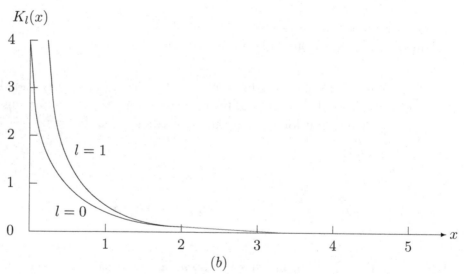

(b)

Figure 4-7   (a) Bessel functions of the first kind (J_l)
            (b) Modified Bessel function of the first kind (K_m)

Obviously, H_z should have the similar form as E_z except the amplitude of H_z is different from that of E_z.

Now we have obtained E_z and H_z. And then E_r and E_ϕ, H_r and H_ϕ can be obtained by substitution of E_z and H_z into Maxwell equation (4.140) to (4.143) in Section 4.7.5.2 this Chapter. Which is very tediously duty. The readers may find the solutions from [65].

How to make single mode fiber for operation at a given λ

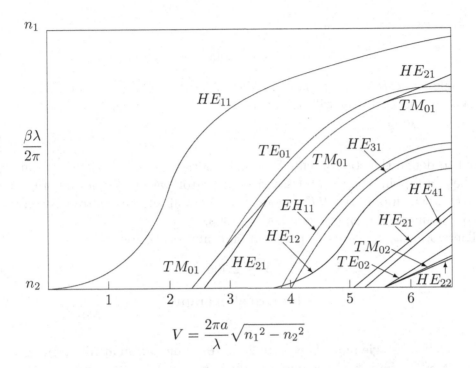

$$V = \frac{2\pi a}{\lambda} \sqrt{n_1{}^2 - n_2{}^2}$$

Figure 4-8   Normalized propagation constant versus "V number" for the first few orders of fiber modes. (From [66])

As we discussed before, the single mode fiber has been used in a long haul optical fiber communication network in the world.

How to make single mode fiber operating at a given λ?

(1) We may consider that it depends on the diameter of the core $2a$. The larger the core diameter $2a$, the more the number of modes propagating in the fiber.

(2) In fact, it also depends on the difference between n_1 and n_2 as indicated in Figure 4-8. In which, the abscissa is the dimensionless V-number parameter

$$V = a\sqrt{U^2 + W^2} = \frac{2\pi a}{\lambda}\sqrt{n_1{}^2 - n_2{}^2} = \frac{2\pi a}{\lambda}NA \qquad (4.33)$$

and NA is numerical aperture of the fiber, given by (4.10). The ordinate is also dimensionless normalized propagation constant $\dfrac{\beta\lambda}{2\pi}$.

Thus, to make single mode fiber operating at a given λ,

(1) A small radius a has to be used,

(2) As well as a small difference $n_1 - n_2$ (the indices difference of core and cladding) has to be used.

For instance, from Figure 4-8, if

$$V < 2.405 \tag{4.34}$$

only single mode HE_{11} is available in fiber. All other modes are cutoff, means that no energy of other modes will propagation in the fiber. Where the condition of

$$\beta = 0 \tag{4.35}$$

is used to define the state of cutoff for the modes in fiber. Obviously, from Figure 4-8, the $\beta = 0$ points on the abscissa gives the cutoff wavelength for all of the modes in fiber, in which only the HE_{11} mode has no cutoff and ceases to exist only when the core diameter $2a$ goes to zero if $V < 2.405$.

Therefore, the single mode propagation condition in fiber is

$$V \leq 2.405 \tag{4.36}$$

Numerical example

For a silicon single mode fiber with $2a = 10\mu m$ operation at $\lambda = 1.5\mu m$, fine the difference of the indices of core and cladding $(n_1 - n_2)$.

Taking the data ($\lambda = 1.5\mu m$ and $2a = 10\mu m$) above, if

$$V \leq 2.405 = \frac{2\pi a}{\lambda}\sqrt{n_1 2 - n_2{}^2} \tag{4.37}$$

we have

$$\sqrt{n_1 2 - n_2{}^2} = \sqrt{(n_1 + n_2)(n_1 - n_2)} \approx \sqrt{2n_1(n_1 - n_2)} \leq \frac{V\lambda}{2\pi a} = 0.0574 \tag{4.38}$$

where $n_1 = 1.45$ for silicon is known, then we have

$$(n_1 - n_2) \leq \frac{(0.0574)^2}{2n_1} = 0.0011367 \tag{4.39}$$

that means that n_1 is larger than n_2 but is very close to n_2 for a single mode fiber.

In this case, from (4.235) in Section 4.7.5.2 this Chapter, the critical angle at the core-cladding interface θ_c is

$$\theta_c = \sin^{-1}\sqrt{\frac{n_2}{n_1}} = \sin^{-1} 0.9921607 = 82.82^0 \tag{4.40}$$

and then, the incident angle into the interface of core-cladding should be

$$\theta_i > \theta_c = 82.82^0 \tag{4.41}$$

so as to keep propagating power in the core and an evanescent wave in the cladding.

Mode pattern in optical fiber

Figure 4-8 indicated that for a step indexes optical fiber:

(1) It can only support lowest mode EH_{11} when $V < 2.405$. In this case, all of modes have been cutoff except EH_{11}.

(2) It can support multi-modes when $V > 2.405$, and the number of the modes, that can be excited in a multi-mode fiber is approximately [67]

$$M = \frac{V^2}{2} \tag{4.42}$$

for large V, if we count the two degenerate states of the same mode as separate.

The solution of E_z in (4.32) and then the solutions of E_r and E_ϕ from Maxwell equations, similarly, the solution of H_z and H_r and H_ϕ from Maxwell equations, can provide us the mode pattern in the fiber as indicated in Figure 4-9.

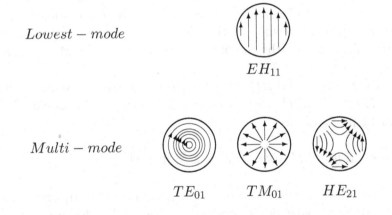

Figure 4-9   Electric field pattern for the first few orders modes

From which, we understand that:

(1) As we discussed in Section 4.7.5.2 this Chapter, TEM wave can present in a free space ($\mu = \mu_0, \varepsilon = \varepsilon_0$) since the medium is uniform any where in the free space. Where, in TEM wave, both electric field E_x and magnetic field H_y are transverse to the propagating direction z. And then, TEM mode can not propagating in optical fiber since optical fiber is not uniform medium, any wave propagating in optical fiber

should match the boundary condition at the interface of core-cladding as shown in Figure 4-4, namely, stand wave in the core and evanescent wave in the cladding.

(2) The simplest mode and the lowest mode in the optical fiber is the HE_{11} mode, this mode is caused by a TEM wave incidence almost pointing to the direction z of the fiber and bounce back at the interface of the core-cladding.

(a) This wave has no cutoff and ceases to exist only when the core diameter $2a$ goes to zero. Which is similar to TEM wave.

(b) However, this wave has nonzero components E_z and H_z so as to match the boundary condition at the interface of core-cladding.

(3) TE_{01} mode as well as TM_{01} mode are also can exist in optical fiber since both of them can match the boundary condition at the interface of core-cladding. Where, TE_{01} mode has nonzero component H_z. Therefore, it is called transverse electric wave. TM_{01} mode has nonzero component E_z. Therefore, it is called transverse magnetic wave.

(4) Most modes in optical fiber are $HE_{l,m}$ modes and and $EH_{l,m}$ modes. Where,

(a) Both l and m are integers. In a cross section of core, $2l$ is the number of maximum value in a 2π of azimuthal angle ϕ. In other words, l is the number of maximum value in a π of ϕ angle. m is the number of the maximum value along the r direction from the center of the core to the interface of core-cladding. For instance, for HE_{21} mode, the maximum number is 2 in π of ϕ angle, and is 1 in r-direction.

(b) Both HE modes and EH modes possesses nonzero $z-$components. In which, it is called HE modes if H_z-component predominates and it is called EH modes if E_z-component predominates.

(5) In optical fiber, Maxwell equations tell us that:

(a) Changing electric fields produce magnetic fields (**Ampere's law**),
(b) Changing magnetic fields produce electric fields (**Faraday's law**).
(c) In which, the electric field **E** and the magnetic field **H** are perpendicular to each other.
(d) Especially, the transverse fields $\mathbf{E}_t$ and $\mathbf{H}_t$ of each guiding mode are in phase so as to propagating power along the fiber.

From (a) to (d), we may imagine the magnetic field **H** from the available **E** of each mode in Figure 4-9. Where

$$\mathbf{S} = \mathbf{E}_t \times \mathbf{H}_t \tag{4.43}$$

is the energy density of each mode in fiber and $\mathbf{E}_t$ and $\mathbf{H}_t$ are the transverse electric field and transverse magnetic field, respectively in the core of the fiber.

4.2.2 Mode analysis of single mode optical fiber

As we mentioned before, if the diameter of the fiber $2a$ is small enough so as $V \leq 2.405$, the optical frequencies of all modes except HE_{11} are below cutoff and the HE_{11} mode has no cutoff and ceases to exist only when the core diameter $2a$ goes to zero. Where the pattern of HE_{11} mode is shown in Figure 4-9. In which, the electric field E is consist of E_y and E_z to form coils inside the core to match the boundary condition at the interface of core-cladding: stand wave in the core and evanescent wave in the cladding. Similarly, the magnetic field H is consist of $-H_x$ and H_z to form coils inside the core to match the boundary condition: stand wave in the core and evanescent wave in the cladding.

The solution of E_y for single mode fiber

We will following the way by D. Marcuse [68] to find the solution of EH_{11} mode in single mode fiber.

Single mode fiber is a weakly guiding fiber (since $n_1 > n_2$ and $n_1 \approx n_2$), and, from numerical example in (4.41) (that $\theta_i > 82.82^0$ if $V \leq 2.405$) and Figure 4-6, the guiding mode of the fiber is very nearly transverse and linearly polarized. Which means the electric field vector of the input field for single mode fiber consists likewise of one dimensional transverse component, say E_y.

Assume that the input field of a single mode fiber is a linearly polarized field

$$E_y = E_y(r)e^{-jl\phi}e^{-j\beta z}, \quad l = 0, \pm 1, \pm 2, \cdots \qquad (4.44)$$

As (4.23), the E_y should obeys a wave equation as follows:

$$\frac{\partial^2 E_y}{\partial r^2} + \frac{1}{r}\frac{\partial E_y}{\partial r} + \frac{1}{r^2}\frac{\partial^2 E_y}{\partial \phi^2} + \frac{\partial^2 E_y}{\partial z^2} + n^2 k_0{}^2 E_y = 0 \qquad (4.45)$$

By substitution of (4.44) into (4.45), we obtain

$$\frac{d^2 E_y(r)}{dr^2} + \frac{1}{r}\frac{d E_y(r)}{dr} + \left(n^2 k_0{}^2 - \beta^2 - \frac{l^2}{r^2}\right)E_y(r) = 0 \qquad (4.46)$$

Which is suitable not only for core $(n = n_1)$

$$\frac{d^2 E_y(r)}{dr^2} + \frac{1}{r}\frac{d E_y(r)}{dr} + \left(n_1{}^2 k_0{}^2 - \beta^2 - \frac{l^2}{r^2}\right)E_y(r) = 0 \qquad (4.47)$$

but also for cladding $(n = n_2)$

$$\frac{d^2 E_y(r)}{dr^2} + \frac{1}{r}\frac{d E_y(r)}{dr} + \left(n_2{}^2 k_0{}^2 - \beta^2 - \frac{l^2}{r^2}\right)E_y(r) = 0 \qquad (4.48)$$

Both of (4.47) and (4.48) are Bessel functions.

Now, the wave guiding condition in fiber is that it should be propagating power in the core and be an evanescent wave in the cladding. Which can be down by the definition of

$$U = (n_1{}^2 k^2 - \beta^2)^{1/2} a \tag{4.49}$$

$$W = (\beta^2 - n_1{}^2 k^2)^{1/2} a \tag{4.50}$$

and then (4.47), (4.48) might ne rewritten as

$$\frac{d^2 E_y(r)}{dr^2} + \frac{1}{r}\frac{dE_y(r)}{dr} + \left[\left(\frac{U}{a}\right)^2 - \frac{l^2}{r^2}\right]E_y(r) = 0 \tag{4.51}$$

and

$$\frac{d^2 E_y(r)}{dr^2} + \frac{1}{r}\frac{dE_y(r)}{dr} - \left[\left(\frac{W}{a}\right)^2 + \frac{l^2}{r^2}\right]E_y(r) = 0 \tag{4.52}$$

Obviously, (4.51) is a Bessel function of the first kind and order l and (4.52) the modified Bessel function of the first kind and order l.

The solution of the Bessel function (4.51) is for core ($r < a$) and the solution of the modified Bessel function (4.52) is for cladding ($r > a$). Combining both of them gives a completed solution of E_y in single mode fiber as follows:

$$E_y \propto \begin{cases} J_0\left(U\dfrac{r}{a}\right) & r \leq a \\ \dfrac{J_0(U)}{K_0(W)}K_0\left(W\dfrac{r}{a}\right) & r \geq a \end{cases} \tag{4.53}$$

Obviously, this solution matches the boundary condition at the interface between core and cladding ($r = a$) as indicated in Figure 4-9. Where $l = 0$, meaning no change along $\mathbf{a}_\phi$, has been considered since E_y, along $\mathbf{a}_y$, is the only direction for electric field in the transverse of single mode fiber.

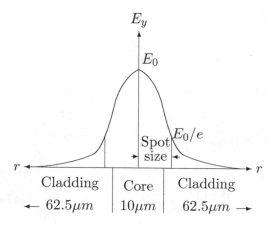

Figure 4-10 EH_{11} Field distribution in single mode fiber

As indicated in Figure 4-7, the Bessel function $J_0(x)$, for $r \leq a$, looks somewhat like decaying sinusoidal function, while the modified Bessel function $K_0(x)$, for $r \geq a$, look like decaying exponent function. Which is exactly what we would expect the wave guiding condition in fiber, namely, it should be propagating power in the core and be an evanescent wave in the cladding.

Recall Figure 4-10,

(1) From the field point of view, the solution of E_y matches the pattern of EH_{11} in Figure 4-9, which is helpful for us to understand that how to lunch the EH_{11} mode in single mode fiber and how to detect the signal from the EH_{11} mode in single mode fiber.

(2) From the light point of view, the solution of E_y also show us the spot size of the light as indicated in Figure 4-10. Which means that the light presents a spot size propagating in the single mode fiber. Here the "spot size" is defined as the distance between the center of the fiber and the point, at which the field is decayed to E_0/e or $0.37E_0$.

4.2.3   The conception of LP_{mn} modes (Linearly Polarized modes)

From the light point of view, the modes in optical fiber can be described by the pattern of the "spot".

As shown in Figure 4-10, LP mode is linearly polarized mode and LP mode numbering is different from TE and TM mode numbering. LP modes are designed LP_{lm} modes, in which m is the number of the maxima along a, the radius of the fiber, l is the number of maxima around half circumference. According to this definition of LP_{lm},

(1) The fundamental mode in optical fiber is referred to in different ways: LP_{01}, HE_{11} or TEM_{00}.

(2) Thus, TE, TM, HE and EH modes can all be summarized and explained using only a single set of LP modes according to the number of the nulls as indicated in Figure 4-11. Which is useful when we discuss multi-modes propagation in the multi-mode fiber.

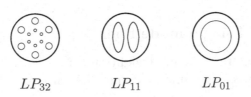

LP_{32} $\qquad$ LP_{11} $\qquad$ LP_{01}

Figure 4-11   Energy distribution of LP_{mn} modes

4.2.4   The characteristics of single mode fiber

Now we will make some conclusions for single mode fiber.

(1) Cutoff wavelength λ_c:

The cutoff wavelength λ_c is the shortest wavelength at which the fiber will be single-mode. Wavelengths shorter than the cutoff wave, $(\lambda < \lambda_c)$, will travel in multi-modes whereas wavelengths longer than cutoff wavelength, $\lambda > \lambda_c$, will travel in a single mode only. Where the lowest mode HE_{11} has no cutoff and ceases to exist only when the core diameter goes to zero. However, if $\lambda < \lambda_c$ the single mode fiber will become a multi-mode fiber. Thus, *the single mode condition of the single mode fiber is that the operation wavelength λ must be larger than the cutoff wavelength λ_c,* namely

$$\lambda > \lambda_c \qquad\qquad (4.54)$$

(2) Mode field of HE_{11} is indicated in Figure 4-9, in which the field is stand wave in the core and is evanescent wave in the cladding so as to propagation the power in the core of single mode fiber.

(3) The characteristics of single mode fiber ware specified by the International Telecommunications Union (ITU) in the 1980's as follows:

- Cladding diameter $= 125\mu m$.
- Mode field diameter $= 9 - 12\mu m$ at $1300nm$ wavelength.
- Maximum cutoff wavelength $= 1260nm$.
- Bend loss (at $1550nm$) must be less than $1dB$ for through 100 turns of fiber wound on a spool of $7.5cm$ diameter.
- Dispersion in the $1300nm$ band ($1285 - 1330nm$) must be less than $3.5ps/nm$ [3]. At wavelength around $1550nm$ dispersion should be less than $20ps/nm/km$.
- The dispersion slope must be less than $.095ps/nm/km$.

Obviously, for a long-haul communication system, the most important characteristics of single mode fiber are:

(1) The attenuation of single mode fiber. The smaller the attenuation the longer the distance between two optical repeater, called repeater spacing.

(2) The dispersion of single mode fiber. The smaller the dispersion the wider the available bandwidth of high speed digital signal.

Therefore, we will concentrate ourself to the attenuation and dispersion of single mode fiber in the following discussion.

4.2.5   Band loss of single mode fiber

As indicated in Figure 4-12, in straight single mode fiber, the ray is incident onto the interface between core and cladding with an incident angle

$$\theta_1 > \theta_c \qquad\qquad (4.55)$$

---

[3] Picoseconds of dispersion, per nanometer of signal bandwidth, per kilometer of distance traveled.

so as the field is stand wave in the core and is evanescent wave in the cladding to propagation the power in the core of single mode fiber.

However, in banding single mode fiber, the condition of $\theta_1 > \theta_c$ is not necessary to be held everywhere. If the ray is incident onto the interface between core and cladding with an incident angle

$$\theta_1 < \theta_c \tag{4.56}$$

part of light will be leaky into the cladding. This light will leave the core and be lost.

In fact,

(1) If the radius of a bend is relatively large, say $10cm$ or more, there will be almost no loss of light.

(2) However, if the band radius is very tight, say $1cm$, then some light will be lost.

Therefore, it has been specified by the International Telecommunications Union (ITU) that the bend loss (at $1550nm$) must be less than $1dB$ for through 100 turns of fiber wound on a spool of $7.5cm$ diameter.

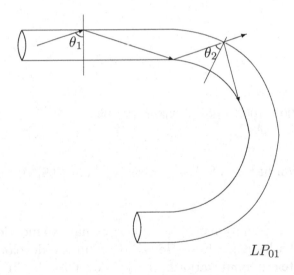

Figure 4-12   Conception of band loss in single mode fiber

4.3   Attenuation in optical fiber

Till now the optical fiber has been considered to be a lossless medium. In this case, the wave number is a real number k.

In fact, the wave propagating in optical fiber will suffer attenuation since optical fiber is made of silicon, which is a glass with loss. In this case, the wave number in optical fiber become a complex number

$$k = \beta + j\alpha \tag{4.57}$$

Where β is called the phase constant and α is called the attenuation constant.

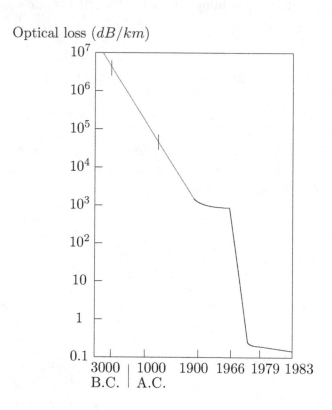

Figure 4-13   Attenuation history of glass materials (From ([69]),@ 1988 IEEE

When we discuss the attenuation of optical fiber, we have to mention "The father of optical fiber" Dr. Charles Kuen Kao. He is pioneer in the development and use of fiber optics in the telecommunications. Kao, known as the "Father of Fiber Optical Communications", was awarded half of the 2009 Nobel Prize in Physics for **"ground-breaking achievements concerning the transmission of light in fibers for communication."**

In 1966, Charles K. Gao and George Hockham proposed optical fiber at **STC Laboratories (STL)** at Harlow, England, when they showed that the losses of $1000dB/km$ in existing glass (compared to $5 - 10dB/km$ in coaxial cable) was due to contaminants, which could potentially be removed. This even has been recorded in Figure 4-13 by ([69]),@ 1988 IEEE.

In 1985, as a specialist in China, I have been so luck to have a opportunity to meet Dr. kao when he spent one year in West Germany, at SEL Research Center. During the discussion, as a consultant of SEL, he introduced his invention of optical fiber with his preform of optical fiber in his hand:

(1) At beginning (1966), the loss of glass is $1000dB/km$. However, he conformed that this was due to contaminants, which could be removed so as to reach $20dB/km$ in an optical fiber. Therefore, he persuaded Corning Glass Works to remove the contaminants from the preform of optical fiber.

(2) In deed, optical fiber was successfully developed in 1970 by Corning Glass Works, with attenuation $20dB/km$. Which is low enough for communication purposes. At the same time, GaAs semiconductor laser were developed, which is suitable for transmitting light through optical fiber cable for long distance.

(3) After then, Dr. Gao considered that the attenuation of optical fiber ($20dB/km$) can be reduced by removing iron (Fe), chromium (Cr) and nickel (Ni) in the fiber. In deed, after removed them by Corning Glass Works, the attenuation of fiber was further reduced from $20dB/km$ to $3dB/km$ as shown in Figure 4-13.

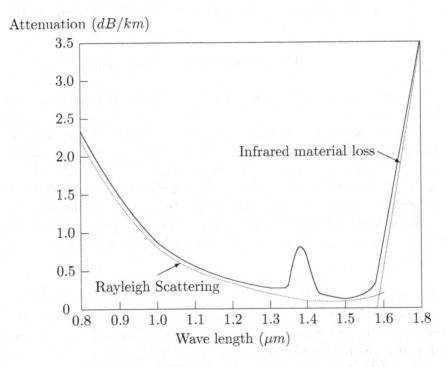

Figure 4-14   Typical attenuation of silicon fiber vs. wavelength

The typical attenuation of silicon single mode fiber is shown in Figure 4-14. In which, the single mode fiber is made of a glass containing about 4% of germanium dioxide (GeO_2) in the core.

Where, the silicon single mode fiber can be operating in several operation windows:

(1) $\lambda = 1.5\mu m$ window: At $1.5\mu m$ the attenuation is $0.12 dB/km$. In the vicinity of $1.5\mu m$, there is a window about $\triangle \lambda = 200nm$ wide , which equal to

$$\triangle f = \left(\frac{c}{\lambda}\right)\left(\frac{\triangle \lambda}{\lambda}\right) = 25,000 GHz \qquad (4.58)$$

wide having the attenuation less than $0.24 dB/km$.

(2) $\lambda = 1.3\mu m$ window: At $1.3\mu m$ the attenuation is about $0.27 dB/km$. In the vicinity of $1.3\mu m$, there is another window $\triangle f = 25,000 GHz$ wide.

Both of $1.3\mu m$ and $1.5\mu m$ windows can be operating with the the available laser based on Indium and Phosphide materials and the photodiode based on germanium or InGaAsP.

(3) $\lambda = 0.85\mu n$ widow: At $0.85\mu m$ the attenuation is about $2dB/km$. In this region, the laser can be easy to make by the same material system, Gallium and Arsenide, and the material of photodiode can be silicon. This region can be used in the short distance communication system with $25,000 GHz$ bandwidth.

The sources of the attenuation

The attenuation of the single mode fiber as indicated in Figure 4-14 is involved several sources:

As discussion above, at beginning (1966 to 1975), the **impurities** in the glass were the chief source of attenuation. Iron (Fe), chromium (Cr) and nickel (Ni) can cause significant absorbtion even in quantities as low as $1/bilillion$.

Today, techniques of purifying have improved to the point that the impurities are no longer a significant concern. The **dopants**, such as germanium dioxide (GeO_2) in the core, have the unwanted increasing of the absorption. This is the one reason that the attenuation of multi-mode optical fiber is higher than that of the single mode fiber.

Recently, the sources of the attenuation in silicon optical fiber are as follows:

(1) Most of the attenuation in fiber is caused by light scattered by minute variation in the density or composition of the glass. This is called "**Rayleigh Scattering**" as indicated in Figure 4-14. In fiber, Rayleigh scattering is inversely proportional to λ^4. This presents at short wavelength end, the shorter the wavelength the higher the attenuation. This is an intrinsic losses, we can't do a lot about Rayleigh scattering by improving the manufacture techniques. We understand that the sunset is red and the sky is blue. Which is also caused by Rayleigh Scattering.

There is another scattering called "**Mie scattering**" caused by the imperfection in the fiber of the size compared with the wavelength. This is not a significant concern with modern fibers as recent improvement of manufacturing techniques have all but eliminated this problem.

(2) At the long wavelength end, there is **infrared material loss** as shown in Figure 4-14. The tail of the absorption-loss spectrum consist of peaks up to $10^{10}dB/km$ at wavelength beyond $9\mu m$. These peaks are due to excitation of molecular vibrations by the incident photons.

(3) In the middle of the spectrum, a peak attenuation shown in Figure 4-14 is centered at $1338\mu m$. This absorption is caused by the presence of -OH atom band, or the water. The bond is resonant at the wavelength of $1338\mu m$.

(4) The incident photons can also stimulate electron transition if the photons are in ultraviolet, therefore this **ultraviolet material** loss also contributes to the overall loss spectrum. Which is much lower than the Rayleigh scattering, we didn't put in Figure 4-14.

4.4 Group velocity and dispersion in optical fiber

4.4.1 Phase velocity v_p and group velocity v_g

The phase velocity v_p is defined as the velocity of the phase surface along propagation direction. Which can be obtained by the solution of the wave equation as follows:[4]

$$E(z,t) = \exp\left[-j(\omega t - kz)\right] = e^{-\alpha z}\exp\left[-j(\omega t - \beta z)\right] \qquad (4.61)$$

and the phase velocity is the velocity of the phase surface $\omega t - \beta z = constant$ propagating along z direction, i.e.

$$v_p = \frac{\omega}{\beta} \qquad (4.62)$$

Where we only consider a single frequency signal propagating in the single mode fiber.

In fact, the optical signal in fiber is a light carrying on a digital signal. Which can be seen as a series of "light on" and "light off". The "light on" is correspondent to "1" and the "light off" is corespondent to "0" in a digital signal. This signal is no long a single frequency signal but a narrow bond signal. This signal possesses not only a phase velocity v_p but also a group velocity v_p. Which can be proved by following example.

For the sack of simplification, we assume that the signal at $z = 0$ is consist of

---

[4]   The wave equation in a homogeneous non-dispersive medium is

$$\frac{\partial^2 E}{\partial z^2} - \left(\frac{k}{\omega}\right)^2\frac{\partial^2 E}{\partial t^2} = 0 \qquad (4.59)$$

and the solution of the wave equation is

$$E(z,t) = \exp\left[-j(\omega t - kz)\right] \qquad (4.60)$$

two sinusoidal waves at slight different angular frequency $(\omega + \triangle\omega)$ and $(\omega - \triangle\omega)$ as follows:

$$E(0, t) = \exp\left[-j(\omega - \triangle\omega)\right]t + \exp\left[-j(\omega + \triangle\omega)\right]t \qquad (4.63)$$

At point z, we have

$$E(0, t) = \exp\left[-j(\omega - \triangle\omega)t - (\beta - \triangle\beta)z\right] + \exp\left[-j(\omega + \triangle\omega)t - (\beta + \triangle\beta)z\right]$$
$$(4.64)$$

Where, from (4.62), k is considered to be a linear function of ω and then at $(\omega \pm \triangle\omega)$ the propagation constant shall be $(\beta \pm \triangle\beta)$. And the equation (4.64) can be rewritten as

$$E(z, t) = \exp[-j(\omega t - \beta z)]\{\exp[j(\triangle\omega t + \triangle\beta z)] + \exp[-j(\triangle\omega t - \triangle\beta z)]\}$$

or

$$E(z, t) = 2\exp[-j(\omega t - \beta z)][cos(\triangle\omega t - \triangle\beta z)] \qquad (4.65)$$

This is a signal of high frequency ω with amplitude varies at a low frequency $(\triangle\omega t - \triangle\beta z)$.

(1) The high frequency $\exp[-j(\omega t - \beta z)]$ is a light wave, and the frequency ω is called the carrier frequency. And the velocity corresponding to the light wave is

$$v_p = \frac{\omega}{\beta}$$

Which is a phase velocity of the light wave.

(2) The amplitude with a low frequency $[cos(\triangle\omega t - \triangle\beta z)$ is an amplitude modulation signal, which is an information wave modulated on the carrier frequency wave. And the velocity of information wave is

$$v_g = \frac{d\omega}{d\beta} \qquad (4.66)$$

Which is so called the group velocity of the information wave.

As indicated in Figure 4-15, the sum of two sinusoidal wave is a group signal. Which means that:

(1) The information wave is a modulated wave, which is a light wave modulated by a slow sinusoidal wave.

(2) The velocity of information wave is no long the phase velocity v_p but the group velocity v_g.

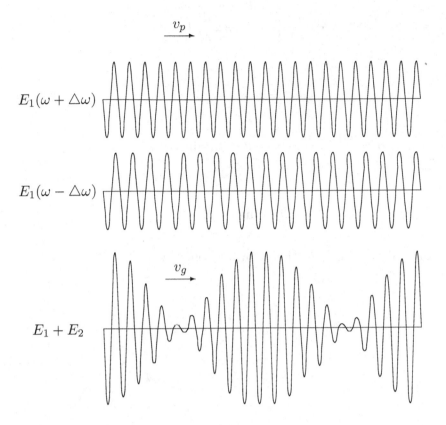

Figure 4-15   The sum of two sinusoidal signal

Phase velocity v_p

In optical fiber, the phase velocity is

$$v_p = \frac{\omega}{\beta}$$

or

$$v_p = \frac{c}{n} \tag{4.67}$$

since

$$\beta = \omega\sqrt{\mu_0\varepsilon_0}\sqrt{\varepsilon_r} = \frac{\omega n}{c} \tag{4.68}$$

Equivalent to (4.67), we may define the group velocity v_g as

$$v_g = \frac{c}{n_g} \tag{4.69}$$

Where

$$n_g = \frac{c}{v_g} = c\frac{d\beta}{d\omega} = c\frac{d}{d\omega}\left(\frac{\omega n}{c}\right)$$

or

$$n_g = n + \omega\frac{dn}{d\omega} = n + \omega\frac{dn}{d\lambda}\frac{d\lambda}{d\omega} = n + \omega\frac{dn}{d\lambda}\left(-\frac{2\pi c}{\omega^2}\right) = n - \lambda\frac{dn}{d\lambda}$$

or

$$n_g = n - \lambda\frac{dn}{d\lambda} \qquad (4.70)$$

Where $\lambda = \frac{2\pi c}{\omega}$ and $\beta = \frac{\omega n}{c}$ has been used in driving (4.70).

$$\textbf{Conclusion} \qquad (4.71)$$

The discussion above comes to the following conclusion:

$$v_p = \frac{c}{n} \qquad (4.72)$$

$$v_g = \frac{c}{n_g} \qquad (4.73)$$

Where

$$n_g = n - \lambda\frac{dn}{d\lambda} \qquad (4.74)$$

4.4.2   Dispersion in optical fiber

Dispersion is one of the parameters that limits the bit-rate on a link. **Dispersion is a difference in velocity of propagation across the signal spectrum.** Which will results in the receiving signal suffer a certain amount of time smearing or **inter-symbol interference** and then produces random data-dependent **bit error rate** even though the attenuation of single mode fiber is low enough at some wavelength regions for the long distance transmission.

The dispersion in optical fiber is mainly caused by [70][71][72]:
(1) Material (Chromatic) dispersion.
(2) Waveguide dispersion.
(3) Polarization dispersion.
When we discuss the dispersion in optical fiber, we assume that the optical fiber is lossless.

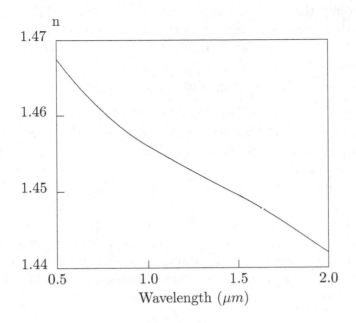

Figure 4-16   Relative index in silicon fiber vs wavelength [73]

Material (Chromatic) dispersion

Since $v_g = \dfrac{c}{n_g}$ and $n_g = n - \lambda\dfrac{dn}{d\lambda}$, **material dispersion** in single mode fiber is caused by the fact that relativity index n of the glass of silicon varies with the wavelength as indicated in Figure 4-16. In this case, some of wavelength has higher group velocity and then travel faster than others. Since every pulse signal is consist of a range of wavelengths, it will spread out to some degree during the long distance travel. Since the material dispersion is caused by the fact that the reflective index of glass varies with the wavelength, therefore, it is also called the **chromatic dispersion**.

All optical signals consist of a range of wavelength. For example, the optical signals are mostly digital signals such as digital TV, digital internal and digital telephone signals. Typically, the range of wavelength in optical pulses used in single mode fiber communication systems are from $.2nm$ wide to $5nm$ wide.

Waveguide dispersion

The profile of relative index in fiber has a very significant effect on the group velocity v_g.

(1) First, considering that the optical fiber should be propagating power in the core and be an evanescent wave in the cladding. In this case, the cladding-side of the boundary will present an evanescent wave, which is a wave with no

power transmission into the cladding and the field strength in cladding-side decays exponentially with distance $(r - a)$ away from the boundary in cladding-side, which can be expressed as $e^{-\alpha(r-a)}$ and, from (4.236) in Section 4.7.5.2 this Chapter, the attenuation constant

$$\alpha = \frac{2\pi}{\lambda}\sqrt{n_1{}^2\sin^2\theta_i - n_2{}^2},$$

From which we understand that the amount of the field overlap between core and cladding depends on strongly the wavelength λ. The longer the wavelength λ the smaller the attenuation constant α and then the further that the electromagnetic wave extends into the cladding.

(2) The RI (reflective index) experienced by the wave is an average of RI of core and cladding depending on the relative proportion of the wave that travels there. Thus, according to the discussion in (1), at shorter wavelength, a greater proportion of the wave is confined within the core, and then, the shorter wavelength "see" a higher RI than do longer wavelength. According to (4.73), the shorter wavelength tend to travel more slowly than longer ones. Thus dispersion occurs. Which is called the waveguide dispersion.

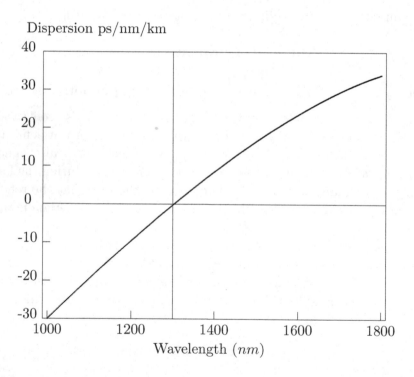

Figure 4-17   Dispersion of "standard" single mode fiber [74]

Cancelation of Two Dispersions

Now the material dispersion and the waveguide dispersion have opposite signs so they tends to counteract one anther. Figure 4-17 shows the wavelength dependent dispersion characteristic of "standard" single mode fiber. It is shown that the two forms of dispersion cancel one anther at wavelength $\lambda = \textbf{1310 nm}$.

Thus if the signal is sent at $\lambda = 1310nm$, dispersion will be minimized in "standard" single mode fiber.

In single mode fiber, the dispersion is usually quoted in picoseconds of dispersion per nanometer of spectrum width per kilometer distance ($ps/nm/km$).

Polarization Mode Dispersion

As shown in Figure 4-9, in single mode fiber, the polarization of electric field on cross section can be E_y or E_x for the lowest mode EH_{11}. Both of them are orthogonal one another and then without any interaction one another in a normal single mode fiber. In this case, both of them have the same characteristics including the propagation parameters along the fiber. This phenomena is called birefringence and two modes are called the birefringence modes.

In normal single mode fiber, a signal is consist of both polarizations. However, two polarization states are not maintained in standard single mode fiber. During its journey light is coupling from one polarization to the other randomly. Which happens at some place in the core where some very slight difference in reflective index exists for different polarization. The difference in reflective index will make the difference in group velocity v_g for different polarization, which will result in the **polarization mode dispersion** in single mode fiber. In which, power of the light is not lost but the polarization and the orientation of electric fields constantly changes. If there is a polarization sensitive device in the circuit , that has significantly higher losses in one polarization than the other, then the signal with the sum of two polarizations will change polarization constantly, which results in the change in total signal power. These changes result in the birefringence noise.

The polarization mode dispersion is also called "**Polarization Mode Noise**" in some publications since the mechanism is quite similar to the mechanism for model noise.

Non-zero Dispersion-Shifted Fiber

As we discussion above, the " **standard single mode fiber**" has its null dispersion wavelength at $\lambda = 1310nm$. At this wavelength, the material dispersion and the waveguide dispersion has been canceled one another. Therefore, at beginning, a lot of single mode fiber system operated at this wavelength since this networks can provide a very wide bandwidth for the information transmission.

However, there are a lot of good reasons that we would like to operate in the $1550nm$ band. For example:

(1) It can provide much lower attenuation for the long-haul transmission, $0.2dB/km$.

(2) It can provide Erbium doped fiber amplifier (**EDFAs**), which operates at

$1550nm$ band.

(3) **WDM** (wavelength division modulation) systems requires a wider bandwidth. Therefore, we would like to use the $1550nm$ window and **EDFAs**.

To this end, we have to face the high dispersion (around $17ps/nm/km$) at $1550nm$ in standard single mode fiber. This is very significant problem for long-haul transmission system. Which has led to development of special fiber that has minimum dispersion at $1550nm$ band.

The bast way to manufacture such special fiber is to modify the core profile to introduce dispersion in the opposite direction to cancel the material dispersion at $1550nm$.

Some successful special fiber are as follows:

(1) **Single mode dispersion shifted fiber** as indicated in Figure 4-18(a). The dispersion of this single mode fiber is minimum at $1550nm$. Which is down by manipulating a core profile to introduce dispersion in opposite direction from the operating direction of the material dispersion. This single mode fiber is good for a non-linear effect called "four wave mixing effect" since it provide the same speed for four wave. However, the signals stay in-phase over a long distance to make near-end crosstalk or interference between optical channels, which is no good for WDM transmission system.

(2) **Single mode flat dispersion fiber** as indicated in Figure 4-18(b). The dispersion over the whole range from $1300nm$ to $1700nm$ is less than $3ps/nm/km$. However, the attenuation of this fiber is high up to $2dB/km$.

(3) **Non-Zero Dispersion-shifted Fiber** as indicated in Figure 4-18(c). This fiber has a dispersion of around $4ps/nm/km$ in the $1530 - 1570nm$ band. In this range of $1530 - 1570nm$, this low but positive dispersion figure not only minimizes dispersion of the signal but also avoids the unwanted effect of four-wave mixing between WDM channels. AT&T calls its dispersion optimized fiber **"Tru-Wave"** and Corning called its **"SMF-LS"**. Both of them are similar one another in characteristic.

The penalty for obtaining the **Non-Zero Dispersion-shifted Fiber** is that the effective cross-section area of the core (A_{eff}) is only $55\mu m^2$ while the standard single mode fiber has A_{eff} of $80\mu m^2$. Which will results in the handling power of Non-Zero Dispersion-shifted Fiber relative lower than that of the standard single mode fiber.

The maximum power handling capability of a fiber is extremely important in high bit-rate systems and WDM systems. However, the smaller effect area A_{eff} will limits the maximum power handling capability of a fiber because of the **"Nonlinear-effect in single mode fiber"**.

In fact, the light almost confined in the core of a fiber and the intensity of the light is proportional to the reflective index n in the core. All of the non-linear effects are almost solely dependent on the intensity of light in the fiber core. The smaller the effect area A_{eff} , the higher the intensity of the light in fiber core. Therefore, the maximum power handling capability requires a lager A_{eff} .

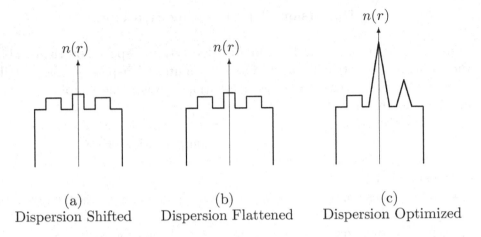

(a)
Dispersion Shifted

(b)
Dispersion Flattened

(c)
Dispersion Optimized

Figure 4-18   Profile of "Special" single mode fiber

Large Effective-Area Fiber (LEAF)

To this end

4.4.3   Control of dispersion in single mode fiber

Dispersion will broadens a pulse to a unacceptable amount. In practical optical fiber communication system, dispersion will become a problem for a receiver when it exceeds about 20% of the pulse length. Hence the higher the data rate, the more important to control the dispersion.

(1) Consider that the material dispersion and the the waveguide dispersion in single mode fiber are wavelength dependence, the best way to control the dispersion is to use a narrow spectral width laser. For instance, in a typical optical single mode fiber link, if the spectral width of laser is $5nm$, the total dispersion of the link is five times lager than that when we use a spectral width of $1nm$ laser.

(2) Waveguide dispersion can be manipulated so that it acts in the opposite direction to the material dispersion. In the later of 1980's the single mode fiber can be adjusted so that both of the material dispersion and the waveguide dispersion are canceled out at wavelength of $1310nm$. However, the attenuation of single mode fiber at $1310nm$ is double of that at $1550nm$. Meanwhile, Er-doped fiber amplifiers (EDFAs) has been developed and it only work at $1500nm$ band.

(3) Therefore, if we would like to use Er-doped fiber amplifiers (EDFAs), we have to work at $1550nm$ band. In this case, we can do alot work to reduce the waveguide dispersion such as to varying the reflective index profile of the core and cladding and changing the geometry of the fiber. Now it is possible to balance the two form of dispersion at $1550nm$ band. This fiber is called dispersion shifted fiber and the

more useful one is the **Non-Zero Dispersion-shifted Fiber**. Therefore, almost all of new long-haul single mode fiber system were been installed at $1500nm$ band.

Dispersion Compensating Strategies

The performance of optical communication system depends on the combination of the transmission attributes as well as the tributes of dispersion fiber attributes. In the first several years, low cost installation drives some simple compensation strategies. For instance:

(1) Transmitter-based electro-optical pre-compensation for dispersion.

(2) Receiver-based electronic dispersion compensation (**EDC**).

(3) Use of grating-based dispersion compensation modules for cost-effective long haul transmission.

(4) Intensive use of digital signal processing and coherent technology to extend un-compensation research at high data rates.

Recently, a novel compensation strategies have been proposed by Srikanth Raghavan etc. In which the **Non-Zero Dispersion-shifted Fiber** with $1 - 6ps/nm/km$ dispersion in C-band, has been used in two systems:

(1) One system, the Corning's **LEAF** optical fiber has been used to mitigate the nonlinear impact of dispersion compensation fiber.

(2) Another system, a combination of optical duobinary modulation format, LEAF optical fiber, and received-based **EDC** has been used to demonstrate a practical and flexible system that requires no in-line dispersion compensation.

Dispersion Compensating Fiber

It is possible to design a fiber profile where the total dispersion is $100ps/nm/km$ in the opposite direction to the dispersion caused by material. This is called the **Dispersion Compensating Fiber**. Which can be placed in series with an existed fiber link to "un-dispersion" the signals.

Now dispersion compensation fiber is commercially available with the following attributes:

(1) The dispersion is $-100ps/nm/km$.

(2) The attenuation is $0.5dB/km$.

(3) Typically, the dispersion compensation fiber has a much narrower core than standard single mode fiber. This will cause the optical power to be more tightly confined and then has a more sensitive to the nonlinear effect in single mode fiber.

Now a single mode fiber link using **Dispersion Compensating Fiber** is indicated in Figure 4-19. In which, the dispersion compensation single mode fiber is placed on a drum and is placed inside the room with the EDFA amplifier. In whole system, the $100km$ standard single mode fiber provide $1700ps/nm/km$ dispersion, which has been canceled out by the total dispersion of $-1700ps/nm/km$ provided by the dispersion compensation single mode fiber. And then the total dispersion of the compensation link is zero for the transmission signal.

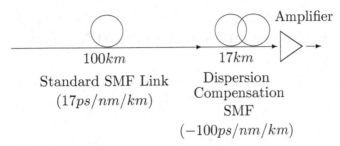

Figure 4-19   Dispersion compensation in SMF link

4.5   Polarization effects in single mode fiber

Polarization effects

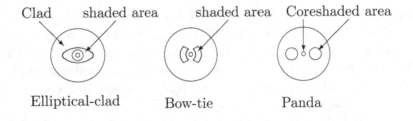

Figure 4-20   Polarization maintaining (PM) fiber

When we set up an optical fiber telecommunication system, the polarization effect of a single mode fiber has to be taken account. The reason is that:

(1) The standard single mode fiber doesn't maintain the polarization state during light traveling,

(2) However, semiconductor laser diodes emit light having a specific polarization direction, and various forms of tunable filters, amplifiers, modulators and receivers are all polarization sensitive components in optical fiber telecommunication system.

As we understand that in single mode fiber, the polarization of the lowest mode, EH_{11}, can be vertical polarization state (E_y) or horizontal polarization state (E_x). Both of them have the same propagation characteristic such as the same group velocity. When the signal travels along the standard single mode fiber, power will couple between the polarization modes more or less randomly as minor variation in the geometry of the fiber.

Polarization maintaining (PM) fibers are designed to maintain the signal polarization state as it was when it is lunched into the fiber. There are three types of polarization maintaining fiber as indicated in Figure 4-20.

Principle of polarization maintaining (PM) fibers

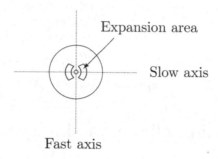

Figure 4-21   Fast axis and slow axis

Polarization maintenance is achieved by making the fiber high birefringent, namely it has a large difference in reflective index, (and then in group velocity), for two orthogonal polarizations. This is achieved by making the fiber asymmetric in profile. The difference in group velocity between the modes minimizes the possibility of coupling so as two modes cannot stay resonant long enough to take place of significant coupling.

In Figure 2-20, the Elliptical-clad polarization maintaining (PM) fiber is the simplest one. The "Bow-tie" and "Panda" PM fiber have the same principle to maintain the signal polarization state in the fibers. In each polarization maintaining fiber, the core is flanked by areas of high expansion and doped glass that shrink-back more than the surrounding silica as the fiber is drawn and freeze the core in tension. This tension induces birefringence and then creates two different refractive indices: a higher index parallel to the applied stress and a lower index perpendicular to the applied stress as indicated in Figure 4-21. The fast axis is the one that the light experiences a lower refractive index than that on the slow axis. Therefore the fast axis is the one that the light has a higher group velocity than that on the low axis. Which is very similar to the phenomena that it will create visible interference fringes when transparent plastics are stressed.

Considering that the light doesn't travel in the expansion area, the unwanted addition attenuation is minimized. However, the attenuation of polarization maintaining fiber still is high to $2dB/km$ at $1550\mu m$, which is much high than that of the standard single mode fiber $0.3dB/km$ at $1550\mu m$.

Therefore, the polarization maintaining fiber is not suitable to the long haul optical fiber communication system but suitable to some components, such as semiconductor laser diodes emit light having a specific polarization direction, and various forms of tunable filters, amplifiers, modulators and receivers are all polarization sensitive components in optical fiber telecommunication system. All of them need the polarization maintaining fiber to provide a good performance.

4.6 Nonlinearity in single mode fiber

After the discussion of attenuation and dispersion in single mode fiber, the nonlinearity in single mode fiber provide the next major limitation on long haul optical fiber transmission systems.

The discussion of attenuation and dispersion in single mode fiber is the linear problem if the light power in single mode fiber is small. However, the power of light in long haul optical fiber system can not small since the light traveling in a long haul optical fiber communication system is multi-channel signal such as wavelength division multiplexing (WDM) signal, in which the power of light is limited by the nonlinearity in single mode fiber for some reasons:

(1) The nonlinearity in single mode fiber will cause the cross-talk from neighborhood channel.

(2) The nonlinearity in single mode fiber will cause the signal distortion.

To discuss the nonlinearity in single mode fiber, we have to mention the nonlinear wave equation.

Nonlinear wave equation

Linear case

Till now the single mode fiber is considered to be a linear dispersion medium, in this case, the wave equation is,

$$\nabla^2 \mathbf{E} + n^2 k_0{}^2 \mathbf{E} = 0 \qquad (4.75)$$

where the refractive index in single mode is

$$n = n_1 \qquad (4.76)$$

This is valid only if the power of light is small.

Nonlinear case

However, if the power of light is over a threshold point, the single mode fiber has to be considered as a nonlinear dispersion medium. In this case, the refractive index of single mode fiber become

$$n = n_1 + \triangle n = n_1 + \left(\frac{n_2}{2n_1}\right)^2 |\mathbf{E}|^2 + \cdots \tag{4.77}$$

Which means that in nonlinear case, we have to consider the interaction between the different light wave $|\mathbf{E}|$ and the material n.

Substitution of (4.77) into (4.76) give us the nonlinear wave equation as follows:

$$\nabla^2 \mathbf{E} + \left(n_1 + [\frac{n_2}{2n_1}]^2 |\mathbf{E}|^2 + \cdots\right)^2 k_0{}^2 \mathbf{E} = 0 \tag{4.78}$$

The solution of the light wave in single mode fiber has to satisfy the nonlinear wave equation (4.78). And different nonlinearity phenomenon has to satisfy different nonlinear equation. [52]

At beginning, we'd better to point out some nonlinear phenomena in single mode fiber as follows:

For example, when light travels in a single mode fiber:

(1) Stimulated Brillouin scattering (**SBS**) has no measurable effect to a signal of $3mW$. However, a significant effect can be measured if the power of the signal is increased to $6mW$. Where the signal to be considered is one channel signal. That means that the power of one channel signal need to be kept below $3mW$ if SBS is to be avoided.

(2) Stimulated Raman Scattering (**SRS**) is not an issue in single-channel systems. However, it can be a significant problem in wavelength division multiplexing (WDM) systems.

(3) Four-wave mixing (FWM): This is one of the biggest problem in WDM system. Four wave mixing occurs when two of more wave propagate in the some direction in single mode fiber. The signals are mixed to produce new signals at wavelengths which are spaced at the same intervals as the mixing signals.

(4) Nonlinear Kerr Effect: When the light acts on the atoms and molecules of the silicon (in the core of single mode fiber), the light in the core will cause a tiny change in the refractive index. This effect is called the **Kerr Effect**.

(a) At very high powers, Kerr nonlinearities can be used to balance the effect of chromatic dispersion in the fiber, and a soliton is formed in single mode fiber. The soliton transmission can keep the shape of the light pulse in a long-haul data transmission links.

(b) At low power levels the results of Kerr Effect are "self-phase" modulation". and "cross-phase modulation".

4.6.1   Stimulated Raman Scattering

This subsection is going to discuss Stimulated Raman Scattering in a long-haul optical fiber communication system studied by author and his students [53][54][55][?].

The Stimulated Raman Scattering (**SRS**) is caused by the interaction between the light and the molecular variation under an injection of a large pumping power. Scattering light can appear in both of the forward and backward directions. In a single-channel system, the "Raman Threshold" (the power level at which Raman Scattering begins to take place) is very high.

The problem and the benefit of SRS

(1) Stimulated raman Scattering is not an issue in single-channel systems.

(2) However, Stimulated raman Scattering can be a significant problem in WDM systems. When multiple channels are present, power is transferred from shorter wavelength to longer ones.

(3) This can be a useful effect for building up an optical amplifier based on Raman Scattering. This is called "**Raman Scattering Amplifier**". In Raman Scattering Amplifier, power has been transferred from (high power) shorter wavelength to (low power) longer wavelength. This results in additive noise at longer wavelength and subtractive noise at the shorter wavelength.

The important characteristics of SRS

The important characteristics of the Stimulated Raman Scattering (SRS) are:

(1) SRS can take affect over about $40THz$ (a very wide range) below the higher frequency (short wavelength) involved. That means it can extend over a range of wavelengths of about $300nm$ longer than the shortest wavelength involved. The effect is maximized when the two frequencies are $13.2THz$ apart.

(2) SRS increases exponentially with increased power. When the power is very high, it is possible that all of the signal power is transferred to the Stocks Wave.

(3) In a 10-channel WDM system with $1nm$ channel spacing, the power level must be less than $3mW$ (per channel) if SRS is to be avoided.

Nonlinear wave equations

In the nonlinear dispersion medium such as an optical single mode fiber, the Maxwell equation is able to be simplified to

$$\nabla^2 \mathbf{E} - \left(\frac{1}{c^2}\right)\frac{\partial^2}{\partial t^2}(\mathbf{E} + 4\pi\mathbf{P}) = \left(\frac{4\pi}{c^2}\right)\frac{\partial^2}{\partial t^2}\mathbf{P}^{(NL)} \tag{4.79}$$

Where $\mathbf{P}$ and $\mathbf{P}^{(NL)}$ denotes the linear polarization vector and the nonlinear polarization vector, respectively.

For the sake of simplification, it is reasonable to assume that the input signal in a single-mode fiber has an uniform distribution of linear polarization states over the transverse cross section. And the direction of the polarization is x, the propagation direction is z. Then

$$P = \chi^{(1)} E(r, \theta, z; t) \tag{4.80}$$

$$P^{(NL)} = \chi^{(3)} E^3(r, \theta, z; t) \tag{4.81}$$

Note, the single-mode fiber is made of the crystal and the crystal is an antisymmetrical material. Therefore, the second order of electrical susceptibility $\chi^{(2)} = 0$.

Substituting (4.80) and (4.81) into (4.79), we have

$$\frac{d^2}{dz^2} E(r, \theta, z; t) - \left(\frac{n_L^2}{c}\right)^2 \frac{d^2}{dt^2} E(r, \theta, z; t) = \frac{16\pi\chi^{(3)}}{c^2} \frac{d^2}{dt^2} [E^3(r, \theta, z; t)] \tag{4.82}$$

Where $\dfrac{d}{dx} = \dfrac{d}{dy} = 0$ since it is assumed that the input signal in single-mode fiber has an uniform distribution of linear polarization states over the transverse cross section.

It is easy to understand that the linear polarization vector $\mathbf{P}$ will cause the wave dispersion, and the nonlinear polarization vector $\mathbf{P^{(NL)}}$ will cause the new frequency components. And then the field in the single-mode fiber is able to be expressed as

$$
\begin{aligned}
|E(r, \theta, z; t)| &= \sum_m \psi_m(r, \theta) |A_m(z, t)| \cos(k_m z - \omega_m t + \phi_m) \\
&\quad + \sum_{m'} \psi_{m'}(r, \theta) |A_{m'}(z, t)| \cos(k_{m'} z - \omega_{m'} t + \phi_{m'})
\end{aligned}
$$

$$\tag{4.83}$$

Where m denotes the wave propagating in the forward direction with a frequency ω_m, and m' denotes the wave propagating in the backward direction with another frequency $\omega_{m'}$. Both of them belong to the single transverse waves in a single-mode fiber .

In the nonlinear fiber optics, the Stimulated Raman Scattering (SRS) is an important phenomenon since *the Raman Scattering Amplifier* is one amplification scheme that can provide a broad and relatively flat gain profile over a wider wavelength range than EDFA (*Erbium Doped Fiber Amplifier*) amplification techniques. The latter has been conventionally used in optical communication systems. Therefore, we would like to concentrate to analyze the Stimulated Raman Scattering as follows.

Assume that the power of the pump wave (p) injected into the single-mode fiber is beyond *the threshold power*, and then the pump wave will stimulate the first order of stocks wave (s) and the first order of anti-stocks wave (a), however, it is not large enough to stimulate the higher order of stocks waves. In this case,

$m = p, s, a; \quad m' = s, a.$

Next, we may consider that the field amplitude

$$F_m(z, t) = N_m A_m(z, t) \tag{4.84}$$

so that the power of the field will be

$$P_m(z, t) = |F_m(z, t)|^2$$

And then the normalization condition gives

$$N_m^2 = \frac{n_{mL} c}{8\pi} \int_0^\infty \int_0^{2\pi} \psi_m^2(r, \theta) r\, dr\, d\theta \tag{4.85}$$

Where n_{mL} is the linear refraction index at the frequency ω_m.

Substituting (4.84) into (4.83), we have

$$
\begin{aligned}
E(r, \theta, z; t) &= \frac{1}{2} \sum_m \psi_m(r, \theta) \Big[\frac{F_m(z, t)}{N_m} e^{i(k_m z - \omega_m t)} \\
&\quad + \frac{F_m^*(z, t)}{N_m} e^{-i(k_m z - \omega_m t)} \Big] \\
&\quad + \frac{1}{2} \sum_{m'} \psi_{m'}(r, \theta) \Big\{ \frac{F_{m'}(z, t)}{N_{m'}} e^{i[k_{m'}(L-z) - \omega_{m'} t]} \\
&\quad + \frac{F_{m'}^*(z, t)}{N_{m'}} e^{-i[k_{m'}(L-z) - \omega_{m'} t]} \Big\}
\end{aligned} \tag{4.86}
$$

Where L is the length of the single-mode fiber from the inject point of the pump wave to the end point of the single-mode fiber.

For the sake of simplification, we may consider the forward waves alone without considering the backward waves for the time been, i.e. let $m' = 0$ in (4.86). And then substituting (4.86) into (4.82), collect the same frequency terms in both sides, we may obtain the coupling wave equation for the Stocks wave as following:

$$
\begin{aligned}
\frac{ik_s \psi_s}{N_s} \Big(\frac{\partial F_s}{\partial z} + \frac{n_{sL}}{c} \frac{\partial F_s}{\partial t} \Big) &= \\
&\quad -\frac{16\pi \omega_s^2}{c} \Big[\chi_{spps}^{(3)} \frac{6\psi_p^2 \psi_s}{8 N_p N_s} |F_p|^2 F_s \\
&\quad + \chi_{spps}^{(3)} \frac{3\psi_p^2 \psi_a}{8 N_p^2 N_a} |F_a|^2 F_a^* e^{i(2k_p - k_a - kc)z} \\
&\quad + \chi_{ssss}^{(3)} \frac{3\psi_s^3}{8 N_s^3} |F_s|^2 F_s
\end{aligned}
$$

$$+\chi_{saas}^{(3)}\frac{6\psi_s\psi_a^2}{8N_sN_a^2}|F_a|^2F_s\bigg] \qquad (4.87)$$

Where

$$\chi_{spps}^{(3)} = \chi^{(3)}(-\omega_s, -\omega_p, \omega_p, -\omega_s) \qquad (4.88)$$

$$\chi_{ssss}^{(3)} = \chi^{(3)}(-\omega_s, -\omega_s, \omega_s, -\omega_s) \qquad (4.89)$$

$$\chi_{saas}^{(3)} = \chi^{(3)}(-\omega_s, -\omega_a, \omega_a, -\omega_s) \qquad (4.90)$$

Of course, we understand that the forward waves contains not only the Stocks wave F_s but also the pumping wave F_p and the anti-Stocks wave F_a. And the coupling wave equations for the pumping wave F_p or the anti-Stocks wave F_a is similar to (4.87). Therefore, we will concentrate to deal with the equation (4.87) first.

Now, to eliminate the transverse distribution function ψ_a and ψ_p etc. from (4.87), it is able to multiply (4.87) by ψ_m and then take a integral over the transverse cross section, and then make a definition

$$< ijkl >= \frac{\int_0^{2\pi}\int_0^{\infty}\psi_i\psi_j\psi_k\psi_l\,rdrd\theta}{N_iN_jN_kN_l} = A_e^{-1} \qquad (4.91)$$

we have

$$\frac{\partial F_s}{\partial z} + \frac{n_{sL}}{c}\frac{\partial F_s}{\partial t}$$

$$= i\frac{3\omega_s}{4}\bigg[2\chi_{spps}^{(3)} < spps > |F_p|^2F_s$$

$$+\chi_{spps}^{(3)} < spap > |F_p|^2F_a^*e^{i(\Delta kz+2\phi_p)}$$

$$+\chi_{ssss}^{(3)} < ssss > |F_s|^2F_s$$

$$+2\chi_{saas}^{(3)} < saas > |F_a|^2F_s\bigg] \qquad (4.92)$$

Similarly, we have

$$\frac{\partial F_p}{\partial z} + \frac{n_{pL}}{c}\frac{\partial F_p}{\partial t}$$

$$= i\frac{3\omega_p}{4}\bigg[\chi_{pppp}^{(3)} < pppp > |F_p|^2F_p$$

$$+2\chi_{pssp}^{(3)} < pssp > |F_s|^2F_p$$

$$+2\chi^{(3)}_{paap} < paap > |F_a|^2 F_p \Big] \qquad (4.93)$$

and

$$\frac{\partial F_a}{\partial z} + \frac{n_{aL}}{c}\frac{\partial F_a}{\partial t}$$

$$= i\frac{3\omega_a}{4}\Big[2\chi^{(3)}_{appa} < appa > |F_p|^2 F_a$$

$$+\chi^{(3)}_{apsp} < apsp > |F_p|^2 F_s^* e^{i(\Delta kz + 2\phi_p)}$$

$$+\chi^{(3)}_{aaaa} < aaaa > |F_a|^2 F_a$$

$$+2\chi^{(3)}_{assa} < assa > |F_s|^2 F_a \Big] \qquad (4.94)$$

Where

$$\Delta k = 2k_p - k_s - k_a$$

$$2\omega_p = \omega_s + \omega_a$$

And in $\chi^{(3)}(-\omega_i, -\omega_j, \omega_k, -\omega_l)$, $\quad (i, j, k, l, = p, s, a)$, we have

$$-\omega_i = -\omega_j + \omega_k - \omega_l$$

$$\mathbf{Im}\Big[\chi^{(3)}(-\omega_i, -\omega_j, \omega_k, -\omega_l)\Big] = -\mathbf{Im}\Big[\chi^{(3)}(\omega_i, \omega_j, -\omega_k, \omega_l)\Big]$$

It is worth to point out, from (4.91) to (4.93), that the amplitude coupling between the pumping wave F_p, Stocks wave F_s and the anti-Stocks wave F_a is caused by the imaginary part of $\chi^{(3)}$, which results in the Stimulated Raman Scattering, and the phase coupling between the pumping wave F_p, Stocks wave F_s and the anti-Stocks wave F_a is caused by the real part of $\chi^{(3)}$, which results in the self-phase modulation phenomenon.

Stimulated Raman Scattering (SRS)

From wave point of view, when the pumping power is beyond a threshold value in a single-mode fiber, it will stimulate the forward Stock wave and the anti-Stocks wave. This phenomenon is said to be *the Stimulated Raman Scattering*.

Step 1: Nonlinear coupling wave equation:

The nonlinear coupling wave equation described the Stimulated Raman Scattering in a single-mode fiber is able to be obtained from (4.93) and (4.92) as follows:

$$\frac{\partial F_p}{\partial z} + \frac{n_{pL}}{c}\frac{\partial F_p}{\partial t} = -\frac{3\omega_p}{2}|\mathbf{Im}[\chi^{(3)}_{pssp}]| < pssp > |F_s|^2 F_p \qquad (4.95)$$

$$\frac{\partial F_s}{\partial z} + \frac{n_{sL}}{c}\frac{\partial F_s}{\partial t} = \frac{3\omega_s}{2}|\mathbf{Im}[\chi^{(3)}_{spps}]| < spps > |F_p|^2 F_s \qquad (4.96)$$

Where F_p and F_s represent the amplitude factor of the forward pumping wave and the forward Stocks wave, respectively.

Multiplying (4.95) and (4.96) by F_p^* and F_s^*, respectively, considering that $F_p \cdot F_p^* = P_P$, $F_s \cdot F_s^* = P_s$, we have the coupling wave equations of the Stimulated Raman Scattering as follows:

$$\frac{\partial P_p}{\partial z} + \frac{1}{v_p}\frac{\partial P_p}{\partial t} = -\frac{g_p}{KA_e}P_s P_p - \alpha_p P_p \qquad (4.97)$$

$$\frac{\partial P_s}{\partial z} + \frac{1}{v_s}\frac{\partial P_s}{\partial t} = \frac{g_s}{KA_e}P_p P_s - \alpha_s P_s \qquad (4.98)$$

Where

$$g_p = 3\omega_p|\mathbf{Im}[\chi^{(3)}(-\omega_p, -\omega_s, \omega_s, -\omega_p)]|$$

$$g_s = 3\omega_s|\mathbf{Im}[\chi^{(3)}(-\omega_s, -\omega_p, \omega_p, -\omega_s)]|$$

are the gain coefficients of the pumping wave and the Stocks wave, respectively. And

$$v_p = c/n_{pL}, \quad v_s = c/n_{sL}$$

are the propagation velocity of the pumping wave and the Stocks wave, respectively. K is the polarization factor. For a single-mode fiber with a holding polarization, $K = 1$. For a single-mode fiber with non-holding polarization, $K = 2$, which is the worst case. Therefore, $1 \leq K \leq 2$. The definition of A_e is in (4.91).

Step 2: Parameter transformation:

Now we will find the solution of (4.97) and (4.98). Normally, the single-mode fiber is a low dissipation fiber. In this case, it is reasonable to consider that the linear attenuation coefficient for both the pumping wave and the Stocks wave are approximately the same, therefore, it is reasonable to assume that $\alpha_p = \alpha_s = \alpha$. In this case, set a parameter transformation as follows [5]

$$z' = z, \quad t'_p = t - z/v_p, \quad t'_s = t - z/v_s \qquad (4.99)$$

---

[5] This is from the contour method (or the characteristic curve method).

And then, in (4.97) and (4.98),

$$\frac{\partial}{\partial z} = \frac{\partial}{\partial z'}\frac{\partial z'}{\partial z} + \frac{\partial}{\partial t'}\frac{\partial t'}{\partial z} = \frac{\partial}{\partial z'} + \frac{\partial}{\partial t'}\left(-\frac{1}{v}\right)$$

$$\frac{1}{v}\frac{\partial}{\partial t} = \frac{1}{v}\left(\frac{\partial}{\partial z'}\frac{\partial z'}{\partial t} + \frac{\partial}{\partial t'}\frac{\partial t'}{\partial t}\right) = \frac{1}{v}\left(\frac{\partial}{\partial t'}\right)$$

Therefore,

$$\frac{\partial}{\partial z} + \frac{1}{v}\frac{\partial}{\partial t} = \frac{\partial}{\partial z'}$$

And then (4.97) and (4.98) become

$$\frac{\partial \tilde{P}_p}{\partial z'} = -\frac{g_p}{KA_e}\tilde{P}_p\tilde{P}_s - \alpha\tilde{P}_p \tag{4.100}$$

$$\frac{\partial \tilde{P}_s}{\partial z'} = \frac{g_s}{KA_e}\tilde{P}_p\tilde{P}_s - \alpha\tilde{P}_s \tag{4.101}$$

Where

$$\tilde{P}_p = P_p(z', t_p'), \quad \tilde{P}_s = P_s(z', t_s')$$

And the boundary conditions are

$$\tilde{P}_p(z', t_p')|_{z'=0} = P_p(0, t_p')$$

$$\tilde{P}_s(z', t_s')|_{z'=0} = P_s(0, t_s')$$

Obviously, the parameter transformation (4.99) transforms the partial differential equations (4.97) and (4.98) into the ordinary differential equations (4.100) and (4.101).

Step 3: Find the solution:

To fine the solution of (4.100) and (4.101), suppose that

$$\tilde{P}_p(z', t_p') = C_p(z', t_p')e^{-\alpha z'} \tag{4.102}$$

$$\tilde{P}_s(z', t_s') = C_s(z', t_s')e^{-\alpha z'} \tag{4.103}$$

Substituting (4.102) and (4.103) into (4.100) and (4.101), we have

$$\frac{\partial C_p(z', t_p')}{\partial z'} = -\tilde{g}_p C_p(z', t_p')\, C_s(z', t_s')\, e^{-\alpha z'} \tag{4.104}$$

$$\frac{\partial C_s(z', t_s')}{\partial z'} = \tilde{g}_s C_p(z', t_p') \, C_s(z', t_s') \, e^{-\alpha z'} \qquad (4.105)$$

Where $\tilde{g}_p = g_p/KA_e$, $\tilde{g}_s = g_s/KA_e$. And the boundary conditions are

$$C_p(z', t_p')|_{z'=0} = P_p(0, t_p')$$

$$\qquad (4.106)$$

$$C_s(z', t_s')|_{z'=0} = P_s(0, t_s')$$

From (4.104) and (4.105), we have

$$\frac{\partial}{\partial z'}(\tilde{g}_s C_p + \tilde{g}_p C_s) = 0 \qquad (4.107)$$

Therefore,

$$\tilde{g}_s C_p + \tilde{g}_p C_s = G(t') \qquad (4.108)$$

Where $G(t')$ is an integral constant in the integral of (4.107) with respect to z', and then $G(t')$ is independent of z'. Therefore, we may consider that

$$G(t') = G(t')|_{z'=0} = [\tilde{g}_s C_p(z', t_p') + \tilde{g}_p C_s(z', t_s')]_{z'=0}$$

and then

$$G(t') = \tilde{g}_s P_p(0, t_p') + \tilde{g}_p P_s(0, t_s') \qquad (4.109)$$

Substituting (4.109) into (4.108), we have

$$C_p = \frac{G(t')}{\tilde{g}_s} - \frac{\tilde{g}_p}{\tilde{g}_s} C_s$$

$$\qquad (4.110)$$

$$C_s = \frac{G(t')}{\tilde{g}_p} - \frac{\tilde{g}_s}{\tilde{g}_p} C_p$$

Substituting (4.110) into (4.104) and (4.105), we have

$$\frac{\partial C_p}{\partial z'} = -G(t') \, C_p \, e^{-\alpha z'} + \tilde{g}_s \, C_p^2 \, e^{-\alpha z'}$$

$$\frac{\partial C_s}{\partial z'} = G(t') \, C_s \, e^{-\alpha z'} - \tilde{g}_p \, C_s^2 \, e^{-\alpha z'}$$

Or

$$\frac{\partial}{\partial z'}\Big(\frac{1}{C_p}\Big) - G(t')\Big(\frac{1}{C_p}\Big)e^{-\alpha z'} = -\tilde{g}_s \, e^{-\alpha z'}$$

$$\qquad (4.111)$$

$$\frac{\partial}{\partial z'}\Big(\frac{1}{C_s}\Big) - G(t')\Big(\frac{1}{C_s}\Big)e^{-\alpha z'} = \tilde{g}_p \, e^{-\alpha z'}$$

These are an ordinary differential equations with the form of

$$Y' + P(x)Y = Q(x)$$

and the solution is

$$Y = \exp\left\{-\int P dx\right\}\left(\int Q \exp\left\{\int P dx\right\} dx + C\right)$$

Following this equation, considering the boundary conditions (4.106) and (4.109), we have

$$C_p(z', t'_p) = \frac{G(t')}{\tilde{g}_s}\frac{1}{1 + H(z', t')} \tag{4.112}$$

$$C_s(z', t'_s) = \frac{G(t')}{\tilde{g}_p}\frac{H(z', t')}{1 + H(z', t')} \tag{4.113}$$

Where

$$H(z', t') = \frac{\omega_p}{\omega_s}\frac{P_s(0, t'_s)}{P_p(0, t'_p)}\exp[G(t')L_e] \tag{4.114}$$

$$L_e = [1 - e^{-\alpha z'}]/\alpha$$

$$G(t') = \tilde{g}_s[P_p(0, t'_p) + (\omega_p/\omega_s)P_s(0, t'_s)]$$

$$t'_p = t - \frac{n_p L}{c}z, \qquad t'_s = t - \frac{n_s L}{c}z \tag{4.115}$$

Substituting (4.112) and (4.115) into (4.102) and (4.103), considering, from (4.99), that $z' = z$, we have

$$P_p\left(z, t - \frac{z}{v_p}\right) = P_0(0, t')\exp(-\alpha z)\frac{1}{1 + H} \tag{4.116}$$

$$P_s\left(z, t - \frac{z}{v_s}\right) = \left(\frac{\omega_s}{\omega_p}\right)P_0(0, t')\exp(-\alpha z)\frac{1}{1 + H} \tag{4.117}$$

Where

$$H = H(z', t')$$

$$P_0(0, t') = P_p\left(0, t - \frac{z}{v_p}\right) + \frac{\omega_p}{\omega_s}P_s\left(0, t - \frac{z}{v_s}\right) \tag{4.118}$$

(4.116) and (4.117) are *the transient-state solution* of the coupling wave equations (4.97) and (4.98) under the condition of $\alpha_p = \alpha_s = \alpha$.

Now we will consider two situations:

(1) Quasi-steady-state solution:

For a pulse signal in a single-mode fiber, if the duration of the pulse signal is larger than the time delay [6] for several times, the dispersion in the single-mode fiber is able to be ignored, in this case, the transient-state solution is said to be *the quasi-steady-state solution*. And then,

$$v_p = v_s = v, \qquad \alpha_p = \alpha_s = \alpha$$

and the Stimulated Raman Scattering coupling wave equations (4.97) and (4.98) become

$$\frac{\partial P_p}{\partial z} + \frac{1}{v}\frac{\partial P_p}{\partial t} = -\frac{g_p}{KA_e}P_pP_s - \alpha P_p \qquad (4.119)$$

$$\frac{\partial P_s}{\partial z} + \frac{1}{v}\frac{\partial P_s}{\partial t} = \frac{g_s}{KA_e}P_sP_p - \alpha P_s \qquad (4.120)$$

The solutions of (4.119) and (4.120) can be obtained by substituting the following transformation

$$t' = t - \frac{z}{v_p} = t - \frac{z}{v_s} = t - \frac{z}{v} \qquad (4.121)$$

into (4.116) and (4.117) to get

$$P_p\left(z, t - \frac{z}{v}\right) = P_0\left(0, t - \frac{z}{v}\right)\exp(-\alpha z)\frac{1}{1+H} \qquad (4.122)$$

$$P_S\left(z, t - \frac{z}{v}\right) = \frac{\omega_s}{\omega_p}P_0\left(0, t - \frac{z}{v}\right)\exp(-\alpha z)\frac{H}{1+H} \qquad (4.123)$$

Where $P_0\left(0, t - \frac{z}{v}\right)$, H are able to be obtained by substituting (4.121) into (4.118) and (4.114).

(2) Steady-state solution:

For a continuous wave in a single-mode fiber, the solution of the Stimulated Raman Scattering is said to be *the steady-state solution*. In this case, the coupling wave equations are

$$\frac{\partial P_p}{\partial z} = -\frac{g_p}{KA_e}P_pP_s - \alpha P_p \qquad (4.124)$$

$$\frac{\partial P_s}{\partial z} = \frac{g_s}{KA_e}P_pP_s - \alpha P_s \qquad (4.125)$$

---

[6]   The time delay is caused by the wave traveling the total length L of the single-mode fiber.

Where the time factor $\exp(i\omega t)$ has been considered in (4.86). And the $\partial/\partial t$ term is absent since the wave amplitude is invariant along the single-mode fiber and then it is said to be *the steady-state solution*. The solution is able to be obtained by substituting

$$t = z/v \qquad (4.126)$$

into (4.122) and (4.123) to get

$$P_p(z) = P_0(0)\exp(-\alpha z)\frac{1}{1+H} \qquad (4.127)$$

$$P_s(z) = \frac{\omega_s}{\omega_p}P_0(0)\exp(-\alpha z)\frac{H}{1+H} \qquad (4.128)$$

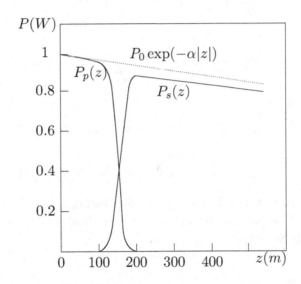

Figure 4-22   The steady-state solution of SRS [53]

Where H is able to be obtained by substituting (4.126) into (4.114). $P_0(0)$ is able to be obtained by substituting (4.126) into (4.118) to get

$$P_0(0) = P_p(0) + \frac{\omega_p}{\omega_s}P_s(0)$$

The steady-state solution (4.127) and (4.128) describes the transformation process between the pumping power $P_p(z)$ and the Stocks power $P_s(z)$ as shown in Figure 4-22.

4.6.2   Stimulated Brillouin scattering

Stimulated Brillouin Scattering is caused by the interaction between the optical fiber mode and the acoustic wave under an injection of a narrow pumping source. And the line-width of the pumping source is about 100MHz. The Brillouin frequency shift is about $1cm^{-1}$. Under this condition, the maximin gain of the Stimulated Brillouin Scattering is two orders large than the maximin gain of the Stimulated Raman Scattering. Since the matching condition of the wave vector, the Brillouin amplification occurs only in the backward direction. Which is called the "**Stock Waves**".

Therefore, substituting $m = p$, (means pumping wave), $m' = s = B$, (means stock wave "$s = B$"), into (4.86), we have the coupling wave equations for the Stimulated Brillouin Scattering as follows:

$$\frac{\partial P_p}{\partial z} + \frac{n_{pL}}{c}\frac{\partial P_p}{\partial t} = \frac{g_p}{KA_e}P_B P_p - \alpha_p P_p$$

$$\frac{\partial P_B}{\partial z} - \frac{n_{sL}}{c}\frac{\partial P_B}{\partial t} = -\frac{g_s}{KA_e}P_p P_B + \alpha_s P_B$$

(4.129)

Where

$$g_p = 3\omega_p \left|Im[\chi^{(3)}(-\omega_p, -\omega_s, \omega_s, -\omega_p)]\right|$$

$$g_s = 3\omega_s \left|Im[\chi^{(3)}(-\omega_s, -\omega_p, \omega_p, -\omega_s)]\right|$$

That means that the Stimulated Brillouin Scattering is also caused by the imaginary part of $\chi^{(3)}$. The process to find the solution is almost the same as that in the Stimulated Raman Scattering .

The problem caused by SBS

In most current system, SBS has not been much of problem for the following reason:

(1) Direct modulation of the transmit laser's inject current produces a chirp and broadens the signal. This significantly reduces the impact of SBS.

(2) The effect is less in $1300nm$ system than in $1500nm$ systems because of the higher attenuation in single mode fiber.

(3) At speeds of below $2.4GHz$ it has not been necessary to use either very high power or very narrow line-width lasers.

(4) SBS effects is decreased when the speed is increased because of the signal broadening affect of the modulation.

However, SBS can be a major problem in following situations:

(1) In a long haul optical fiber transmission system where the span between amplifiers is great and the bit rate is low (below about $2.5Gbps$).

(2) In WDM system (up to about 10Gbps) where the spectral width of the signal is very narrow.

4.6.3 Nonlinear Kerr Effect and Soliton

Nonlinear Kerr Effect

The light traveling in the fiber will cause a tiny change in refractive index (RI). This is because the electromagnetic field acts on the atom and molecules that make up the glass. This phenomenon is called "**Kerr Effect**". At low intensity of light the effect is linear, that means that amount of RI change varies linearly with the intensity of the light.

At high intensities the effect is highly nonlinear, the RI change is much greater than that in the linear case. This is called the "**Nonlinear Kerr Effect**".

At very high powers **Kerr** effect can be used to balance the effect of chromatic dispersion in single mode fiber and a **soliton** is formed.

As well known, in a long-haul optical fiber communication system, the nonlinearity in a single-mode fiber will narrow the width of the pulse in the transmission. However the dispersion in a single-mode fiber will broad the width of the pulse in the transmission. If the narrowing effect and the broadening effect are in balance in the transmission, the pulse shape will be preserved in the transmission. This transmission is called the **soliton transmission**. The equation to describe this phenomenon is called *the KdV equation*, i.e.

$$\frac{\partial u}{\partial t} + u\frac{\partial u}{\partial x} + \mu\frac{\partial^3 u}{\partial x^3} = 0 \qquad (4.130)$$

Where the second term is the nonlinear and the third term is the dispersion in single mode fiber. In which, μ is a constant and it can be positive or negative. If $\mu > 0$, the solution is a soliton. The process to obtain the solution is similar to that of the Burgers equation. Now we will use it to discuss the KdV equation for far field in a long-haul optical fiber communication system.

Let

$$\xi = x - vt, \quad u = u(\xi)$$

then (4.130) becomes

$$-vu_\xi + uu_\xi + \mu u_{\xi\xi\xi} = 0 \qquad (4.131)$$

Where u_ξ represents the derivative of u with respect to ξ. Let

$$\bar{u} = u - v$$

then (4.131) becomes an original differential equation

$$\bar{u}\bar{u}_\xi + \mu\bar{u}_{\xi\xi\xi} = 0 \qquad (4.132)$$

Taking an integral of (4.132) with respect to ξ from ∞ to ξ we have

$$\frac{1}{2}(\bar{u}^2 - \bar{u}_\infty^2) + \mu\bar{u}_{\xi\xi} = 0 \qquad (4.133)$$

Where $\lim_{\xi\to\infty}\bar{u} = \bar{u}_\infty = constant$.

Multiplied both sides of (4.133) by $\bar{u}_\xi$, and then taking the integral from ∞ to ξ, in which

$$\int_\infty^\xi (\bar{u}^2 - \bar{u}_\infty^2)\bar{u}_\xi d\xi = \int_\infty^\xi (\bar{u}^2 - \bar{u}_\infty^2)d\bar{u} = \left[\frac{1}{3}\bar{u}^3 - \bar{u}_\infty^2\bar{u}\right]\Big|_\infty^\xi = \frac{1}{3}\bar{u}^3 - \bar{u}_\infty^2\bar{u} + \frac{2}{3}\bar{u}_\infty^3$$

and

$$\int_\infty^\xi \bar{u}_{\xi\xi}\bar{u}_\xi d\xi = \int_\infty^\xi \frac{\partial}{\partial\xi}(\bar{u}_\xi)\bar{u}_\xi d\xi = \frac{1}{2}\int_\infty^\xi \bar{u}_\xi^2 d\bar{u}_\xi = \frac{1}{2}\bar{u}_\xi^2\Big|_\infty^\xi = \frac{1}{2}\bar{u}_\xi^2$$

and then we have

$$\frac{1}{6}\bar{u}^3 - \frac{1}{2}\bar{u}_\infty^2\bar{u} + \frac{1}{3}\bar{u}_\infty^3 + \frac{\mu}{2}\bar{u}_\xi^2 = 0$$

Where

$$\frac{1}{6}\bar{u}^3 - \frac{1}{2}\bar{u}_\infty^2\bar{u} = \frac{1}{6}(\bar{u} - \bar{u}_\infty)^2(\bar{u} - \bar{u}_\infty + 3\bar{u}_\infty)$$

Taking the variables separation of (4.134) and then integral, we have

$$\int \frac{d\bar{u}}{(\bar{u} - \bar{u}_\infty)[-1/3(\bar{u} - \bar{u}_\infty) - \bar{u}_\infty]^{1/2}} = \frac{\xi}{\sqrt{\mu}} + C \qquad (4.134)$$

Let $C = 0$, $\bar{u}_\infty \le 0$, the integral (4.134) is in form of [7]

$$\int \frac{dx}{x\sqrt{z}} = \frac{1}{\sqrt{a}}\ln\frac{\sqrt{z} - \sqrt{a}}{\sqrt{z} + \sqrt{a}} \qquad (4.135)$$

Where

$$x = (\bar{u} - \bar{u}_\infty),$$
$$b = -\frac{1}{3},$$
$$a = -\bar{u}_\infty$$
$$z = a + bx = -\bar{u}_\infty - \frac{1}{3}(\bar{u} - \bar{u}_\infty)$$

and then the integral (4.134) becomes

$$\frac{1}{\sqrt{-\bar{u}_\infty}}\ln\frac{[-1/3(\bar{u} - \bar{u}_\infty) - \bar{u}_\infty]^{1/2} - [-\bar{u}_\infty]^{1/2}}{[-1/3(\bar{u} - \bar{u}_\infty) - \bar{u}_\infty]^{1/2} + [-\bar{u}_\infty]^{1/2}} = \frac{\xi}{\sqrt{\mu}}$$

---

[7]   I.S. Gradshteyn/I.M. Ryzhik, "Table of Integrals, and products." 1980. pp.73, 2.224,5

After rearranged, we have

$$\bar{u} = \bar{u}_\infty - 3\bar{u}_\infty \, sech^2[1/2(-\bar{u}_\infty/\mu)^{1/2}\xi] \qquad (4.136)$$

Where $\bar{u} = u - v$, $\bar{u}_\infty = u_\infty - v$, $\xi = x - vt$. Let $-3\bar{u}_\infty = \varepsilon$, then (4.136) becomes

$$u = u_\infty + \varepsilon \, sech^2\left\{\left(\frac{\varepsilon}{12\mu}\right)^{1/2}\left[x - \left(u_\infty + \frac{\varepsilon}{3}\right)t\right]\right\} \qquad (4.137)$$

This is the soliton solution of the KdV equation . This solution leads us to some important conclusions as follows:

(1) If the amplitude of the wave envelope $\varepsilon = constant$, then the soliton will be transmission forward with a constant velocity of $v = u_\infty + \varepsilon/3$ and then the soliton will preserves its identity.

(2) The effect of the dispersion factor μ is to broaden the soliton since the width of the waveform of the soliton (4.137) is proportional to $\sqrt{\mu}$.

(3) Meanwhile, the width of the waveform of the soliton (4.137) is proportional to $1/\sqrt{\varepsilon}$, that means the higher the ε the narrower the width of the wave. Which means the nonlinearity of single mode fiber will narrow the width of the wave since ε is the amplitude of the wave envelope, the higher the ε means the stronger the nonlinearity.

(4) If $u_\infty = \lim_{x\to\pm\infty} u = 0$, then the solution of the soliton becomes

$$u = \varepsilon \, sech^2\left[\left(\frac{\varepsilon}{12\mu}\right)^{1/2}\left(x - \frac{\varepsilon}{3}t\right)\right] \qquad (4.138)$$

In this case, the propagation velocity of the soliton $v = \varepsilon/3$ is proportional to the amplitude ε, namely, the higher the wave amplitude ε the higher the wave velocity. This phenomena is valid either for (4.137).

4.6.4 Four-wave mixing

One of the biggest problem in WDM systems is called the "Four-wave mixing" (FWM). Which can be illustrated by a simple picture as indicated in Figure 4-23.

Signals (Two Channels)

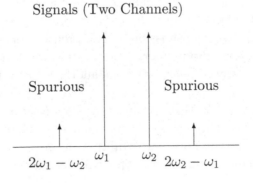

Figure 4-23   Four wave mixing

Where ω_1 and ω_2 are the carrier frequency of two channels, respectively. And $2\omega_1 - \omega_2$ and $2\omega_2 - \omega_1$ are the spurious side modes produced by the nonlinearity of refractive index in single mode fiber when the power in single mode fiber is high enough.

This is easy to be understood. We are familiar with the mixer in the circuit, in which, when two signals with frequency ω_1 and ω_2 are input into a mixer, the mixer will produce ω_1, ω_2, $2\omega_1 - \omega_2$, $2\omega_2 - \omega_1$ and etc. output. Similarly, in single mode fiber, the nonlinearity of the refractive index will produce $2\omega_1 - \omega_2$, $2\omega_2 - \omega_1$ when two light signals ω_1, ω_2 are traveling in the same direction in single mode fiber. Where, ω_1 and ω_2 are the frequencies of two neighborhood channels in a WDM, and $2\omega_1 - \omega_2$, $2\omega_2 - \omega_1$ are the spurious side modes traveling in the same direction with the signals ω_1 and ω_2 in single mode fiber. Which is called the "Four Wave Mixing" in single mode fiber.

The charateristic of Four Wave Mixing (4.139)

Four Wave Mixing (FWM) is a big problem in a WDM transmission system since the spurious side modes will become added noise (crosstalk) in a channel signal.

(1) First, the effect of FWM is relating to the channel spacing. The effect of FWM becomes greater when the channel spacing is reduced. The smaller the channel spacing the greater the effect of FWM. In conception, this is the same as that in circuit mixer.

(2) Secondary, the effect of FWM is relating to the signal power in single mode fiber. As signal power increase the effect of FWM increase exponentially.

(3) Third, the effect of FWM is strongly influenced by the chromatic dispersion. In fact FWM is caused only if the signals stay in phase with one another **over a significant distance**. The lasers produce light with large "coherence length" and then a number signals will stay in phase over a long distance if there is no chromatic dispersion in single mode fiber. Now the chromatic dispersion in single mode fiber will destroy "in phase transmission". The greater the dispersion the smaller the effect of FWM in single mode fiber. Because chromatic dispersion ensures that different signals do not stay in phase with one another over a long distance in single mode fiber.

(4) Forth, the effect of FWM is relating to the spacing distribution of signals in WDM. If the signals are evenly spaced, then the new spurious signals will appear in signal channels and cause the noise (crosstalk). One method to reduce the effect of FWM is to space the channel unevenly. This will mitigate the problem of crosstalk in unrelated channels.

4.7 Basic knowledge

4.7.1 Maxwell equation

Maxwell equation is a basic knowledge not only in the discussion of optical fiber but also in the semiconductor laser. In fact, a light propagating in an optical fiber is exactly an electromagnetic field propagating in an optical fiber. Meanwhile, a light propagating in a semiconductor laser is also an electromagnetic field propagating in a laser waveguide structure. Meanwhile, in the discussion of semiconductor laser, we need **quantum-mechanics** as the basic.

The classical Maxwell equation

$$\nabla \times \mathbf{E} = -\frac{\partial \mathbf{B}}{\partial t} \tag{4.140}$$

$$\nabla \times \mathbf{H} = \mathbf{J} + \frac{\partial \mathbf{D}}{\partial t} \tag{4.141}$$

$$\nabla \cdot \mathbf{D} = \rho \tag{4.142}$$

$$\nabla \cdot \mathbf{B} = 0 \tag{4.143}$$

Where we use boldface to denote a vector, use "$\times$" to denote cross product, use "$\cdot$" to denote dot product.

The classical Maxwell equation is a conclusion based on a long times of experimental works, which dominates the electromagnetic field in the macroscopic world, such as the free space, the dielectric medium, the magnetic medium, the isotropic and anisotropic medium, the linear and nonlinear medium etc in the world:

(1) **Faraday's law** (4.140).

Faraday's law is the basic law in the electric industry, which says that changing magnetic fields $\frac{\partial \mathbf{H}}{\partial t}$ produce electric fields $\mathbf{E}$.

(2) **Ampere's law** (4.141).

Ampere's law is another basic law in the electric industry, which says that moving charges produce magnetic fields, changing electric field produce magnetic fields as well. Where J is the current density and $\partial D/\partial t$ is the displacement current density.

The displacement current $\partial D/\partial t$ in Ampere's law is initial in predicating the existence of propagating electromagnetic wave and is amended by Maxwell. This contribution is so important that it results in the IT industry in the world.

(3) (4.142) is **Gauss' law** for the electric field.

Gauss' law says that electric charges produce electric fields. Where ρ is volume electric charge density.

(4) (4.143) is **Gauss' law** for the magnetic field.

Which says that any lines of magnetic flux are continuous since no magnetic charge is available.

4.7.2 Material equations

In a given material, **D** and **E**, **B** and **H** are related by the following equations:

$$\mathbf{D} = \varepsilon_0 \mathbf{E} + \mathbf{P} \tag{4.144}$$

$$\mathbf{H} = \frac{\mathbf{B}}{\mu_0} - \mathbf{M} \tag{4.145}$$

and ε_0 and μ_0 are the permittivity and permeability in the free space. **P** is the polarization and **M** is magnetization.

In isotropic linear medium, **P** is proportional to **E** and **M** is proportional to **B**, namely,

$$\mathbf{P} = \varepsilon_0 \chi_e \mathbf{E} \tag{4.146}$$

$$\mathbf{M} = \chi_m \mathbf{H} \tag{4.147}$$

Substitution of (4.146), (4.147) into (4.144) (4.145) gives us

$$\mathbf{D} = \varepsilon \mathbf{E} \tag{4.148}$$

$$\mathbf{B} = \mu \mathbf{H} \tag{4.149}$$

and

$$\varepsilon = \varepsilon_0 (1 + \chi_e) = \varepsilon_0 \varepsilon_r \tag{4.150}$$

$$\mu = \mu_0 (1 + \chi_m) = \mu_0 \mu_r \tag{4.151}$$

Where,

χ_e is the electric susceptibility.

χ_m is the magnetic susceptibility.

ε_r is the relative permittivity.

μ_r is the relative permeability.

Therefore, **D** and **E**, **B** and **H** are related by the parameters of the material.

In free space,

$$\mathbf{D} = \varepsilon_0 \mathbf{E} \tag{4.152}$$

$$\mathbf{B} = \mu_0 \mathbf{H} \tag{4.153}$$

In general case,

$$\mathbf{D} = \varepsilon\mathbf{E} \tag{4.154}$$

$$\mathbf{B} = \mu\mathbf{H} \tag{4.155}$$

or

$$\mathbf{D} = \varepsilon_0\varepsilon_r\mathbf{E} \tag{4.156}$$

$$\mathbf{B} = \mu_0\mu_r\mathbf{H} \tag{4.157}$$

$$\varepsilon_r = 1 + \chi_e \tag{4.158}$$

$$\mu_r = 1 + \chi_m \tag{4.159}$$

4.7.3 Wave equation

In far field, $\rho = 0$, $J = 0$, $D = \varepsilon E$ and $B = \mu H$, the Maxwell equation become

$$\nabla \times \mathbf{E} = -\mu\frac{\partial\mathbf{H}}{\partial t} \tag{4.160}$$

$$\nabla \times \mathbf{H} = \varepsilon\frac{\partial\mathbf{E}}{\partial t} \tag{4.161}$$

Take the curl of Faraday's law (4.160), we have

$$\nabla \times (\nabla \times \mathbf{E}) = -\mu\frac{\partial(\nabla \times \mathbf{H})}{\partial t} \tag{4.162}$$

Substitute Ampere's law (4.161), we have the wave equation as follows:

$$\nabla^2\mathbf{E} - \mu\epsilon\frac{\partial^2\mathbf{E}}{\partial t^2} = 0 \tag{4.163}$$

This is the three dimensional wave equation in vector form.

4.7.4 Plane wave

A sinusoidal wave in antenna

$$E = E_0\sin(\omega t + \phi) \tag{4.164}$$

become a sinusoidal plane wave at the far field

$$E_\theta = E_0\sin(\omega t - kr + \phi) \tag{4.165}$$

$$H_\phi = -H_0\sin(\omega t - kr + \phi) \tag{4.166}$$

as indicated in Figure 4-24.

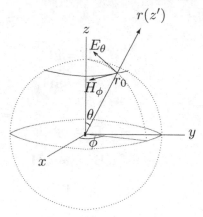

Figure 4-24   The far field of a dipole antenna.

Obviously, the far field E_θ and H_ϕ is a spherical surface wave, that means that the far field (E_θ and H_ϕ) are lying on a spherical surface ($r = r_0$).

For the sack of simplification, physicists found that the far fields (4.165) and (4.166) can be viewed approximately as a plane wave as follows:

$$E_{x'} = E_0 \sin(\omega t - kz' + \phi) \tag{4.167}$$

$$H_{y'} = H_0 \sin(\omega t - kz' + \phi) \tag{4.168}$$

as indicated in Figure 4-25. This wave can be viewed as a far field far away from the source of the field.

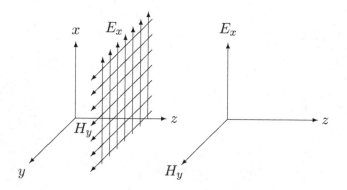

Figure 4-25   A plane wave

Definition of plane wave

As shown in Figure 4-25, a plane wave is an idea wave, which obeys the following definition:

(1) A plane wave is a transverse electromagnetic wave (or TEM wave), that both of **E** and **H** are lying on a transverse plane (x, y) corresponding to the propagation direction z.

(2) Meanwhile, the amplitude of both **E** and **H** are uniform in the transverse plane (x, y).

Of course, both of (4.167) and (4.168) have to obey the Maxwell equation and wave equation (4.163).

Substitution (4.167) into (4.163), considering that $E_y = E_z = H_x = H_z = 0$ and $\frac{\partial}{\partial x} = 0$, $\frac{\partial}{\partial y} = 0$, we have from (4.163) that

$$\frac{\partial^2 E_x}{\partial z^2} - \mu\varepsilon\frac{\partial^2 E_x}{\partial t^2} \tag{4.169}$$

This is an one-dimensional wave equation and the solution is

$$E_x = f\left(t - \frac{z}{v}\right) + f\left(t + \frac{z}{v}\right) \tag{4.170}$$

Where

$$v = \frac{1}{\sqrt{\mu\varepsilon}} \tag{4.171}$$

and $f\left(t \pm \frac{z}{v}\right)$ is an arbitrary function depending on the initial condition.

For a normal sinusoidal oscillation, we have

$$E_x = E_0 \sin[\omega(t - \frac{z}{v})] + E_0 \sin[\omega(t + \frac{z}{v})] \tag{4.172}$$

Now we will investigate the first term first.

The initial wave at $t = t_1$ and the second wave at $t = t_2$ are illustrated in Figure 4-26.

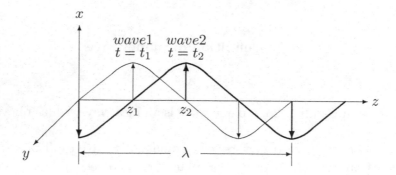

Figure 4-26   Propagation of a plane wave

Obviously, the amplitude of the plane wave at $t = t_1, z = z_1$ and $t = t_2, z = z_2$ should be equal to each other, namely

$$E_0 \sin \omega(t_1 - \frac{z_1}{v}) = E_0 \sin \omega(t_2 - \frac{z_2}{v}) \qquad (4.173)$$

And then

$$(t_1 - \frac{z_1}{v}) = (t_2 - \frac{z_2}{v}) \qquad (4.174)$$

From which, we have

$$v = \frac{z_2 - z_1}{t_2 - t_1} \qquad (4.175)$$

Therefore,

(1) "v" is the phase velocity of the plane wave, which means the phase variation in unit length along "+z". And, from (4.171), which depends on the material parameter (μ, ε) in the medium.

(2) " $-$ " in first term of (4.172) means that the first term is a plane wave propagating along the " $+ z$" direction.

(3) "$+$" in second term means that the second term is a plane wave propagating along the " $- z$" direction.

Obviously, the plane wave propagating along " $+ z$" direction is

$$E_x = E_0 \sin(\omega t - \frac{\omega}{v} z) \qquad (4.176)$$

which is the same as (4.167), in which,

$$k = \frac{\omega}{v} \qquad (4.177)$$

is so called the phase constant of the plane wave, which is the phase shift per unit length along z.

The phasor representation of a plane wave

For a time-harmonic field being the amplitude A and frequency ω time-invariant, the sine wave also can be expressed as a phasor (or phasor vector). Which has been used in electric engineering to calculate the current I, voltage V and power P in a complex circuit.

Phasors decompose the behavior of a sinusoidal wave into three independent factor: amplitude A, frequency ω, and phase ϕ. Since the frequency ω is often common to all the component of a linear combination of sinusoidal wave, in this case, phasor allow this common feature to be factored out, leaving the time-independent amplitude A and phase information, which can be combined algebraically rather then trigonometrically. Similarly, linear differential equations can be reduced to algebraic ones when we analyze an electric circuit.

Euler's formula indicates that the sine wave can be expressed as the sum of two complex-value functions. [8] Therefore, we have

$$A \cdot e^{(j\omega t + \phi)} = A \cdot \cos(\omega t + \phi) + jA \cdot \sin(\omega t + \phi) \tag{4.178}$$

and then, a sine wave can be expressed as

$$
\begin{aligned}
i = A \cdot \cos(\omega t + \phi) &= Re\{A \cdot e^{j(\omega t + \phi)}\} \tag{4.179}\\
&= Re\{Ae^{j\phi} \cdot e^{j\omega t}\} \tag{4.180}
\end{aligned}
$$

The phasor of i is

$$I = Ae^{j\phi}e^{j\omega t} \tag{4.181}$$

and then

$$\frac{\partial}{\partial t} = j\omega \tag{4.182}$$

Similarly, for a sine plane wave

$$E_x = E_0 \sin(\omega t - kz + \phi) = E_0 \cos(\omega t - kz) \tag{4.183}$$

where, $\phi = \pi/2$ is the initial phase at $t = 0, z = 0$ and the phasor of the plane wave E_x is

$$E_x = E_0 e^{j(\omega t - kz)} \tag{4.184}$$

The propagation characteristics of plane wave

---

[8] $\sin\theta = \frac{1}{2j}(e^{j\theta} - e^{-j\theta})$, $\cos\theta = \frac{1}{2}(e^{j\theta} + e^{-j\theta})$. and then, $e^{j\theta} = \cos\theta + j\sin\theta$

The discussion above give us two important propagation characteristics.

(1). The phase velocity, from (4.171),

$$v = \frac{1}{\sqrt{\mu\varepsilon}} \tag{4.185}$$

which dependents on the material parameters μ and ε. For instance, in free space,

$$\mu = \mu_0 = 4\pi \times 10^{-7} \quad (H/m) \tag{4.186}$$

$$\varepsilon = \varepsilon_0 = \frac{1}{36\pi} \times 10^{-9} \quad (F/m) \tag{4.187}$$

and then, the phase velocity in free space is

$$v = \frac{1}{\sqrt{\mu_0 \varepsilon_0}} = 3 \times 10^8 \quad (m/s) \tag{4.188}$$

which is just the light velocity c. Obviously, in general case,

$$v \le c \tag{4.189}$$

(2). The phase constant, from (4.177) and (4.188),

$$k = \frac{\omega}{v} = \omega\sqrt{\mu\varepsilon} \tag{4.190}$$

Again, the phase constant k is the phase shift per unit length along z.

(3). The wave length λ of a sinusoidal wave.

As indicated in Figure 4-26, which is the distance over which the wave's shape repeats. In other words, which is the spatial period of the wave or the distance over which the phase of the wave shifts one period (2π). And then,

$$\lambda \times k = 2\pi \tag{4.191}$$

$$\lambda = \frac{2\pi}{k} \quad (m) \tag{4.192}$$

And then,

$$k = \frac{2\pi}{\lambda} = \omega\sqrt{\mu\varepsilon} \tag{4.193}$$

$$\lambda = \frac{2\pi}{\omega\sqrt{\mu\varepsilon}} = \frac{v}{f} \tag{4.194}$$

(4) The wave impedance.

Following the electric impedance $Z = V/I$, the wave impedance of a plane wave η is the radio of the transverse components E_T/H_T, namely,

$$\eta = \frac{E_T}{H_T} = \frac{E_x}{H_y} \tag{4.195}$$

Where H_y can be obtained by substitution of E_x into Maxwell equation

$$\nabla \times \mathbf{E} = -\mu_0 \frac{\partial \mathbf{H}}{\partial t} \tag{4.196}$$

For a plane wave $H = a_y H_y$, $E = a_x E_x$, we have, from (4.196),

$$\mu \frac{\partial H_y}{\partial t} = -\frac{\partial E_x}{\partial z} \tag{4.197}$$

From phasor point of view, (4.197) leads to

$$j\omega\mu H_y = j\frac{\omega}{v} E_x \tag{4.198}$$

and then, we have

$$H_y = \sqrt{\frac{\varepsilon}{\mu}} E_x \tag{4.199}$$

Therefore, from (4.195), the wave impedance

$$\eta = \frac{E_x}{H_y} = \sqrt{\frac{\mu}{\varepsilon}} \tag{4.200}$$

Which means,
   (a) H_y is in phase with E_x in time domain,
   (b) H_y is perpendicular to E_x in spatial domain.
   In other words, that E_x and H_y is in phase propagation in the z direction. And $a_x \times a_y = a_z$ tell us that E_x and H_y are following the right hand law to point to the propagation direction z.
   The wave impedance η of the plane wave depends on the parameters μ and ε in the propagation medium. In free space,

$$\eta = \sqrt{\frac{\mu_0}{\varepsilon_0}} = 120\pi \approx 337 \quad \Omega \tag{4.201}$$

(5)Poynting Vector **S**.

Poynting vector is defined as the directional energy density of an electromagnetic field as follows:

$$\mathbf{S} = \mathbf{E} \times \mathbf{H} \tag{4.202}$$

Where, **S** is pointing to the propagation direction and **E** and **H** are the transverse components of the field (relative to the propagation direction).

For a plane wave, $\mathbf{S}$ is pointing to z and the transverse fields are Ex and H_y. As $a_x \times a_y = a_z$, we have

$$S = E_x \cdot H_y = \sqrt{\frac{\mu}{\varepsilon}} H_y^2 \tag{4.203}$$

Here, from (4.200), we have

$$\frac{\varepsilon E_x^2}{2} = \frac{\mu H^2 y}{2} \tag{4.204}$$

Which means, for a plane wave, the energy is equally distributed into electric field E_x and the magnetic field H_y. And then,

$$S = \frac{1}{\sqrt{\mu\varepsilon}}(\mu H_y^2) = v\left(\frac{\varepsilon E_x^2}{2} + \frac{\mu H^2 y}{2}\right) \tag{4.205}$$

Which gives us two information:

(a) The Poyinting vector S is a energy flux density with a velocity v.

(b) The energy density S is distributed equally into electric field E_x and the magnetic field H_y.

Conclusion

For a plane wave:

(1) Wave velocity $v = \dfrac{1}{\sqrt{\mu\varepsilon}}$

(2) Wave phase constant $k = \omega\sqrt{\mu\varepsilon}$

(3) Wavelength $\lambda = \dfrac{2\pi}{k}$

(4) Wave impedance $\eta = \sqrt{\dfrac{\mu}{\varepsilon}}$

(5) Poyinting vector $\mathbf{S} = \mathbf{E} \times \mathbf{H} = \mathbf{a}_z v\left(\dfrac{\varepsilon E_x^2}{2} + \dfrac{\mu H_y^2}{2}\right)$

(6) Energy density $\dfrac{\varepsilon E_x^2}{2} = \dfrac{\mu H_y^2}{2}$

(7) The plane wave is a TEM wave or a transverse electromagnetic wave.

Figure 4-27 shows us a plane wave propagating along the z direction.

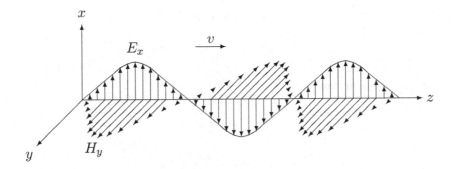

Figure 4-27   A plane wave propagating in a lossless dielectric

4.7.5   Reflection and refraction of plane wave

4.7.5.1   Normal incidence on plane boundaries

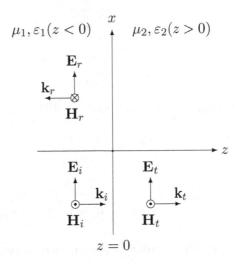

Figure 4-28   Normal incidence on plane boundary

Since plane wave $(\mathbf{E}_t, \mathbf{H}_t)$ is TEM wave, a plane wave normally incident onto plane boundary is similar to the reflection on a transmission lines.

As shown in Figure 4-28, the incident plane wave is:

$$\mathbf{E}_i = \mathbf{a}_x E_{i0} e^{-jk_1 z} \tag{4.206}$$

$$\mathbf{H}_i = \mathbf{a}_y \frac{E_{i0}}{\eta_1} e^{-jk_1 z} \tag{4.207}$$

The reflection wave is

$$\mathbf{E}_r = \mathbf{a}_x E_{r0} e^{jk_1 z} \tag{4.208}$$

$$\mathbf{H}_r = -\mathbf{a}_y \frac{E_{r0}}{\eta_1} e^{jk_1 z} \tag{4.209}$$

and the transmission wave is

$$\mathbf{E}_t = \mathbf{a}_x E_{t0} e^{-jk_2 z} \tag{4.210}$$

$$\mathbf{H}_t = \mathbf{a}_y \frac{E_{r0}}{\eta_2} e^{-jk_2 z} \tag{4.211}$$

Now the boundary condition at boundary plane $z = 0$ is

$$E_{i0}(1 + r) = \tau E_{i0} \tag{4.212}$$

$$\frac{E_{i0}}{\eta_1}(1 - r) = \tau \frac{E_{i0}}{\eta_2} \tag{4.213}$$

Where

$$r = \frac{E_{r0}}{E_{i0}} \tag{4.214}$$

is reflection coefficient.

$$\tau = \frac{E_{t0}}{E_{i0}} \tag{4.215}$$

is transmission coefficient. And then, from (4.212), (4.214), we have

$$r = \frac{\eta_2 - \eta_1}{\eta_2 + \eta_1} \tag{4.216}$$

$$\tau = \frac{2\eta_2}{\eta_2 + \eta_1} \tag{4.217}$$

4.7.5.2 Oblique incidence onto a plane boundaries

The conception of a plane wave obliquely incident onto a plane boundaries is very important for discussion of optical fiber as well as the semiconductor laser.

In fact, the wave propagation in optical fiber can be viewed as a total reflection of a TEM wave obliquely incident onto the boundaries between the core (ε_1) and the cladding (ε_2). And the wave inside the semiconductor laser can be viewed as a TEM wave obliquely incident onto the boundaries between the inside waveguide (ε_1) and the outside waveguide (ε_2).

Polarization of the wave

Before discussion, we have to define the polarization of the wave. Two polarization are defined with respect to **a plane of incident**, which contains both of the normal vector to the boundary, $\mathbf{a}_z$, and $\mathbf{k}$ vector of the incident signal.

Parallel polarization has the E-field lying on the **plane of incident** as shown in Figure 4-29(b). Perpendicular polarization has the E-field perpendicular to the **plane of incident** as shown in Figure 4-29(a).

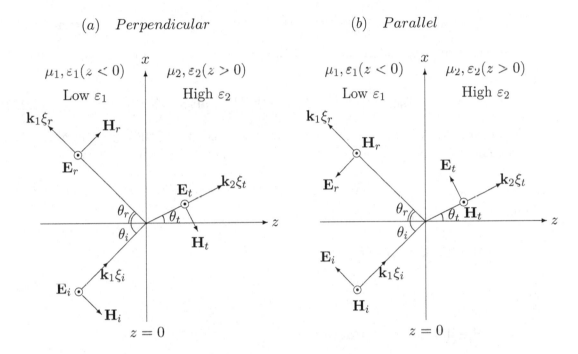

Figure 4-29   Oblique incidence onto a plane boundary

Boundary condition

The next step is to find the solutions of the reflection wave $\mathbf{E}_r, \mathbf{H}_r$ and the refraction wave $\mathbf{E}_t, \mathbf{H}_t$ by the known incident wave $\mathbf{E}_i, \mathbf{H}_i$. The solution can be found by the boundary condition at the boundary plane $z = 0$ as follows.

$$E_{\tau 1} = E_{\tau 2} \tag{4.218}$$

$$H_{\tau 1} = H_{\tau 2} \tag{4.219}$$

Where $E_{\tau 1}$ and $H_{\tau 1}$ are the total tangential component at boundary plane in region 1 ($z < 0$), $E_{\tau 2}$ and $H_{\tau 2}$ are the total tangential component on boundary plane in region 2 ($z < 0$). The boundary condition (4.218) and (4.219) is not only suitable for perpendicular polarization but also suitable for parallel polarization.

For perpendicular wave

For perpendicular wave in Figure 4-29(a),

$$\mathbf{E}_i = \mathbf{E}_{i0}e^{j(\omega t - k_1\xi)}, \quad \mathbf{E}_r = \mathbf{E}_{r0}e^{(\omega t + k_1\xi')}, \quad \mathbf{E}_t = \mathbf{E}_{t0}e^{j(\omega t - k_2\theta_t)} \qquad (4.220)$$

Where ξ_i is the propagation direction for incident wave, ξ_r is for reflection wave and ξ_t is for the transmission wave. They are, from Figure 4-29(a),

$$\xi_i = x\sin\theta_i + z\cos\theta_i, \quad \xi_r = -x\sin\theta_r + z\cos\theta_r, \quad \xi_t = x\sin\theta_t + z\cos\theta_t \qquad (4.221)$$

Considering that all of $\mathbf{E}_i$, $\mathbf{E}_r$ and $\mathbf{E}_t$ are tangential components at the boundary plane, we have, from the boundary condition (4.218),

$$\mathbf{E}_{i0}e^{-jk_1 x\sin\theta_i} + \mathbf{E}_{r0}e^{-jk_1 x\sin\theta_r} = \mathbf{E}_{t0}e^{-jk_2 x\sin\theta_t} \qquad (4.222)$$

Equation (4.222) is valid for all of x, this is possible only if the exponential function are equal to each other for both sides of the equality. Namely, we have,

$$k_1 x\sin\theta_i = k_1 x\sin\theta_r = k_2 \sin x\theta_t \qquad (4.223)$$

Therefore, we have

(a) **Snell's law of reflection:**

$$\theta_i = \theta_r \qquad (4.224)$$

(b) **Snell's law of refraction:**

$$\sqrt{\varepsilon_1}\sin\theta_i = \sqrt{\varepsilon_2}\sin\theta_t \qquad (4.225)$$

Considering that $\varepsilon = \varepsilon_0\varepsilon_r = \varepsilon_0 n^2$, where n **is the index of the dielectric,** the Snell's law of refraction can be writing to another form:

$$n_1\sin\theta_i = n_2\sin\theta_t \qquad (4.226)$$

Now, from the boundary condition (4.219), we have

$$\frac{E_{i0}}{\eta_1}e^{-jk_1 x\sin\theta_i} + \frac{E_{r0}}{\eta_1}e^{-jk_1 x\sin\theta_r} = \frac{E_{t0}}{\eta_2}e^{-jk_2 x\sin\theta_t} \qquad (4.227)$$

Substitution of the Snell's law (4.224), (4.225) into (4.227), we have

$$\Gamma_\perp = \frac{\eta_2\sec\theta_t - \eta_1\sec\theta_i}{\eta_2\sec\theta_t + \eta_1\sec\theta_i} \qquad (4.228)$$

$$\tau_\perp = \frac{2\eta_2 \sec\theta_t}{\eta_2 \sec\theta_t + \eta_1 \sec\theta_i} \tag{4.229}$$

Where $\Gamma_\perp = \dfrac{E_r}{E_i}$ is the reflection coefficient of the electric field, $\tau_\perp = \dfrac{E_t}{E_i}$ is the transmission coefficient of the electric field.

For parallel wave

Following the same process, we will have the following conclusion:

(a) For parallel wave, the Snell's law are the same as that of the perpendicular wave.

(b) For parallel wave, the reflection coefficient $\Gamma = \dfrac{E_{r0}}{E_{i0}}$ and the transmission coefficient $\tau = \dfrac{E_{t0}}{E_{i0}}$ are

$$\Gamma_\| = \frac{\eta_2 \cos\theta_t - \eta_1 \cos\theta_i}{\eta_2 \cos\theta_t + \eta_1 \sec\theta_i} \tag{4.230}$$

$$\tau_\| = \frac{2\eta_2 \cos\theta_t}{\eta_2 \cos\theta_t + \eta_1 \cos\theta_i} \tag{4.231}$$

Two special angles:

Two special angle are very useful:

(a) **Brewster Angle** θ_B: The **Brewster Angle** is an incident angle such that $\Gamma_\| = 0$, provide that $\mu_1 = \mu_2$, $\varepsilon_1 \neq \varepsilon_2$. Obviously, the **Brewster Angle** is for parallel wave only.

Setting the numerator of $\Gamma_\| = 0$ from (4.230) and using the Snell' law of refraction (4.225), gives us the **Brewster Angle** as follows:

$$\tan\theta_B = \sqrt{\frac{\varepsilon_2}{\varepsilon_1}} \tag{4.232}$$

At the **Brewster Angle**, waves with parallel polarization will be completely into the medium 2.

For perpendicular polarization wave, we may find the similar angle such that $\Gamma_\perp = 0$, however, this would require materials with different permeability and identical permittivity, which does not often occur in nature.

(b) **Critical Angle** θ_c: **Critical Angle** is an incident angle such that $\theta_t = 90^0$ and total reflection occurs for parallel polarization wave. Which occurs if the wave travels from a medium with **higher** permittivity (ε_1) into a medium with **lower** permittivity (ε_2). In this case, if the incident angle

$$\theta_i \geq \theta_c, \tag{4.233}$$

total reflection occurs and the transmission angle

$$\theta_t = 90^0 \tag{4.234}$$

in this case, there is no transmission signal into region 2 (ε_2). Which can be proved by setting $\theta_t = 90^0$ in (4.230), (4.231) to get $\Gamma_{\parallel} = -1$, $\tau_{\parallel} = 0$.

The critical angle can be found by Snell's law of refraction (4.225) with $\theta_t = 90^0$, namely,

$$\theta_c = \sin^{-1} \sqrt{\frac{\varepsilon_2}{\varepsilon_1}} \tag{4.235}$$

Obviously, (4.235) is valid only if $\varepsilon_1 > \varepsilon_2$ since $\sin \theta_c \leq 1$.

Physically, there is still some field penetration into region 2, however, the field strength decays exponentially away from the boundary, this wave is so called the evanescent wave.

Evanescent wave

As we discussion above, when $\theta_i > \theta_c$, even though there is complete reflection on n_1-side [9] of the boundary, some field on n_2-side of the boundary must be present in order to satisfy the boundary condition of (4.218) and (4.219), i.e. $E_{\tau 1} = E_{\tau 2}$ and $H_{\tau 1} = H_{\tau 2}$. The total reflection on n_1-side results in a standing wave in n_1-side of the boundary and an evanescent wave in n_2-side of the boundary as shown in Figure 4-30.

The evanescent wave is a wave with no power transmission into the n_2-side, which means that in evanescent wave, **E** field and **H** field are 90^0 out of phase and then the Poynting vector $\mathbf{E} \times \mathbf{H}$ of the evanescent wave is a pure imaginary vector, which is parallel to the boundary in the plane of incidence. The field strength decays exponentially with distance z away from the boundary on n_2-side, which can be expressed as $e^{-\alpha z}$ as shown in Figure 4-30. Where the attenuation constant α is

$$\alpha = \frac{2\pi}{\lambda} \sqrt{n_1{}^2 \sin^2 \theta_i - n_2{}^2} \tag{4.236}$$

Which can be proved by the fact, that the wave number in n_2-side is

$$k_z = \frac{2\pi n_2}{\lambda} \cos \theta_t \tag{4.237}$$

---

[9] $\quad n = \sqrt{\varepsilon_r}, \quad \varepsilon = \varepsilon_r \varepsilon_0$

Considering the Snells' law of refraction (4.225), (4.237) becomes

$$k_z = \frac{2\pi n_2}{\lambda}\sqrt{1 - \left(\frac{n_1}{n_2}\right)^2 \sin^2 \theta_i} \tag{4.238}$$

When $\theta_i > \theta_c$, we have $\sin \theta_i > \dfrac{n_2}{n_1}$, and then k_z becomes imaginary, or $k_z = j\alpha$ where α is given by (4.236).

Conclusion

Therefore, from (4.238)

(1) When $\theta_i > \theta_c$, total reflection occurs in n_1-side of the boundary and then the n_2-side of the boundary presents an evanescent wave as shown in Figure 4-30.

(2) When $\theta_i < \theta_c$, k_z = real number in n_2-side of the boundary. The wave will be passed into n_2-side of the boundary and lost.

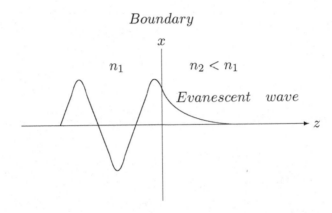

Figure 4-30   Wave propagation in an optical fiber

Application example

The light traveling in an idea optical fiber can be explained by the critical angle θ_c as indicated in Figure 4-31. Where an idea optical fiber contains a core (n_1) and a cladding (n_2) with $n_1 > n_2$ and $n_1 \approx n_2$, and the end of the fiber is covered by air.

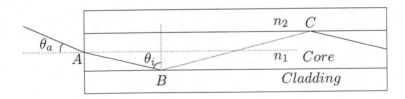

Figure 4-31   Wave propagation in an optical fiber

Now, when a light with a certain acceptance angle θ_a is entering a fiber, it must fall in order to be totally reflected inside the fiber. If light comes in along a ray path at an angle smaller than θ_a, then when the ray path hits the core-cladding boundary at point B, **it must be total reflected** and bound back to point C under the condition of

$$\theta_i \geq \theta_c \tag{4.239}$$

where θ_c is the critical angle. If

$$\theta_i \leq \theta_c \tag{4.240}$$

the light will be passed into the cladding and lost.

Chapter 5

Optical source - Semiconductor Laser

5.1 The fundamental of a Laser

5.1.1 Introduction

We may consider a semiconductor laser as a device that it can provide a light with frequency ν_{12} by injection a bias current into the laser. Therefore, the semiconductor laser is E to O converter or E/O converter. Now, what is the mechanism of the semiconductor laser? The semiconductor laser is a device involved in the interaction of light with the laser medium or the interaction between the photon and electrons, meanwhile, it is an energy transformation process between the photons and the electrons.

In the interaction between the photon and the electron , we have to face a word which is so called the "**population inversion**".

This is a fundamental important word to deal with the semiconductor laser by means of the quantum mechanics and the theory of the electromagnetic field.

To understand "population inversion", we have to deal with the semiconductor laser as an atom system from the quantum mechanics point of view. Then discuss the interaction between the photon and electron by the quantum mechanics and the theory of electromagnetic field.

In physics, a population inversion occurs when a system (such as a group of atoms or molecules) exists in state with more members in an exited state than in lower energy states. This concept is of fundamental important in semiconductor laser since the production of a population inversion is a necessary step to obtain the light amplification in the laser.

5.1.2 The energies model of atoms

In a material such as a crystal, the energy levels of electrons are indicated in Figure 5-1. In which, at temperature $T = 0K$, the Fermi energy E_F is the highest energy level occupied by the electrons, namely,

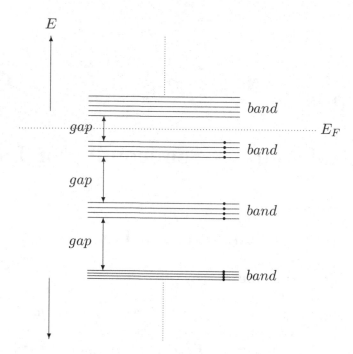

Figure 5-1   The energy levels of electrons in crystal

† The energy levels are empty if the energy of the electrons $E > E_F$.

† The energy levels are occupied by the electrons if the energy of the electrons $E < E_F$.

† The ground level is the lowest level of the electrons.

† The first upper level is the excited level 1.

† The second upper level is the excited level 2, $\cdots$ etc.

† Between the ground level and the excite level 1 is the energy gap E_g.

When we discuss the interaction of the photon and the electron, we may find tow process:

† A photon is incident into the crystal to give its energy $h\nu$ to the electron, which will make the electron transit from the ground level to the excited level according to $E_2 - E_1 = h\nu$. Which is so called the "absorption". Where E_1 is the ground level and E_2 is the excited level.

† An electron at excited level E_2 falls down to the ground level E_1, gives its energy to a photon according to $E_2 - E_1 = h\nu$. Which is called the "emission" and ν is the frequency of the photon.

A semiconductor is a device to convert the electric energy into the photon energy. The electric energy can be obtained by the bias current, which means that the semiconductor is

† Using the electric energy to excited the electrons transiting from the ground level E_1 to the excited level E_2.

† A photon emission occurs when the electron at excited level E_2 fall down to

the ground level E_1 according to $E_2 - E_1 = h\nu$. Thus, to make the photon emission continuously, we need $N_2 > N_1$, which is so called the population inverse. Where N_2 is the number of the electrons at the excited level E_2 and N_1 is the number of the electrons at the ground level.

5.1.3 Boltzmann distributions and thermal equilibrium

To understand the concept of a population inverse, we have to understand some thermodynamics and the light interaction with the laser medium. To this end, we may consider a simple assembly of atoms forming a laser medium.

Assume there are a group of N atoms, each of which is capable of being in one of two energy states, either

† The ground state with energy E_1; or

† The excited state with energy E_2 as shown in Figure 5-2. Where

$$E_2 > E_1 \tag{5.1}$$

The number of these atoms is noted by N_1 for which in the ground state and is noted by N_2 for which in the exited state and the total number

$$N = N_1 + N_2 \tag{5.2}$$

The energy different between the exited state and the ground state is

$$\triangle E_{12} = E_2 - E_1 = h\nu_{12} \tag{5.3}$$

Which can be used to determines the frequency ν_{12} of the light, that will interact with the atom. Where h is the planck's constant.

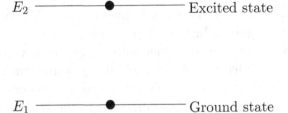

Figure 5-2   Two energy levels of electrons in laser medium

In thermal equilibrium

If the group of the atoms is **in thermal equilibrium**, it is proved from thermodynamics that the ratio of the number of atoms in each state is giver by a Boltzmann distribution as follows:

$$\frac{N_2}{N_1} = \exp\frac{-(E_2 - E_1)}{kT} \tag{5.4}$$

Where T is the temperature of the group of the atoms and k is Boltzmann's constant.

For instance, at room temperature ($T \approx 300^0 K$), for a frequency corresponding to a visible light $\nu \approx 5 \times 10^{14}$ Hz, the difference of the energy

$$\triangle E = E_2 - E_1 \approx 2.07 eV \tag{5.5}$$

and

$$kT \approx 0.026 eV \tag{5.6}$$

Obviously,

$$E_2 - E_1 >> kT \tag{5.7}$$

it follows that the argument of the exponential in (5.4) is a large negative number, and then

$$N_2 << N_1 \tag{5.8}$$

that means that there is almost no atoms in the excited state E_2 **in thermal equilibrium**.

When the temperature T increases, the number of electrons N_2 in the higher energy state E_2 increases but N_2 never exceeds N_1 for a system at thermal equilibrium, rather, when $T \to \infty$, the population N_2 and N_1 become equal.

Conclusion

The population inverse

$$N_2 > N_1 \tag{5.9}$$

can never be achieved at thermal equilibrium. This is a very important conclusion.

Thus, to achieve population inverse ($N_2 > N_1$), we have to push the system into a thermal non-equilibrium state.

For instance, the semiconductor laser can never be possible to provide a laser without any injection of external energy source such as bias current. To push the atoms system of semiconductor laser into the population inverse ($N_2 > N_1$), it is necessary to inject the bias current into the laser since the bias current will make the atoms transit from the ground state E_1 into the excited energy state E_2, so as to get ($N_2 > N_1$) in semiconductor laser.

5.1.4   The interaction of electrons and photons

Now we will discuss the interaction of electrons and photons in direct semiconductor.

As shown in Figure 5-3, a photon is incident into a laser medium with excited level E_2 and ground level E_1. The laser medium can be a semiconductor laser medium with conduction band (as the excited level E_2) and the valence band (E_1).

The interaction of the photon and electron in this laser medium is involved in three process as follows:

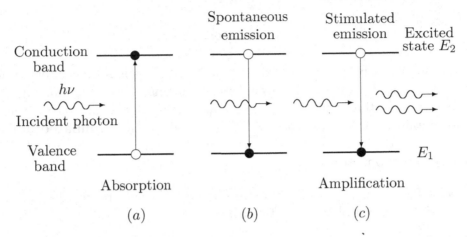

Figure 5-3   The interaction of photon and electron in semiconductor

Absorption

As shown in Figure 5-3 (a), an electron transits from the ground state into the excited state, which is induced by an incident of a photon, as the result the photon is absorbed. This is so called "absorption".

In fact, this is a energy conversion from a photon ($h\nu$) into the electron, used to make the electron transited from the ground state (E_1) into the excited state (E_2) based on

$$E_2 - E_1 = h\nu \tag{5.10}$$

Spontaneous emission

As shown in Figure 5-3 (b), an electron at excited state returns from the excited state into the ground state emitting a photon with

$$h\nu = E_2 - E_1 \tag{5.11}$$

Which is so called **"spontaneous emission"**.

This process can be mathematically modeling by

$$-\frac{dN_2}{dt} = A_{21}N_2 \equiv \frac{N_2}{\tau_{21}} \tag{5.12}$$

means that the spontaneous rate $\frac{dN_2}{dt}$ is proportional to the number of the atoms at the exited level N_2, which lead to a solution of

$$N_2(t) = N_2(0)e^{-A_{21}t} \tag{5.13}$$

Where

$$A_{21} = 1/(t_{spont})_{21} \tag{5.14}$$

is the spontaneous transition rate. And $(t_{spont})_{21} \equiv A_{21}^{-1}$ is called the spontaneous lift-time associated with the transition $2 \to 1$.

Stimulated emission

As shown in Figure 5-3 (c), an external photon causes that an electron returns from the excited state into the ground state emitting a second photon. As the result, an external photon incident causes two photons output, which means the light has been amplification in this process. This is so called **the stimulated emission**.

Look at the process carefully, we found that the amplification of the light is depends on two factors:

- The frequency of the external photon, ν, should be equal to the frequency of the second photon, ν_{21} following

$$h\nu_{21} = E_2 - E_1 \tag{5.15}$$

Therefore, the second photon is a coherent light with the external photon, which is a necessary result of the stimulated emission.

- The population distribution of N_2 and N_1 should be

$$N_2 > N_1 \tag{5.16}$$

Which is another condition for the light amplification. This condition is so called **"population inverse ($N_2 > N_1$)"**. In other words, that the necessary condition of the light amplification is the **"population inverse ($N_2 > N_1$)"**. This is a fundamental important conclusion.

Stimulated emission can be modeled mathematically by considering an atoms that stay at one of lower level state E_1 or the excited state E_2.

If the atom is in the excited state E_2, it may decay into the lower level state E_1 by the process of spontaneous emission, releasing the difference in energies between the two state as a photon, which has the frequency ν_{21} and energy $h\nu_{21}$ as follows:

$$E_2 - E_1 = h\nu_{21} \tag{5.17}$$

where h is Planck's constant.

Alternatively, if the atom at the excited state E_2 is perturbed by an electric field of frequency ν_{21}, it may emit an additional photon of the same frequency and in phase, leaving the atom in the lower energy state E_1. This process is known as stimulated emission.

If the number of atoms in the excited state E_2 is given by N_2, the rate of the stimulated emission is given by

$$R_{21st} = \frac{\partial N_2}{\partial t} = -\frac{\partial N_1}{\partial t} = B_{21}\rho(\nu)N_2 \qquad (5.18)$$

means that the stimulation rate R_{21st} is proportional not only to the number of the atoms at the exited level N_2 but also to $\rho(\nu)$, the radiation density of the incident field at frequency ν. Where the proportionality constant B_{21} is known as the Einstein B coefficient for that particular transition, and $\rho(\nu)$ is the radiation density of the incident field at frequency ν [1].

At the same time, there well be a process of atomic absorption, which removes the energy from the field while raising electrons from the lower state E_1 into the upper state E_2. The absorption rate is given by

$$R_{abs} = \frac{\partial N_2}{\partial t} = -\frac{\partial N_1}{\partial t} = -B_{12}\rho(\nu)N_1 \qquad (5.19)$$

Which means that:

† The absorption rate R_{abs} is proportional to the the number of the atoms in lower state N_1.

† The absorption rate R_{abs} is proportional to the radiation density of the incident field $\rho(\nu)$ as well.

Einstein showed that the coefficient for this transition B_{12} must be identical to the coefficient for the stimulated emission transition B_{21}, i.e.

$$B_{12} = B_{21} \qquad (5.20)$$

Thus absorption and stimulated emission are reverse process at somewhat different rates.

When we consider the light amplification, we have to take both of the absorption and the stimulated emission as a single process. In this case, the net rate of the transition from E_2 to E_1 due to this combined process can be found by adding the stimulated emission rate and the absorption rate as follows:

$$R_{st} = R_{21st} - R_{abs} = \frac{\partial N_2(net)}{\partial t} = -\frac{\partial N_1(net)}{\partial t} = B_{21}\rho(\nu)(N_2 - N_1) \qquad (5.21)$$

Thus a net power P is released into the electric field, which is given by

$$P = R_{st}h\nu = B_{21}\rho(\nu)(N_2 - N_1) \qquad (5.22)$$

which means:

---

[1] Actually, the incident field is the incident photons.

- The condition to obtain a positive net power is

$$N_2 > N_1 \qquad (5.23)$$

which is known as a *population inverse*. Physically, in order to obtain a positive net power P, there must be more atoms in excited state E_2 than in lower level state E_1. Otherwise there is net absorption and the power of the wave is reduced during passage through the medium.

- The notable characteristic of stimulated emission compared to the spontaneous emission is that the emission photons have the same frequency, phase, polarization, and direction of propagation as the incident photons. In other words, both of the stimulated emitted photons and the incident photons are mutual coherent.

Line shape

The above rate equation refers only to excitation at the particular optical frequency ν_0 corresponding to the energy of the transition. At frequency offset from ν_0 the strength of stimulation (or spontaneous) emission will be decreased according to the "line shape".

Now consider only homogeneous broadening affecting, an atom or molecular resonance, the spectral line shape function is described as a Lorentzian distribution:

$$g'(\nu) = \frac{1}{\pi} \frac{(\Gamma/2)}{(\nu - \nu_0)^2 + (\Gamma/2)^2} \qquad (5.24)$$

Where Γ is the full width at half maximum bandwidth. The peak value of Lorentzian line shape occurs at the line center, $\nu = \nu_0$. And the normalization of the line shape function is

$$\int_{-\infty}^{\infty} g'(\nu) d\nu = 1 \qquad (5.25)$$

In this case of a Lorentzian, we have

$$g(\nu) = \frac{g'(\nu)}{g'(\nu_0)} = \frac{(\Gamma/2)^2}{(\nu - \nu_0)^2 + (\Gamma/2)^2} \qquad (5.26)$$

Thus stimulated emission at frequencies away from ν_0 is reduces by the factor $g(\nu)$.

In practice there may be also be broadening of the line due to inhomogeneous broadening, most notably due to Doppler effect resulting from the distribution of velocities in a gas at certain temperature. This has a Gaussian shape and reduces the peak strength of the line shape function. Therefore optical amplification will add power to an incident optical field at frequency ν a rate given by

$$P = h\nu g(\nu) B_{21} \rho(\nu)(N_2 - N_1) \qquad (5.27)$$

5.1.5 Optical amplification

Under certain conditions, stimulated emission can provide a physical mechanism for optical amplification.

A group of atoms can be excited into the excited state E_2 from the ground state E_1 so as to make the number of atoms at excited state N_2 is lager than the number of atoms at ground N_1, creating a population inverse.

When a light of frequency ν corresponding to

$$h\nu = E_2 - E_1 \tag{5.28}$$

passes through the population inverse medium, the photons stimulated the excited atoms to emit additional photons of the same frequency, phase, and direction,, resulting in an amplification of the intensity of the input light.

The population inversion, in units of atoms per cubic meter, is

$$N_2 - N_1 = \left(N_1 - \frac{g_2}{g_1}N_1\right) \tag{5.29}$$

where g_1 and g_2 are the degeneracies of energy levels 1 and 2, respectively.

Small signal gain equation

The intensity (in watts per square meter) of the stimulated emission I is governed by the following differential equation:

$$\frac{dI}{dz} = \sigma_{21}(\nu)(N_2 - N_1)I(z) \tag{5.30}$$

as long as the intensity $I(z)$ is small enough so that it does not have a significant effect on the magnitude of the population inverse. The equation may be simplified as

$$\frac{dI}{dz} = \gamma_0(\nu)I(z) \tag{5.31}$$

where

$$\gamma_0(\nu) = \sigma_{21}(\nu)(N_2 - N_1) \tag{5.32}$$

is **small-signal gain coefficient** (in unit of radians per meter).

Which means that the intensity of the stimulation emission $I(z)$ is increased along z in a population inverse $N_2 - N_1 > 0$ region $0 \to z$.

Now, we can solve the differential equation using separation of variables:

$$\frac{dI}{I(z)} = \gamma_0(\nu)dz \tag{5.33}$$

Integrating, we have

$$\ln\left(\frac{I(z)}{I_{in}}\right) = \gamma_0(\nu)z \tag{5.34}$$

or

$$I(\nu) = I_{in} \exp\{\gamma_0(\nu)z\}$$ (5.35)

and the small signal gain is:

$$G = \frac{I(z)}{I_{in}} = \exp\{\gamma_0(\nu)z\}$$ (5.36)

Where $I_{in} = I(z = 0)$ is the optical intensity of the input signal (in watts per square meter).

Saturation intensity

The saturation intensity I_s is defined as the input intensity at which the gain of the optical amplifier drops to exactly half of the small-signal gain.

We can compute the saturation intensity as

$$I_s = \frac{h\nu}{\sigma(\nu)\tau_s}$$ (5.37)

Where:

  † h: is planck's constant.

  † τ_S: is the saturation time constant, which depends on the spontaneous emission lifetimes of the various transitions between the energy levels related to the amplification.

  † ν: is the frequency in Hz.

General gain equation

The general form of the gain equation, which applies regardless of the input intensity, drive from the general differential equation for the intensity I as a function of position z in the gain medium:

$$\frac{dI}{dz} = \frac{\gamma_0(\nu)}{1 + \bar{g}(\nu)\dfrac{I(0)}{I_s}}I(z)$$ (5.38)

where I_s is the saturation intensity.

This equation can be solved by using the separate variable method as follows:

$$\frac{dI}{I(z)}\left[1 + \bar{g}(\nu)\frac{I(z)}{I_s}\right] = \gamma_0(\nu)dz$$ (5.39)

Integrating both side, we have

$$\ln\left(\frac{I(z)}{I_{in}}\right) + \bar{g}(\nu)\frac{I(z) - I_{in}}{I_s} = \gamma_0(\nu)z \tag{5.40}$$

or

$$\ln\left(\frac{I(z)}{I_{in}}\right) + \bar{g}(\nu)\frac{I_{in}}{I_s}\left(\frac{I(z)}{I_{in}} - 1\right) = \gamma_0(\nu)z \tag{5.41}$$

The gain of the amplifier is defined as the optical intensity I at position z divided by the input intensity:

$$G = G(z) = \frac{I(z)}{I_{in}} \tag{5.42}$$

Substituting this definition into (5.41), we found **the general gain equation**:

$$\ln(G) + \bar{g}(\nu)\frac{I_{in}}{I_s}(G - 1) = \gamma_0(\nu)z \tag{5.43}$$

Small signal approximation

In case of

$$I_{in} << I_s \tag{5.44}$$

the general gain equation (5.43) become

$$\ln(G) = \ln(G_0) = \gamma_0(\nu)z \tag{5.45}$$

or the gain become

$$G = G_0 = \exp\{\gamma_0(\nu)z\} \tag{5.46}$$

which is identical to the small signal gain (5.36).

Large signal asymptotic behavior

In case of

$$I_{in} >> I_s \tag{5.47}$$

the gain approaches unit

$$G \to 1 \tag{5.48}$$

and the general gain equation approaches a linear asymptote, from (5.45):

$$I(z) = I_{in} + \frac{\gamma_0(\nu)z}{\bar{g}(\nu)}I_s \tag{5.49}$$

5.1.6   Pumping

As we discussed before, we can never get "population inverse $(N_2 > N_1)$" in temperature equilibrium condition.

How to get "population inverse"?

By "pumping"

One way of the pumping is injection of a bias current into the semiconductor laser. When a bias current injects into the semiconductor, the electrons at the valence band (or the ground state) will be transited into the conduction band (or the excited state) to set up the population inverse $(N_2 > N_1)$.

In fact, we are more interested in the operation of the semiconductor laser, which is involved in three processes as we discussed before:

† Initially, atoms are energized from the ground state E_1 to the excited state E_2 by a process of **pumping**.

† Some of these atoms (in the excited state) decay via **spontaneous emission**, releasing incoherent light as photons of frequency ν. These photons are fed back into the laser medium. Some of these photons are absorbed by the atoms in the ground state. And the photons are lost to the laser process.

† However, some photons cause **stimulated emission** in excited state atoms, resulting another coherent photon [2], this result in optical amplification.

If the number of photons being amplified per unit time is greater than the number of photons being absorbed, than the net result is a continuously increasing number of photons being produced, and the laser medium is said to have a gain of great than unit.

At this point we have to point out that:

† The rate of the absorption process are proportional to the the number of atoms N_1 in the ground state E_1 (or valence band),

† The rate of the amplification process are proportional to the number of the atoms N_2 in the excited state E_2 (or conduction band),

† Thus, if $N_1 > N_2$, the process of the **absorption** will dominates and therefore is a net attenuation of the photon.

† If $N_1 = N_2$, the rate of absorption of light exactly balance the rate of emission, the medium is then said to be **optically transparent**.

† If $N_2 > N_1$, then the emission process dominates, and the light in this system undergoes a net increase in light intensity or the laser medium has a **gain** of great

---

[2]   Coherent photon: that means the another photon is in frequency and in phase with the initial photon.

than unit.

Obviously, it is required to have the population inverse $N_2 > N_1$ in the semiconductor laser operation, which can be obtained by, for instance, an injection of the bias current. Now we will use two real model of laser medium, three-level lasers and four-levels laser, to explain that how to obtain the population inverse as follows.

5.1.7 Three-level lasers

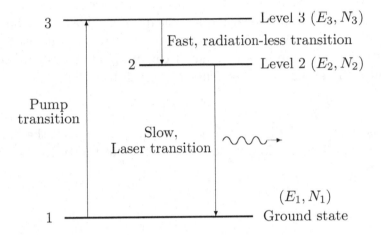

Figure 5-4   Three-levels energy diagram

As shown in Figure 5-4, consider a group of N atoms with each atom able to exit in any of three energy-level state 1, 2 and 3 with energies E_1, E_2 and E_3 and populations N_1, N_2 and N_3, respectively. Where

$$E_3 > E_2 > E_1 \tag{5.50}$$

That means the level 2 is lies between level 1 and level 3.

Initially, the system of atoms is at thermal equilibrium, and the majority of the atoms will be in the ground state, i.e.

$$N_1 \approx N \tag{5.51}$$

$$N_2 \approx N_3 \approx 0 \tag{5.52}$$

Now, suppose that a light of frequency

$$\nu_{12} = h^{-1}(E_3 - E_1) \tag{5.53}$$

is incident, which induces an electron transiting from ground state (E_1) into the level 3 (E_3) and results in a light absorption. The process of light absorption will excite the atoms from ground state into the level 3. This is so called the "pumping transition" as shown on the left side of Figure 5-4. The other method, such as injection of bias current into the semiconductor laser or chemical reaction, also may be used to the "pumping transition" for the atoms at level 3. If we continue pumping the atoms, we will excite an appreciable number of atom into level 3, such that

$$N_3 > 0 \tag{5.54}$$

Then, the excited atoms N_3 at level 3 will quickly decay to level 2. The energy released in this transition may be emitted as a photon (spontaneous emission), however, the transition ($3 \rightarrow 2$) is a fast radiation-less transition, with the energy being converted to vibrational motion (heat) of the host material surrounding the atoms. Therefore, there is no any generation of a photon in this transition.

After then, an atom in level 2 may decay by spontaneous emission to the ground state, releasing a photon of frequency ν_{12}, given by

$$E_2 - E_1 = h\nu_{12} \tag{5.55}$$

, which is shown as the laser transition on the right side. If the lifetime of the laser transition, τ_{12}, is much longer than the lifetime of the transition from $3 \rightarrow 2$, τ_{32}, namely, if

$$\tau_{12} >> \tau_{32} \tag{5.56}$$

we have

$$N_3 \approx 0 \tag{5.57}$$

and a population of excited state atoms will be accumulated in level 2, namely,

$$N_2 > 0 \tag{5.58}$$

If over half of the N atoms are accumulated in level 2, we will have

$$N_2 > N_1 \tag{5.59}$$

which means that the population inverse ($N_2 > N_1$) has been achieved between level 2 and level 1 and then we may obtain the optical amplification in this laser medium.   Obviously, at least half population of atoms N must be excited from ground state to level 3 to obtain the population inverse, the laser medium must be very strongly pumped. This make the three-level laser rather inefficient. Therefore, most of lasers are four-level laser.

5.1.8   Four-level lasers

As shown in Figure 5-5, the four-levels energy diagram is consist of four level with energy E_1, E_2, E_3, E_4 and population N_1, N_2, N_3, N_4 in level 1, 2, 3, 4, respectively. Where

$$E_1 < E_2 < E_3 < E_4 \tag{5.60}$$

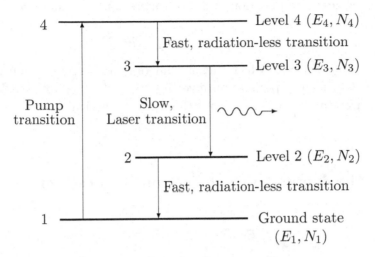

Figure 5-5   Four-levels energy diagram

In this system,

† The pumping transition excites the atoms in the ground state into the level 4 as shown on the left side of Figure 5-5,

† And the atoms at level 4 fast transit from level 4 into level 3 with non-radiation because of the lifetime of this transition τ_{43} is very short. And then

$$N_4 \approx 0. \tag{5.61}$$

† Since the lifetime of this transition, τ_{43}, is much shorter than the lifetime of the laser transition from level 3 to level 2, τ_{32}, i.e. because of

$$\tau_{32} \gg \tau_{43} \tag{5.62}$$

a population accumulates in level 3, which may relax by spontaneous or stimulated emission into level 2. And then this transition is so called the laser transition.

† Then the level 2 also has a fast non-radiation decay into ground level. Therefore, the population N_2 at level 2 is negligible, i.e.

$$N_2 \approx 0 \tag{5.63}$$

This is fair important, since any appreciable population in level 3 will form a population inverse between level 3 and level 2, i.e. the whole process reaches to a population inverse

$$N3 > N2, \tag{5.64}$$

$$N_2 \approx N_4 \approx 0 \tag{5.65}$$

Tush, an optical amplification and laser operation can take place at a frequency of ν_{32} with

$$E_3 - E_2 = h\nu_{32} \tag{5.66}$$

Since $N_2 \approx N_4 \approx 0$, which means that only a few population has to be excited into level 3 to form a population inverse ($N_3 > N_2$), a four level laser is much efficient than a three-level laser. And then most practical lasers are of the four-level laser.

5.1.9   The energy levels of electrons in a crystal

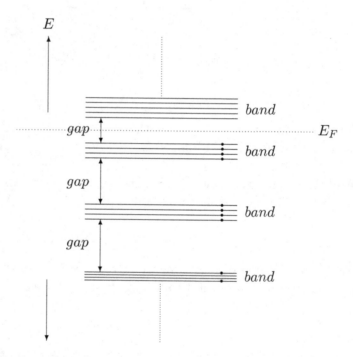

Figure 5-6   The energy levels of electrons in a crystal

In a semiconductor bulk, no two electrons in a crystal can be placed in the same quantum state. This is so-called **Pauli exclusion principle**, which is one of the more important axiomatic foundations of quantum mechanics. Each electron thus must possess a unique spatial wave-function and an associated eigen-energy (the total energy associated with the state).

As shown in Figure 5-6, the energy levels of electrons in a crystal are indicated by horizontal lines. And the energy levels cluster within bands, which are separated by "energy gaps", or so-called the "forbidden gaps". On the uppermost "gap", the lower band, which is so-called valence band, is filled up with electrons while the upper band, which is so-called the conduction band, is completely empty. The gap between them is lager enough, say $\approx 3eV$, so that thermal excitation across the gap is negligible.

If the gap between the conduction band and the valence band is small, say $\approx 2eV$, then the thermal excitation cause partial convert of electron from the valence band to the conduction band and the crystal can conduct electricity. **Such crystals are called the semiconductors.** Their conductivity can be controlled not only by the temperature but also by "doping" them with impurity atoms. The wave function of an electron in a valence band is characterized by a vector **k** and a corresponding wave function

$$\psi_v(\mathbf{r}) = u_{v\mathbf{k}}(\mathbf{r})e^{i\mathbf{k}\cdot\mathbf{r}} \tag{5.67}$$

The function possess the same periodicity as the lattice. The factor $\exp(ik \cdot r)$ is responsible for the wave nature of the electronic motion and is related to the wavelength λ_e of the electron as follows:

$$\lambda_e = \frac{2\pi}{k} \tag{5.68}$$

The vector **k** can only possess a set of value, i.e. it is quantized value, which is obtained by requiring that the total phase shift $\mathbf{k}\cdot\mathbf{r}$ across a crystal with dimension L_x, L_y, L_z be some multiple integer of 2π. Therefore, $k_iL_i = s\ 2\pi$, or

$$k_i = \frac{2\pi}{L_i}s, \quad s = 1, 2, 3, \ldots \tag{5.69}$$

where $i = x, y, z$. Obviously, from (5.69), **the value of k in crystal is a quantized value**.

Thus, we can divide the total volume in k space into cells each with a volume

$$\triangle V_k = \triangle k_x \triangle k_y \triangle k_z = \frac{(2\pi)^3}{L_xL_yL_z} = \frac{(2\pi)^3}{V} \tag{5.70}$$

The energy, measured from the bottom of the conduction band is

$$E_c(\mathbf{k}) = \frac{\hbar^2 k^2}{2m_c} \tag{5.71}$$

where m_c is the effective mass of an electron in the conduction band.

Obviously, the energy $E_c(\mathbf{k})$ depends only on the magnitude k of the electron propagation and not its direction. Therefore, when we need to perform electron counting, we don't need in **k** space but as a function of the energy. **This is an important point**.

5.1.10 Energy band model of direct semiconductor

Figure 5-7 shows the energy-k relationship of a direct gap semiconductor. Where, the conduction band minimum and the valence band maximum space at the same value of k. The dots represent allowed (not necessary occupied) electron energies. And according to (5.69), these energies are spaced uniformly along the k axis.

Again, here the dots represent allowed but not necessary occupied electron energies.

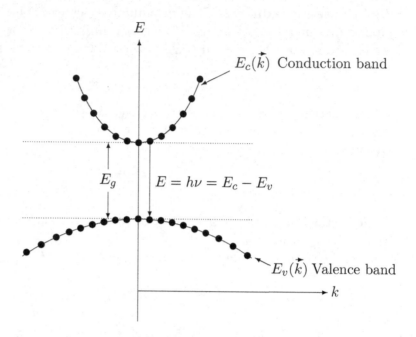

Figure 5-7   Energy band structure of a direct semiconductor with $m_c \leq m_v$

The Fermi-Dirac distribution law

The probability that an electron state energy E is occupied by an electron is given by the Fermi-Dirac distribution law [75]

$$f(E) = \frac{1}{\exp[(E - E_F)/kT] + 1} \tag{5.72}$$

Where E_F is the Fermi energy and T is the temperature.

According to the Fermi-Dirac distribution law:

† When $E = E_F$, $f(E) = \dfrac{1}{2} = 50\%$, which is the definition of the Fermi level, means that the Fermi level E_F is a energy of electron with the Fermi-Dirac distribution function $f(E_F) = 50\%$.

$$f(E) \to 1, \quad when \ E_F - E >> kT \tag{5.73}$$

That means when $E_F - E >> kT$, almost all of the energies below Fermi level are occupied by electrons, while above the Fermi level the energy approaches the

Boltzmann distribution, i.e.

$$f(E) \approx e^{-E/kT}, \quad when \quad E - E_F \gg kT \tag{5.74}$$

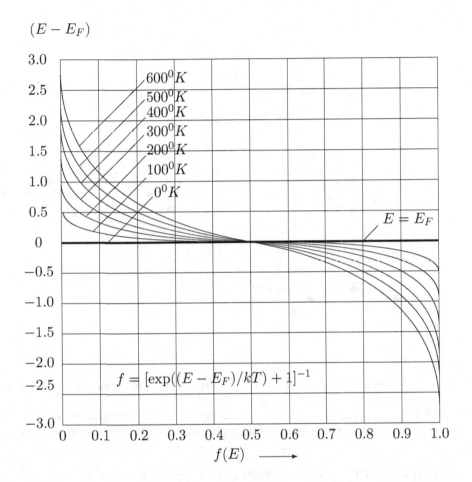

$(E - E_F)$

$$f = [\exp((E - E_F)/kT) + 1]^{-1}$$

$f(E) \longrightarrow$

Figure 5-8   Fermi-Dirac distribution function (After W.W. Gartner)

† Especially, at $T = 0K$

$$f(E) = 1, \quad For \quad E < E_F \tag{5.75}$$

$$f(E) = 0, \quad For \quad E > E_F \tag{5.76}$$

That means at $T = 0K$, the energies are all occupied by electrons below the Fermi level, while the energies are empty above Fermi level E_F.

In fact, the Fermi distribution function $f(E)$ is the function of $(E - E_F)$ with different temperatures T as shown in Figure 5-8.

E_F - **The Fermi energy**

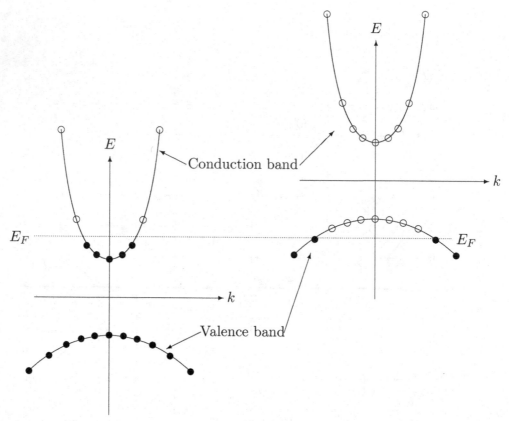

(a)   n-type semiconductor at $T = 0K$ (b)   p-type semiconductor at $T = 0K$

Figure 5-9   (a) Energy band of a degenerate n-type semiconductor at 0 K.
(b) Energy band of a degenerate p-type semiconductor at 0 K.

Now from (5.72), in semiconductor, **the Fermi energy is a energy of electron with the Fermi-Dirac distribution function** $f(E_F) = 0.5$.

Especially, from (5.75) and (5.76), at $T = 0K$, the energies of electrons are occupied by electrons below Fermi energy while the energies of electrons are empty above Fermi energy. Therefore, **at T=0K, the Fermi energy is the highest energy occupied by the electrons.**

Thermal equilibrium in semiconductor

In thermal equilibrium, a single Fermi energy is applied to both the conduction band and the valence band as shown in Figure 5-9.

In very high doped semiconductors, the Fermi level is forced into either the conduction band for donor impurity doping indicated in Figure 5-9(a), or the valence band for acceptor impurity doping indicated in Figure 5-9(b).

At $T = 0K$, the energies are occupied by the electrons below the Fermi level E_F as indicated by dots both in conduction band in Figure 3-9(a) and valence band in Figure 5-9(b), while the energies are empty above the Fermi level E_F both in conduction band in Figure 5-9(a) and valence band in Figure 5-9(b) as indicated by empty circles.

The unoccupied stales in the valence band in Figure 3-9(b) are called holes and they are treated exactly like the electrons except their charge is positive and their energy increase downward in the diagram. The unoccupied states of electrons in Figure 5-9(b) can be viewed as the states of holes in the valence band. The process of exciting an electron from state a to state b can be viewed as a hole is excited from state b to state a.

Non-thermal equilibrium in semiconductor

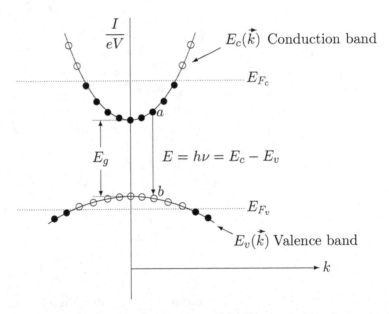

Figure 5-10   Electrons are Injected at a rate of I/eV per unit volume (I=total current) into the conduction band of a semiconductor

However, if the semiconductor is in non-thermal equilibrium, such as injection of electrons and holes by an external bias current $\dfrac{I}{eV}$ or the semiconductor undergoes a strong light incident, the populations (i.e. electrons and holes) will displaced from equilibrium. The population displaced from equilibrium is such that the carrier population can no longer be described by a single Fermi energy level. In this case, we have to use two Fermi energy levels as shown in Figure 5-10. One, E_{F_c}, used for conduction band only and another one, E_{F_v}, used for valence band only. Both of them are called **the quasi-Fermi levels**.

The concept of quasi-Fermi levels is valid based on the fact that carrier scattering time ($\approx 10^{-12}s$) within a band is much shorter than the equilibrium time ($\approx 3 - 4^{-9}s$) between bands. This is usually true at the large carrier densities used in $p - n$ junction laser.

To understand the conception of the quasi-Fermi level in non-thermal equilibrium, we may consider that electrons are excited into the conduction band of a p-type semiconductor (which can be done by injecting electrons into the p region across a $p - n$ junction or by subjecting the semiconductor to an high intensity of light beam with $h\nu > E_g$), so that, for each absorbed photon, an electron is excited into the conduction band from the valence band. This situation is indicated in Figure 5-11.

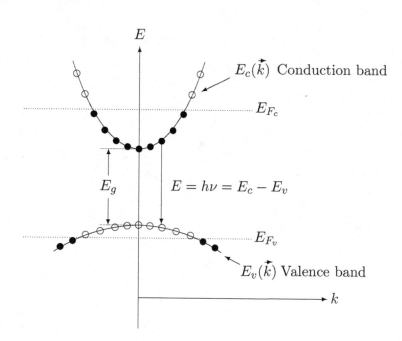

Figure 5-11   Energy band structure of a direct semiconductor

Following the excitation, electrons relax by emitting optical and acoustic photons [3] to the bottom of the conduction band in time of

$$\approx 10^{-12}s \qquad (5.77)$$

---

[3]   In the process of electron transition from valence band to the conduction bane caused by photons, the momentum of the electron can not be changed so that the acoustic photon have to be involved.

while their relaxation across the gap back to the valence band, this process is called the electron-hole recombination with time constant of

$$\tau \approx 3 - 4^{-9}s \qquad (5.78)$$

When a disturbance from a thermal equilibrium situation occurs, the populations of holes and electrons will be changed. If the disturbance is not too great or not to changing too quickly, the populations of electrons and holes each relax to a state of quasi thermal equilibrium. Because the relaxation time for electrons within the conduction band is smaller than across the band gap, we may consider that the electrons are in thermal equilibrium in the conduction band. This also applicable for holes in the valence band. In this case we may define a quasi Fermi level E_{F_c} and temperature due to thermal equilibrium of electrons in conduction band and quasi Fermi level E_{F_v} for holes similarly.

Where the quasi Fermi level E_{F_c} for conduction band is corresponding to

$$f_c(E) = \frac{1}{e^{(E-E_{F_c})/kT} + 1} \qquad (5.79)$$

The quasi Fermi level E_{F_v} for valence band is corresponding to

$$f_v(E) = \frac{1}{e^{(E-E_{F_v})/kT} + 1} \qquad (5.80)$$

Where, the populations in semiconductor with non-thermal equilibrium is a new atoms system. The injection of bias current only cause a new energy of atoms system even though the process of the electrons transition is involved the $p-n$ conjunction. The only different from the thermal equilibrium is that the process of the injection of bias current force that the single Fermi level E_F to be separated into two quasi Fermi levels, E_{F_c} and E_{F_v}.

The quasi Fermi level for conduction band – E_{F_c}

The quasi Fermi level for conduction band E_{F_c} is the energy level in which the probability of occupation of electron is 0.5.

At $T = 0K$, the energies of electrons in conduction band are occupied by electrons below the quasi Fermi energy E_{F_c} while the energies of electrons are empty above the quasi Fermi energy E_{F_c}. Therefore, **the quasi Fermi energy E_{F_c} is the highest energy occupied by the electrons in conduction band when $T = 0K$.**

The quasi Fermi level for valence band – E_{F_v}

The quasi Fermi level for valence band E_{F_v} is the energy level in which the probability of occupation of electron is 0.5.

At $T = 0K$, the energies of electrons in valence band are occupied by electrons below the quasi Fermi energy E_{F_v} while the energies of electrons are empty above the quasi Fermi energy E_{F_v}. Therefore, **the quasi Fermi energy E_{F_v} is the highest energy occupied by the electrons in the valence band when** $T = 0K$.

On the other hand, the quasi Fermi level for valence band E_{F_v} can be viewed as the energy level in which the probability of occupation of **hole** is 0.5. Therefore, E_{F_v} also can be viewed as the hole quasi Fermi level.

At $T = 0K$, the energies of holes in valence band are occupied by hole above the quasi Fermi energy E_{F_v} while the energies of holes are empty below the quasi Fermi energy E_{F_v}. Therefore, **the quasi Fermi energy E_{F_v} is the lowest energy occupied by holes in the valence band when** $T = 0K$. And their energy increase downward in the diagram.

In thermal equilibrium, both the electron quasi Fermi level E_{F_c} and the hole quasi Fermi level E_{F_v} are equal to Fermi level E_F.

5.2 The interaction of electrons and photons in semiconductor

The interaction of electrons and photons in semiconductor is consist of three processes:

† The absorption of photons in semiconductor. In which, the absorption of a photon in semiconductor results in the transition of an electron from valence band into the conduction band.

† The spontaneous emission in semiconductor. In which, the transition of an electron from conduction band into valence band results in a photon emission.

† The stimulated emission in semiconductor. In which, the transition of an electron from conduction band into the valence band caused by a photon incident with frequency

$$\nu = h^{-1}(E_c - E_v) \tag{5.81}$$

results in an emission of the second photons with the same frequency ν, same phase, same direction as the first one. So that the light is been amplified in semiconductor. Where E_c is the energy level in conduction band and E_v is the energy level in valence band.

5.2.1 The spontaneous emission in semiconductor

The process of the spontaneous emission in semiconductor is indicated in Figure 5-12. In which, the transition of an electron from conduction band into valence band results in a photon emission.

The rate of the spontaneous emission R_{spon}

A measurement of the spontaneous emission is the rate of the spontaneous emission R_{spon}.

The rate of the spontaneous emission R_{spon} corresponds to the number emitted per time unit with a certain energy E of a certain oscillation state. Where

$$R_{spon} = A_{21} \cdot \rho(E) \cdot f_c(E)[1 - f_v(E)] \qquad (5.82)$$

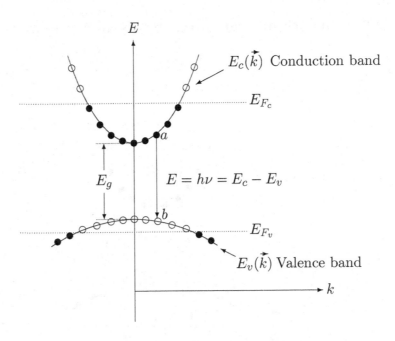

Figure 5-12   Spontaneous emission in semiconductor

In which:

† A_{21} : is the proportionality constant from the excited state 2 (conduction band) to state 1 (valence band).

† $\rho(E)$: is the density of the energy state E in the conduction band and the valence band.

† $f_c(E)$: is the occupation probability of the electrons in the conduction band.

† $f_v(E)$: is is the occupation probability of the electrons in the valence band.

† $(1 - f_v(E))$: is the probability of the free states in the valence band.

The equation (5.82) indicates that:

† The rate of the spontaneous emission R_{spon} is proportional to the density of the energy state $\rho(E)$ in the conduction band and the valence band. This is understandable.

† However, not all of the density of the energy state $\rho(E)$ but only some percentage of $\rho(E)$ joint the transition of electrons from conduction band into the valence

band, this percentage is just $f_c(E)$, the occupation probability of the electrons in the conduction band.

† Not only this, the transition of the electrons from conduction band into valence band has to be accepted by free states in the valence band, this free states in the valence band is just $(1 - f_v(E))$. Which is understandable as indicated in Figure 3-12.

5.2.2 The stimulated emission in semiconductor

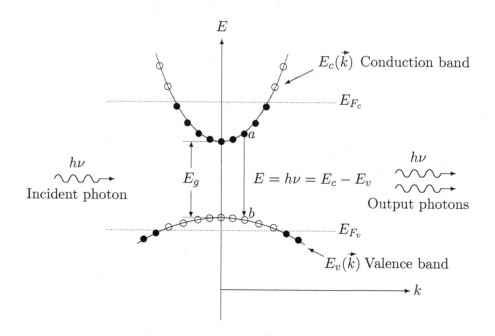

Figure 5-13   Stimulated emission in semiconductor

As shown in Figure 5-13, the stimulated emission in semiconductor occurs at the following condition:

† The semiconductor is pumped [4] and the electrons are in the conduction band (the excited state).

† A photon with frequency $\nu = h^{-1}(E_c - E_v)$ is incident.

† Then the induced transition of electron from conduction band into valence band caused by the photon incident contributes one photon of frequency $\nu = h^{-1}(E_c - E_v)$ to add second photon to the output.

Therefore, only the photon with frequency

$$\nu = h^{-1}(E_c - E_v) \tag{5.83}$$

---

[4]   Such as by injection of the bias current into the semiconductor laser.

can caused the stimulated emission.

The stimulated emission rate R_{stim}

The stimulated emission rate R_{stim} corresponds to the number of emitted photons per unit time and per photon in a certain oscillation state, which corresponds to a frequency $\nu = h^{-1}(E_c - E_v)$. Where

$$R_{stim} = B_{21} \cdot \rho(E) \cdot f_c(E)[1 - f_v(E)] \tag{5.84}$$

In which

† B_{21} : is the proportionality constant from the conduction state 2 to the valance state 1.

† $\rho(E)$: is the density of the energy state E in the conduction band and the valence band.

† $f_c(E)$: is the occupation probability of the electrons in the conduction band.

† $f_v(E)$: is the occupation probability of the electrons in the valence band.

† $(1 - f_v(E))$: is the probability of the free states in the valence band.

Obviously, R_{stim} in (5.84) is the same as R_{spon} in (5.83) in form except that A_{21} in (5.83) is replaced by B_{21} in (5.84).

5.2.3 The absorption in semiconductor

As shown in Figure 5-14, an incident photon may cause the transition of an electron from valence band b to conduction band a by absorption.

The absorption rate R_{abs}

The absorption rate R_{abs} corresponds to the number of absorbed photons per unit time and per photon. Where

$$R_{abs} = B_{12} \cdot \rho(E) \cdot f_c(E)[1 - f_v(E)] \tag{5.85}$$

Where

† B_{12} : is the proportionality constant from the valance state 1 to the conduction state 2.

† $\rho(E)$: is the density of the energy state E in the conduction band and the valence band.

† $f_c(E)$: is the occupation probability of the electrons in the conduction band.

† $f_v(E)$: is is the occupation probability of the electrons in the valence band.

† $(1 - f_v(E))$: is the probability of the free states in the valence band.

Obviously, R_{abs} in (5.85) is the same as R_{spon} in (5.82) in form except that A_{12} in (5.83) is replaced by B_{12} in (5.85).

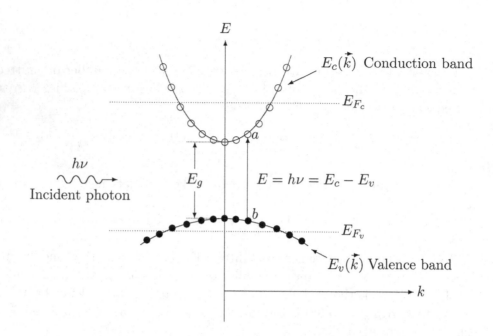

Figure 5-14   Absorption in semiconductor

The condition of absorption

Obviously, from Figure 5-14, the condition of absorption is

$$E > E_c - E_v \tag{5.86}$$

That means the absorbed photon energy E must be larger than the energy level difference between the conduction band and the valence band, $E_c - E_v$, so that the electron will obtain enough energy E to transit from valence band into conduction band.

The Einstein relation

Where the probability constants are given by the Einstein relation:

$$A_{21} = B_{12} = B_{21} \tag{5.87}$$

which has been proved by Einstein.

5.2.4   Condition of optical amplification in semiconductor

Now it is the time to discuss the condition of optical amplification, which corresponds to the net stimulation rate R_{st} in semiconductor as follows:

$$R_{st} = R_{stim} - R_{abs} = B_{21} \cdot \rho(E) \cdot [f_c(E) - f_v(E)] \tag{5.88}$$

The equation (5.88) explain this fact that, when a light beam is incident into the pumped semiconductor laser [5], it will induce electrons transition downward $a \to b$, this transition leads to light amplification; meanwhile, it will induce electrons transition upward $b \to a$, which leads to light absorption. The net amplification of the light beam means that the rate of $a \to b$ exceeds the rate of $b \to a$ transition.

Condition of optical amplification in semiconductor

Now, from (5.88), with $B_{21} > 0$ and $\rho(E) > 0$, we have

$$R_{st} > 0, \quad if \quad f_c(E) > f_v(E) \tag{5.89}$$

Obviously, an external excitation by the injection of bias current increases the electrons number in the conduction band such that the probability of stimulated emission is larger than that of absorption. And then the light amplification occurs in the semiconductor.

It is interested to point out, that
† For a laser medium, the amplification condition of light is "population inverse"

$$N_2 > N_1 \tag{5.90}$$

† In semiconductor, the condition of light amplification is governed by the difference of the occupation probability between conduction band and the valence band as follows:

$$f_c(E) > f_v(E) \tag{5.91}$$

From (5.91), the inverse condition can be written as

$$\frac{1}{f_v(E)} - \frac{1}{f_c(E)} > 0 \tag{5.92}$$

Where, from (5.79), (5.80),

$$\frac{1}{f_v(E)} = [\exp(E_v - E_{F_v}) + 1] \tag{5.93}$$

$$\frac{1}{f_c(E)} = [\exp(E_c - E_{F_c}) + 1] \tag{5.94}$$

and then we have, from (5.92),

$$E_c - E_v < E_{F_c} - E_{F_v} \tag{5.95}$$

---

[5] The pumped semiconductor means that the semiconductor has been injected bias current, which has made the electrons transition from the valence band to the conduction band.

which gives the energy levels necessary to provide optical gain.

The equation (5.95) leads to a conclusion that *only a photon having an energy of E corresponding to*

$$E_c - E_v < E < E_{F_c} - E_{F_v} \tag{5.96}$$

can be amplified.

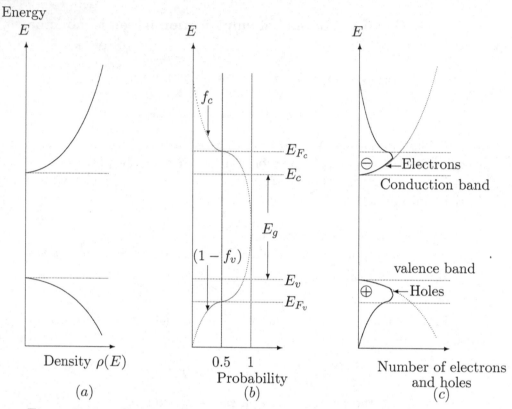

Figure 5-15   Graphical representation of the inversion state

Figure 5-15 may help us to understand the inverse state in semiconductor.

† As we mentioned before, in thermal equilibrium, there is only a Fermi energy level F_F for both of the conduction band and the valence band in semiconductor as shown in Figure 5-15(a).

† If the semiconductor is in non-thermal equilibrium with inject bias current, there are two Fermi energy levels, one is the quasi Fermi level E_{F_c} for the conduction band, another one is the quasi Fermi level E_{F_v} for the valence band as indicated in Figure 3-15(b). The energy levels E_{F_c} and E_{F_v} are determined by the material doping level and the injected carrier density.

† Where E_{F_c} is corresponding to the occupation probability of electrons in the conduction band:

$$f_c(E) = \frac{1}{\exp[(E - E_{F_c})/kT] + 1} \tag{5.97}$$

which means that the energy levels of electrons in conduction band are below the quasi Fermi level F_{F_c} as indicated in Figure 3-15(c).

† And E_{F_v} is corresponding to the occupation probability of electrons in the valence band:

$$f_v(E) = \frac{1}{\exp[(E - E_{F_v})/kT] + 1} \tag{5.98}$$

which means that the energy levels of electrons in valence band are below the quasi Fermi level F_{F_v} as indicated in Figure 3-15(c). In other words, $(1 - f_v(E))$ is the occupation probability of free states. The free states for electrons are just the hole states. Therefore, $(1 - f_v(E))$ is the occupation probability of holes. That means that the energy levels of holes are below $(1 - f_v(E))$ in valence band as indicated in Figure 5-15(c).

Conclusion

The discussion above leads to a conclusion: that when the semiconductor is in non-thermal equilibrium such as the injection of bias current into the $p - n$ junction in semiconductor, the semiconductor shows us an inverse state:

$$f_c(E) > f_v(E) \tag{5.99}$$

as indicated in Figure 5-15(b). This inverse state will result the population inverse as shown in Figure 3-15(c)

$$N_2 > N_1 \tag{5.100}$$

as indicated in Figure 5-15(c) and then results in the light amplification in the semiconductor. Where N_2 is the number of electrons in conduction band and N_1 is the number of electrons in valence band.

5.3 $p - n$ junction in semiconductor

5.3.1   Laser diode structure

Figure 5-16 displays some types of the semiconductor laser diode structures, in which

† (a) is a homostructure p-n conjunction. Where the intersection of p-n junction is the active layer.

† (b) is a double heterstructure with planar active layer.

† (c) is a double heterstructure with ridge waveguide active layer.

And (c) is the most important structure for the semiconductor laser diode used to the optical fiber communications.

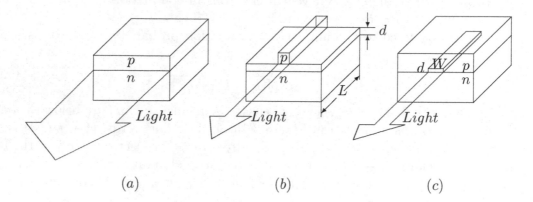

(a) (b) (c)

Figure 5-16   (a) Homostructure
(b) Double heterstructure with planar active region
(c) Buried double heterstructure

5.3.2   The double heterstructure of the semiconductor laser diode

The most important two classes of semiconductor laser diode are both based on III-V semiconductors.

† In first class, the active layer is GaAs or $Ga_{1-x}Al_xAs$. Where x indicates the fraction of the Ga atoms in GaAs, which are replaced by Al. The resulting lasers emit at

$$\lambda = 0.75\mu m \sim 0.88\mu m \qquad (5.101)$$

depends on the active region molar fraction x and its doping. This special region is suitable for the short-haul ($\leq 2km$) optical fiber transmission system since in this region the dissipation loss of the optical fiber is not so small.

† In second class, the active layer is $Ga_{1-x}In_xAl_{1-y}As_y$ as shown in Figure 5-17. The laser emit at

$$\lambda = 1.1\mu m \sim 1.6\mu m \qquad (5.102)$$

depends on the active region molar fraction x and y as well as its doping. Where x indicates the fraction of the Ga atoms in GaIn, which are replaced by In, y indicates the fraction of the Al atoms in AlAs, which are replaced by As.

† The region near $1.55\mu m$ is suitable for the long haul optical fiber telecommunications system since the single mode fiber with the dissipation loss of 0.15dB/km is available. And the laser has been used to the 10Gbps optical fiber transponder for the long-haul optical fiber transmission link. This system has been installed in USA, Canada, China, India, Japan, Germany and worldwide area to transmission of 1000Gbps data, digital TV, telephone an so on.

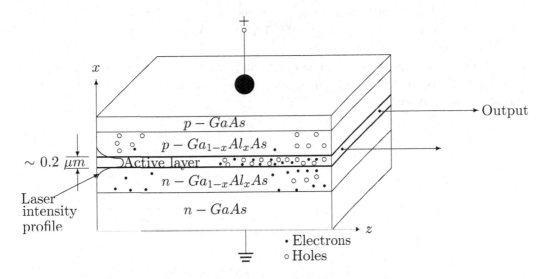

Figure 5-17   A typical double heterstructure GaAs-GaAlAs laser [76]

A typical double heterstructure GaAs-GaAlAs laser is shown in Figure 5-17.

When the bias current are injected into the laser from p-GaAs to n-GaAs, the holes are injected into the active layer from p-GaAlAs and the electrons are injected into the active layer from n-GaAlAs. In this case, the "population inverse" exits in the active layer. And the laser intensity with the frequency near $\nu = E_g/h$ are amplified by the simulating electron-hole recombination. The laser output from the right side of the active layer after the amplification.

In this classes of laser,

† The thin active region is usually undoped while the upside of the $p-Ga_{1-y}Al_yAs$ is doped heavily p-type and the downside is n-type.

† The refraction index of the active layer n_{GaAs} is slight higher than that of the cladding layers. The difference

$$n_{GaAs} - n_{Ga_{1-x}Al_xAs} \simeq 0.62x \qquad (5.103)$$

between the refraction index of active layer $GaAs$ and the cladding layers, $p - Ga_{1-y}Al_yAs$, $n - Ga_{1-x}Al_xAs$, $n_{p-Ga_{1-y}Al_yAs}$, $n_{n-Ga_{1-x}Al_xAs}$, respectively, is used to form a planar waveguide in the active layer. Thus, the laser intensity profile is indicated in Figure 5-17. This laser intensity profile is amplified along the z-direction when the positive bias is applied to the laser.

† The wavelength λ of the laser depends on x and y as shown in Figure 5-18.

† This class of semiconductor laser can provide two important wavelengthes suitable for the long haul optical fiber transmission system:

$$\lambda = 1.3\mu m \qquad (5.104)$$

$$\lambda = 1.55\mu m \qquad (5.105)$$

† The optical single mode fiber transmission will be zero dispersion transmission at $\lambda = 1.3\mu m$ with dispassion loss of 0.3dB/km. Which is suitable for the broadband optical fiber transmission system.

† The dispassion loss of the optical single mode fiber transmission will be minimum (0.15dB/km) at $\lambda = 1.55\mu m$. Which is suitable for the long-haul optical fiber transmission system.

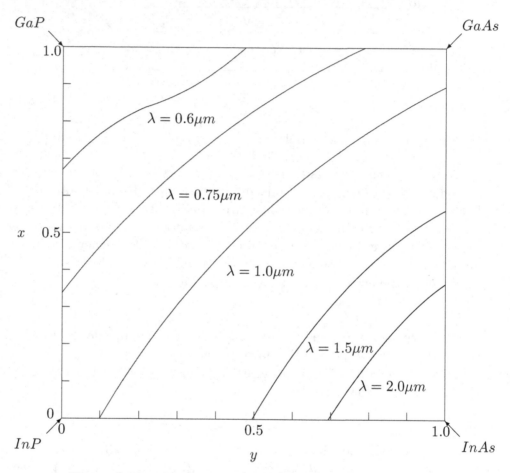

Figure 5-18 λ corresponding to the bandgap energy
for quaternary III-V semiconductor alloys whose
composition in $In_{1-x}As_xAs_{1-y}P_y$ (From [[4]])

Now, from Figure 5-18,

† $\lambda = 1.5\mu m$ when $x = 0$ and $y = 0.5$. At this point, the active layer should be $Ga_1In_0As_{0.5}P_{0.5}$.

† Following the curve of $\lambda = 1.5\mu m$, we may choice a new y corresponding to a new x. Thus, the new active layer is $Ga_{1-x}In_xAs_{1-y}P_y$.

5.3.3 $p - n$ **junction in semiconductor**

Now we will discuss the p-n junction in semiconductor laser so as to understand that how the laser intensity is amplified along the z-direction of the active layer.

Without bias current $I_{bias} = 0$

The energy level distribution of electrons in $p - n$ junction of semiconductor depends on the density of the bias current.

When the bias current $I_{bias} = 0$ or in equilibrium condition, the energy level distribution in $p - n$ junction is indicated in Figure 5-19.

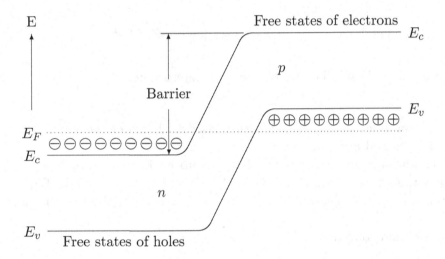

Figure 5-19 $p - n$ junction in semiconduction when $I_{bias} = 0$

Where,

† The energy levels of electrons in n side are below the Fermi level F_F as indicated in Figure 5-19.

† The free states of electrons in p side are above the Fermi level E_F. In other words, the energy levels of holes are above the Fermi level E_F as indicated in Figure 5-19.

Obviously, there is no population inverse or inverse state in the equilibrium condition. Meanwhile, there is no any current through the $p - n$ junction because of the barrier.

Under the forward bias current $I_{bias} > 0$

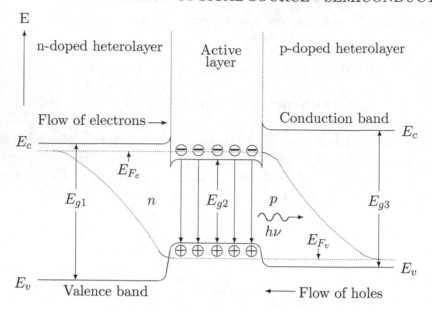

Figure 5-20   Band diagram of a double heterostructure laser diode
under forward bias current

The diagram of a double heterostructure laser diode under the forward bias current I_{bias} is indicated in Figure 5-20.

In which the forward bias current I_{bias} set up an inverse state $E_{F_c} > E_{F_v}$ as well as the population inverse $N_2 > N_1$ in the active layer as shown in Figure 5-20, which will result in the light amplification in the active layer. Where in Figure 5-20,

† The inversion condition

$$E_c - E_v < E_{F_c} - E_{F_v} \tag{5.106}$$

is satisfied in the active layer, and the photon with energy E in

$$E_c - E_v < E < E_{F_c} - E_{F_v} \tag{5.107}$$

can be amplified in the active layer. Where the energy E_c is a solid line from the n-doped heterolayer to the p-doped heterolayer as indicated in Figure 5-20 and the energy E_v is another solid line from the n-doped heterolayer to the p-doped heterolayer.

† Meanwhile, the population inverse

$$N_2 > N_1 \tag{5.108}$$

is satisfied in the active layer, which will results in the photon amplification in the active layer. Where N_2 is the number of electrons in the conduction band of the active layer and N_1 is the number of electrons in the valence band of the active layer and here the number of electrons in conduction band is equal to the number of holes in the valence band.

† The forward bias current I_{bias} will result in a flow of electrons from n-doped heterolayer into the p-doped heterolayer as well as a flow of holes from the p-doped heterolayer into the n-doped heterolayer as indicated in Figure 3-20.

5.3.4   Optical gain and the net rate of stimulated emission

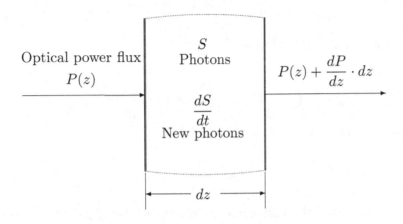

Figure 5-21   Optical gain

Figure 5-21 indicates the process of optical amplification, in which the pumped medium is indicated by a rectangular area. When the input optical power flux $P(z)$ is incident into the pumped medium, the optical beam will be amplified by the stimulated emission inside the pumped medium, and then results in an output optical power flux $P(z) + \dfrac{dP}{dz} \cdot dz$.

In more detail, the photons beam S incident into the pumped medium will result in a new photons per second $\dfrac{dS}{dt}$ as follows:

$$\frac{dS}{dt} = R_{st} \cdot S \tag{5.109}$$

Where R_{st} is the net rate of the stimulated emission, and the optical power travels with a group velocity

$$V_g = \frac{C}{n_g} \tag{5.110}$$

where n_g is the group refractive index.

Now, the optical energy in the rectangular area is

$$E_{opt} = h\nu \cdot S = P(z) \cdot \frac{dz}{V_g} \tag{5.111}$$

Where $dt = dz/v_g$ is the time to travel dz.

Thus, from (5.109) and (5.111), the optical power increment due to the stimulated emission is

$$dP(z) = h\nu \cdot \frac{dS}{dt} = h\nu \cdot R_{st} \cdot S = P(z) \cdot \frac{dz}{V_g} \cdot R_{st} \tag{5.112}$$

And then, the increment of the output power $dP(z)$ in dz of the active layer is

$$\frac{dP(z)}{dz} = P(z) \cdot \frac{R_{st}}{v_g} \tag{5.113}$$

Thus, the solution of this equation is

$$P(z) = P_0 \cdot e^{\frac{R_{st}}{V_g} \cdot z} = P_0 \cdot e^{g_{st} \cdot z} \tag{5.114}$$

Where, the stimulated gain coefficient is

$$g_{st} = \frac{R_{st}}{V_g} \tag{5.115}$$

Figure 5-22 indicates the calculated gain spectral at different injected carrier density for undoped InGaAsP. In which,

† The InGaAsP bulk displays an absorption behavior when the injected carrier density N less than $1 \times 10^{18} cm^{-1}$.

† The InGaAsP bulk displays a gain behavior in a certain bandwidth B when the injected carrier density N large than $1 \times 10^{18} cm^{-1}$.

† The gain of the InGaAsP bulk is increasing and the bandwidth B is widening when the injected carrier density is increasing. The gain spectral displays a resonance characteristic because that only a photon having an energy of E with

$$E_c - E_v < E < E_{f_c} - E_{F_v} \tag{5.116}$$

can be amplified, which has given in (5.107). Where, in (5.116),

$$E = h\nu \tag{5.117}$$

and

$$\lambda = c/n\nu \tag{5.118}$$

where c is the light velocity in the air and n is the index of the reflection in InGaAsP bulk.

Which means that the energy E between $E_c - E_v$ and $E_{f_c} - E_{F_v}$ corresponds to $\lambda_1 < \lambda < \lambda_2$. Which has been indicated on the bottom and top, respectively, in Figure 5-22.

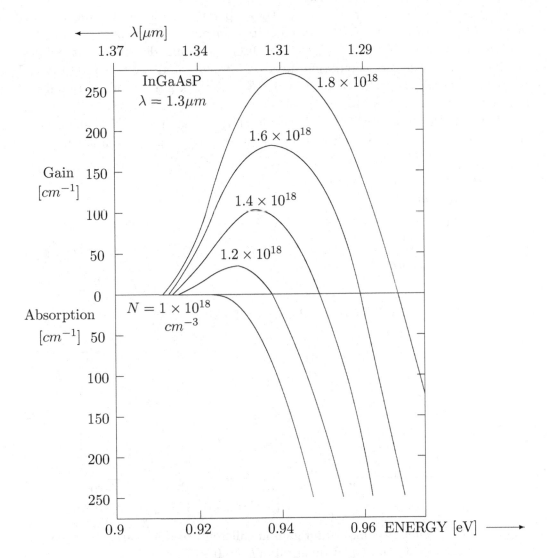

Figure 5-22   Calculated gain spectral at different injected carrier density
for undoped InGaAsP. ($E_g = 0.96eV, \lambda = 1.31\mu m$)
After: Dutta, 1980

5.3.5   Maximum Gain

In InGaAsP bulk, the dependence of peak gain G_{max} on the injected carrier density N is indicated in Figure 5-23. In which the peak gain G_{max} is proportional to the population inverse in the bulk as follows:

$$g_{max} \approx A(N - N_{tr}) \tag{5.119}$$

Where:

A: is the gain coefficient.

N: is the injected carrier density.

N_{tr}: is the injected carrier density at transparency point or the injected carrier density at the start point when the population inverse is available.

Obviously, the peak gain g_{max} of the InGaAsP bulk depends on the injected carrier density N or the population inverse $(N - N_{tr})$ and the temperature T. For a fixed temperature, for instance $T = 100K$, the peak gain is proportional to the population inverse $(N - N_{tr})$.

However, the equation (5.119) is an idea formula without taking the other real situations into account. Which will be discuss step by step in the following.

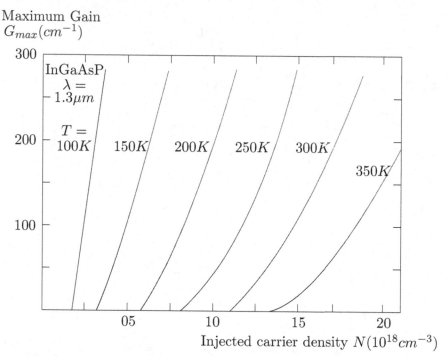

Figure 5-23   Calculated maximum gain at different injected carrier density
for undoped InGaAsP. ($E_g = 0.96eV, \lambda = 1.31\mu m$)
After: Dutta and Nelson, 1982

5.3.6   Linewidth enhancement factor and transverse confinement factor

Plane wave propagation in the positive z of a bulk material

The one of the real situation is that the light wave has to be propagation in a InGaAsP dielectric waveguide.

To discuss the light wave propagation, we may discuss a plane wave propagation in the positive z-direction in a bulk material.

$$\mathbf{E} = \mathbf{E}_0 \exp\left(-jk_0 \cdot z\sqrt{\bar{\varepsilon}}\right)$$

(5.120)

Where

$$\bar{n} = \sqrt{\bar{\varepsilon}} = n' - jn"$$

(5.121)

and **E** and **E**$_0$ are the electric field in vector at $z = z$ and $z = 0$, respectively.

The transmitted power is

$$P \sim |\mathbf{E}|^2$$

(5.122)

or

$$P = P_0 \cdot \exp\{-2n"k_0 \cdot z\} = P_0 \cdot \exp\{+g_{st} \cdot z\}$$

(5.123)

Where

$$n" = -g_{st}/2k_0 = -g_{st} \cdot \lambda/4\pi$$

(5.124)

The polarization **P** of the electric field is defined as

$$\mathbf{P} = \varepsilon_0 \bar{\chi} \cdot \mathbf{E}$$

(5.125)

Where

$$\bar{\chi} = \bar{\chi}_0 + \bar{\chi}_p$$

(5.126)

is the susceptibility. And

$\bar{\chi}_0$: is the susceptibility without external pumping.

$\bar{\chi}_p$: is the susceptibility due to external pumping.

In electromagnetic field theory, the dielectric constant is

$$\bar{\varepsilon} = \varepsilon' + j\varepsilon"$$

(5.127)

or

$$\bar{\varepsilon} = 1 + \bar{\chi} = 1 + \bar{\chi}_0 + \bar{\chi}_p$$

(5.128)

Therefore, in bulk material, we have

$$\bar{\varepsilon} = 1 + \bar{\chi}_0 + \bar{\chi}_p$$

(5.129)

And the complex index of reflection in the bulk material is

$$\bar{n} = n' - jn" = \sqrt{\bar{\varepsilon}}$$

(5.130)

Therefore, from (5.129) and (5.130), we have

$$n' = \sqrt{1 + Re\{\bar{\chi}_0\} + Re\{\bar{\chi}_p\}}$$

(5.131)

$$n" = \frac{-\frac{1}{2}Im\{\bar{\chi}_0 + \bar{\chi}_p\}}{\sqrt{1 + Re\{\bar{\chi}_0\} + Re\{\bar{\chi}_p\}}}$$

(5.132)

Linewidth enhancement factor

A remarkable feature of semiconductor laser is that both of the gain and the refractive index of the active region change with injected carrier density. Which will result in an amplitude-phase noise coupling that effects many laser characteristics such as the linewidth enhancement factor α.

Laser linewidth is the spectral linewidth of a laser beam. Here, two of most distinctive characteristics of laser emission are spatial coherence and spectral coherence. While spatial coherence is related to the beam divergence of the laser, spectral coherence is evaluated by measuring the laser linewidth of the radiation.

As well known, due to the noise from spontaneous emission into resonator modes, a free-running single-frequency laser has a certain finite linewidth. The fundamental limit for the linewidth was calculated by Schawlow and Towmes even before the first laser was experimentally demonstrated. This limit was later shown to be closely approached by a number of soloid-state laser, significantly higher linewidth value were measured for semiconductor laser. It was then laser found by Henry [77] that the increased linewidth result from a coupling between intensity and phase noise, caused by a dependence of the refractive index on the carrier density in the semiconductor laser. Henry then introduced the linewidth enhancement factor α (also called Henry factor or alpha factor) to quantify this amplitude-phase coupling mechanism: α is a proportionality factor relating phase change to change of the gain.

$$\Delta\phi = \frac{\alpha}{2}\Delta g \qquad (5.133)$$

Henry then found that the linewidth of the laser should be increased by the factor $(1 + \alpha^2)$, which turned to be in agreement with experimental data.   The linewidth enhancement factor α result from the coupling of amplitude-phase noise in semiconductor is, from [78],

$$\alpha = \frac{4\pi \cdot dn'/dN}{\lambda \cdot dg_{st}/dN} = \frac{\triangle n'}{\triangle n"} \qquad (5.134)$$

Where N is the carrier concentration.   The linewidth enhancement factor (α-factor) is very important for semiconductor lasers, as it is one of the main feathers distinguishes the behavior of semiconductor laser with respect to other types of laser. The α-factor influences several fundamental aspects of semiconductor lasers, such as linewidth, chirp under current modulation, mode stability, laser dynamics, laser behavior in presence of optical feedback and injection.

The light wave propagation

Now we will discuss the light wave propagation in a three layer slab waveguide as indicated in Figure 5-24. Where, we only show the active layer, in which,

† Carrier inverse in active layer at reasonable current.

† The waveguide design reaches a condition that only two perpendicularly polarized waves (TE and TM) should be able to be propagation in the active layer.

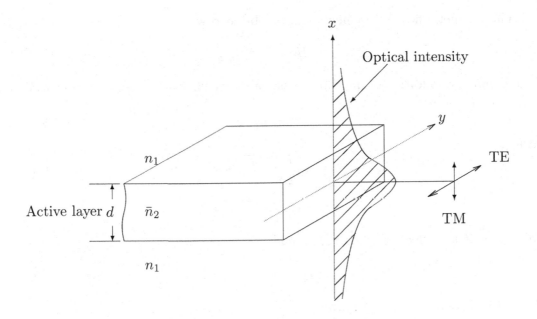

Figure 5-24   Three layer slab waveguide

Where
† For the TE mode: $\vec{E}\ ||y - axis.$
† For the TM mode: $\vec{H}\ ||y - axis.$
The refraction index in three layer are, respectively,:

$$n_1 = real\ \ number \tag{5.135}$$

$$\bar{n}_2 = n_2' - jn_2" \tag{5.136}$$

$$n_2' > n_1 \tag{5.137}$$

And

$$n_2'' = -g_{st}/2k_0 \tag{5.138}$$

$$g = \gamma \cdot g_{st} \tag{5.139}$$

The solution of the wave equation is:

† For the fundamental TE-mode $(H_y = E_x = E_z = 0)$:

$$E_y = A \cdot \cos u \cdot \exp\left[-\frac{2v}{d}\left(x - \frac{d}{2}\right)\right], \quad |x| \geq \frac{d}{2} \tag{5.140}$$

$$E_y = A \cdot \cos \frac{2u}{d} \cdot x, \qquad\qquad |x| < \frac{d}{2} \qquad\qquad (5.141)$$

The magnetic field components can be obtained with

$$\omega \mu_0 H_x = -\beta_{TE} E_y \qquad\qquad (5.142)$$

The boundary condition of the three layer slab waveguide will lead to the eigenvalue equation as follows

$$u \cdot \tan u = v \qquad\qquad (5.143)$$

Where

$$u = k_0 \frac{d}{2} \sqrt{n_2'^2 - n_e^2} \qquad\qquad (5.144)$$

$$v = k_0 \frac{d}{2} \sqrt{n_e^2 - n_1^2} \qquad\qquad (5.145)$$

Where n_e is the effective mode index, and

$$n_2' > n_e > n_1 \qquad\qquad (5.146)$$

From (5.144) and (5.145), we have

$$u^2 + v^2 = k_0^2 \frac{d^2}{4} \left(n_2'^2 - n_1^2 \right) \qquad\qquad (5.147)$$

and

$$\beta_{TE}^2 = k_0^2 n_e^2 = k_0^2 n_2'^2 - 4 \frac{u^2}{d^2} \qquad\qquad (5.148)$$

† For the fundamental TM-mode ($E_y = H_x = H_z = 0$):

The magnetic field components are

$$H_y = B \cdot \cos r \cdot \exp \left[- \frac{2s}{d} \left(x - \frac{d}{2} \right) \right], \quad |x| \geq \frac{d}{2} \qquad\qquad (5.149)$$

$$H_y = B \cdot \cos \frac{2r}{x} \cdot x, \qquad\qquad |x| < \frac{d}{2} \qquad\qquad (5.150)$$

The electric field components can be obtained with

$$\omega \varepsilon_0 n_2'^2 E_x = \beta_{TM} H_y \qquad\qquad (5.151)$$

This equation is from Maxwell equation.

The boundary condition of the three layer slab waveguide leads to the eigenvalue equation for TM-mode as follows:

$$r \cdot \tan r = s \left(\frac{n'^2}{n_1} \right)^2 \qquad\qquad (5.152)$$

Where

$$r = k_0{}^2 \frac{d}{2} \sqrt{n_2'{}^2 - n_e{}^2} \tag{5.153}$$

$$s = k_0{}^2 \frac{d}{2} \sqrt{n_e{}^2 - n_1{}^2} \tag{5.154}$$

$$r^2 + s^2 = k_0^2 \frac{d^2}{4} \left(n_2'^2 - n_1^2\right) \tag{5.155}$$

and

$$\beta_{TM}^2 = k_0{}^2 n_e{}^2 = k_0^2 n_2'^2 - 4\frac{r^2}{d^2} \tag{5.156}$$

Transverse confinement factor

The transverse confinement factor is defined as

$$\Gamma_T = \frac{\displaystyle\int_d \phi^2(x,y)\,dx}{\displaystyle\int_{-\infty}^{+\infty} \phi^2(x,y)\,dx} \tag{5.157}$$

Where
$\phi(x,y)$ is the transverse field distribution for TE-mode or TM-mode in the slab waveguide. $\int_d$ is the integral in the active layer.

Polarization dependence

Obviously, from Figure 5-25, the confinement factor of the TE-mode is different from that of the TM-mode. Where

$$\Gamma_{TE} > \Gamma_{TM} \tag{5.158}$$

Meanwhile, the propagation constant of the TE-mode is different from that of the TM-mode.

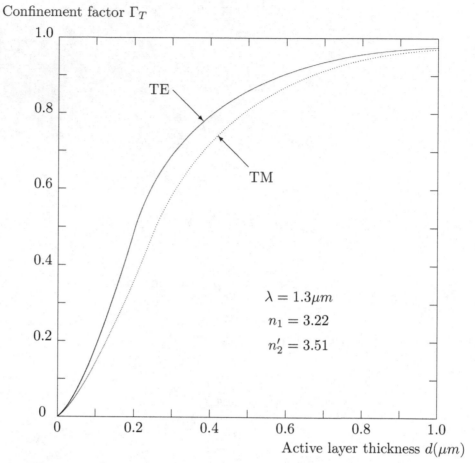

Figure 5-25   Transverse confinement factor as a function
of active thickness for a $1.3\mu m$ InGaAsP laser with InP
cladding layers (After: Agrawal and Dutta,1988)

Summery

† Peak gain coefficient:

$$g_{max} = A(N - N_0) - A_2(\lambda - \lambda_p)^2 \qquad (5.159)$$

Where:

$A \approx 2.7 \cdot 10^{-16} cm^2$

$A_2 \approx 0.15 cm^{-1} \cdot nm^{-2}$

$N_0 \approx 1.1 \cdot 10^{18} cm^{-3}$ is the transparency density

† Peak gain wavelength:

$$\lambda_p = \lambda_0 + A_3(N - N_0) \tag{5.160}$$

Where:

$A_3 \approx 2.7 \cdot 10^{-17} nm \cdot cm^{-3}$

† 3-dB gain curve bandwidth:

$$2 \cdot \triangle\lambda = 2 \cdot \sqrt{\frac{A(N - N_0)}{2A_2}} = 2 \cdot \sqrt{\frac{\ln 2}{A_2 \cdot \Gamma \cdot L}} \tag{5.161}$$

† Net gain per unit length:

$$g = \Gamma \cdot g_{st} - \alpha \tag{5.162}$$

Where

g_{st} :   is the material gain coefficient.

Γ : ≈ 0.2, is the confinement factor.

α : $\approx 10....50 cm^{-1}$, is the loss coefficient.

L :   is the amplifier length.

† Carrier density rate equation:

$$\frac{dN}{dt} = \frac{j}{ed} - R(N) - \frac{\Gamma \cdot g_{st,max}}{h\nu}\left(\beta I_{sp} + I\right) \tag{5.163}$$

Where

j : is the current density.

e : is the electron charge.

d : is the active layer thickness.

$R(N)$: is the recombination rate.

$h\nu$: is the photon energy.

β : is the spontaneous emission efficient.

I_{sp} : is the intensity of the spontaneous emission.

I : is the average signal intensity.

† Spontaneous emission coefficient:

$$\beta = \frac{cA\tau_s}{n'V} \tag{5.164}$$

Where

c: is the velocity of light.

τ_s: is the carrier lifetime.

n':   is the material refractive index.

V: is the active volume.

† Recombination rate:

$$R(N) = R_1 \cdot N + R_2 \cdot N^2 + R_3 \cdot N^3 \tag{5.165}$$

Where, for $\lambda = 1.3\mu m$,
$R_1 = 1 \cdot 10^8 \cdot s^{-1}$
$R_2 = 8 \cdot 10^{-17} m^3 \cdot s^{-1}$
$R_3 = 4 \cdot 10^{-41} m^6 \cdot s^{-1}$
Therefore,

$$R(N) \approx \frac{N}{\tau_s} \tag{5.166}$$

For the steady state, $\dfrac{dN}{dt} = 0$ and $a = 0$, then, from (5.163), we have

$$\frac{j}{ed} = \frac{N}{\tau_s} + \frac{\Gamma \cdot g_{st,max} I}{h\nu} \tag{5.167}$$

$$g_{st,max} = \frac{g_0}{1 + I/I_s} \tag{5.168}$$

$$I_s = \frac{h\nu}{\Gamma A \cdot \tau_s} = \frac{c \cdot h\nu}{\beta \cdot n' \cdot V \cdot \Gamma} \tag{5.169}$$

Where
g_0:   is the unsaturated material gain coefficient.

† Single pass gain:

$$G_s = \exp\left[\left(\frac{\Gamma \cdot g_0}{1 + I/I_s} - a\right)L\right] \tag{5.170}$$

† Phase shift due to n':

$$\phi = \phi_0 + g_0 \frac{L \cdot \alpha}{2}\left[\frac{I}{I + I_s}\right] \tag{5.171}$$

$$\phi_0 = \frac{2\pi}{c}(f - f_0)L \cdot n' \tag{5.172}$$

Where
α: is the linewidth enhancement factor.

† Saturation output power:

$$P_{0s} = \frac{W \cdot d \cdot h \cdot h\nu}{\Gamma \cdot \tau_s \cdot A}\ln 2 \tag{5.173}$$

Where
W: is the width of active layer.

† Output intensity:

$$I_0 = I \cdot g_{st} \cdot L \tag{5.174}$$

5.4 The direct modulation of semiconductor laser

5.4.1 The direct modulation

In 10Gbps optical fiber transponder, the direct modulation of semiconductor laser has been selected as the 10Gbps digital modulation. In which,

† The signal to be transmitted in a long-haul optical fiber link is 10Gbps optical digital signal.

† The direct modulation on the bias current of the semiconductor laser has been selected to transmit 10Gbps digital signal. Which means that the bias current I_{bias} is "high" when the data is "1", and the bias current I_{bias} is "low" when the data is "0".

† Thus,

— The bias current I_{bias} will lunch the "population inverse" $N_2 > N_1$ in the active layer.

— The "population inverse" $N_2 > N_1$ will result in the photon amplification in the active layer.

— After the optical amplification in the semiconductor laser, the output optical power P is "high" when the data is "1", the output optical power P is "low" when the data is "0". In the other words, the input electrical digital signal will become the optical digital signal output. Therefore, the direct modulator of the semiconductor laser is an electric/optical convertor or E/O convertor.

Now we will discuss step by step after then.

5.4.2 Carrier density rate equation

Under the forward bias current $I_{bias} > 0$

The diagram of a double heterostructure laser diode under the forward bias current I_{bias} is indicated in Figure 5-26.

In which the forward bias current I_{bias} set up an inverse state $E_{F_c} > E_{F_v}$ as well as the population inverse $N_2 > N_1$ in the active layer as shown in Figure 5-26, which will result in the light amplification in the active layer. Where in Figure 5-26, the inversion condition

$$E_c - E_v < E_{F_c} - E_{F_v} \tag{5.175}$$

in the active layer is satisfied and the photon with energy E corresponding to

$$E_c - E_v < E < E_{F_c} - E_{F_v} \tag{5.176}$$

can be amplified in the active layer. Where the energy E_c is a solid line from the n-doped heterolayer to the p-doped heterolayer as indicated in Figure 5-26 and the energy E_v is another solid line from the n-doped heterolayer to the p-doped heterolayer.

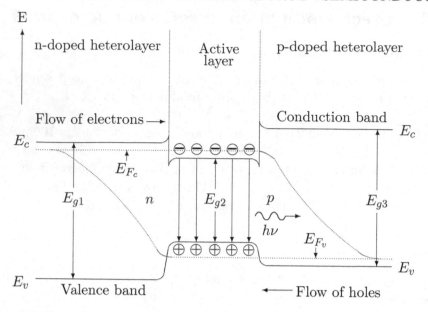

Figure 5-26   Band diagram of a double heterostructure laser diode
under forward bias current

As shown in Figure 5-26, the stimulated emission in semiconductor occurs at the following conditions:

† The semiconductor is pumped [6] and the electrons are in the conduction band (or in the excited state).

† A photon with frequency $\nu = h^{-1}(E_c - E_v)$ is incident, which will induce the transition of an electron from the conduction band into the valence band and contributes one photon of frequency $\nu = h^{-1}(E_c - E_v)$ to add the input photon to the output.

These process may be described by a *carrier density rate equation* as follows:

$$\frac{dN}{dt} = \frac{j}{ed} - \frac{N}{\tau} - \frac{\Gamma \cdot g_{stmax}}{h \cdot \nu}(\beta I_{SB} + I) \tag{5.177}$$

Where

† j : is the injection current density.

† $\frac{dN}{dt}$: is the variation rate of the electrons N in the conduction band (or the holes N in the valence band).

† $\frac{N}{\tau}$: is the recombination rate of the electrons and the holes. τ is the carrier life.

---

[6]   Such as by injection of the bias current into the semiconductor laser.

† In $\dfrac{\Gamma \cdot g_{stmax}}{h \cdot \nu}(\beta I_{SB} + I)$:

$-\ h\nu = \varepsilon$: is the energy of one photon.
$-\ I$: is the average signal intensity.
$-\ I_{SB}$: is the intensity of the spontaneous emission.
$-\ \beta$: is the spontaneous emission coefficient.
$-\ \Gamma$: is the confinement factor of the active dielectric waveguide.

Thus,

$$P = \frac{(\beta I_{SB} + I)}{h\nu} \tag{5.178}$$

is the photon density.

Now, from (5.119),

$$g_{st,max} = A(N - N_{tr}) \tag{5.179}$$

Thus, (5.177) may be rewritten as

$$\frac{dN}{dt} = \frac{j}{ed} - \frac{N}{\tau} - \Gamma A(N - N_{tr})P \tag{5.180}$$

or

$$\frac{dN}{dt} = \frac{j}{ed} - \frac{N}{\tau} - \frac{dP}{dt} \tag{5.181}$$

Where

$$\frac{dP}{dt} = \Gamma A(N - N_{tr})P - \frac{P}{\tau_p} \tag{5.182}$$

and $\dfrac{P}{\tau_p}$ is added after considering the photon lift τ_p, since the increment per second of the photon number $\left(\dfrac{dP}{dt}\right)$ will be decreased some $\left(\dfrac{P}{\tau_p}\right)$ because of the vanish of the photons.

† $\dfrac{dN}{dt} = \dfrac{d(eN)}{dt} \cdot \dfrac{d}{ed}$ is a quantity corresponding to $\dfrac{j}{ed}$.

† Thus, (5.181) points out that the injection carrier density per second $\dfrac{j}{ed}$ will result in the "population inverse" per second $\dfrac{dN}{dt}$ and thephoton emission per second $\dfrac{dP}{dt}$.

† In those process, we need to consider the recombination rate of the electrons and the holes $\dfrac{N}{\tau}$.

5.4.3   The steady-state solution of the carrier density rate equation

In the steady-state [7], the photon density P is a constant, and then, $\dfrac{d}{dt} = 0$. Thus, the equations (5.180) and (5.182) become

$$0 = \frac{j_0}{ed} - \frac{N_0}{\tau} - \Gamma A(N_0 - N_{tr})P_0 \qquad (5.183)$$

$$0 = \Gamma A(N - N_{tr})P_0 - \frac{P_0}{\tau_p} \qquad (5.184)$$

† Equation (5.183) says that the photon density P_0 is proportional to the injection current density j_0. Which has been verified by a test results as shown in Figure 5-27. Where j_{th} is called the threshold current intensity, which means that the photon density P_0 is proportional to the injection current density j_0 after $j_0 > j_{th}$.

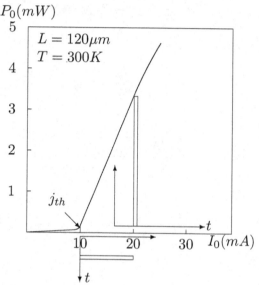

Figure 5-27   Light output power P_0 versus injection current I_0 of a laser
(After Reference[26])

† If a digital pulse is added on the injection current, the output light from the laser will be a light pulse as shown in Figure 5-27.

† Which means that the laser is a converter to convert the electrical digital signal into the light digital signal. *Which is so called the direct modulation of laser.*

† There are two important light modulations: the direct modulation and the phase modulation. The direct modulation is an amplitude modulation on the laser itself. The phase modulation of light has to be modulated on an external phase modulator. However, it can provide more wide modulation bandwidth or higher speed of the data modulation. Thus, the direct modulation of laser is the simplest light modulation. Meanwhile, it is the cheapest light modulation. But, we need to

---

[7]   The steady-state solution is a continuous wave)

investigate the bandwidth of the direct modulation. Which will be discussed after then.

† Equation (5.184) tell us that

$$A(N - N_{tr}) = \frac{1}{\Gamma \tau_p} \tag{5.185}$$

5.4.4 The frequency response of the direct modulation of semiconductor laser

To investigate the bandwidth of the direct modulation of the semiconductor laser, we may consider that the injection current is a sinusoidal current(density) $j_1 e^{i\omega t}$ added to a DC current(density) j_0. Namely,

$$j = j_0 + j_1 e^{i\omega t} \quad Injection \quad current \tag{5.186}$$

$$N = N_0 + n_1 e^{i\omega t} \quad Carrier \quad density \tag{5.187}$$

$$P = p_0 + p_1 e^{i\omega t} \quad photon \quad density \tag{5.188}$$

Where, the injection current(density) j causes the population inverse N and then results in the photon density P.

Using (5.186), (5.187) and (5.188) in (5.180) and (5.182), we have

$$i\omega n_1 = \frac{j_1}{ed} - \left(\frac{1}{\tau} + \Gamma A p_0\right) n_1 - \frac{1}{\tau_p} p_1 \tag{5.189}$$

$$i\omega p_1 = A p_0 \Gamma n_1 \tag{5.190}$$

From (5.190), we have $n_1 = \dfrac{i\omega}{A p_0 \Gamma} p_1$. Then (5.189) becomes

$$\frac{(i\omega)^2}{A p_0 \Gamma} p_1 = \frac{j_1}{ed} - \frac{i\omega}{A p_0 \Gamma}\left(\frac{1}{\tau} + \Gamma A p_0\right) p_1 - \frac{1}{\tau_p} p_1 \tag{5.191}$$

Timing $A p_0 \Gamma$, we have

$$-\omega^2 p_1 = \frac{j_1}{ed} A p_0 \Gamma - i\omega\left(\frac{1}{\tau} + \Gamma A p_0\right) p_1 - \frac{A p_0 \Gamma}{\tau_p} p_1 \tag{5.192}$$

The solution of (5.192) is

$$p_1(\omega) = \frac{-(j_1/ed) A p_0 \Gamma}{\omega^2 - i\omega\left(\dfrac{1}{\tau} + \Gamma A p_0\right) - \dfrac{A p_0 \Gamma}{\tau}} \tag{5.193}$$

Equation (5.193) is just the frequency response of the direct modulation of the semiconductor laser. The amplitude of $p_1(\omega)$, $|p_1(\omega)| \sim \omega$, is plotted in Figure 5-28.

$20 \log(|p_1(\omega)|) dB$

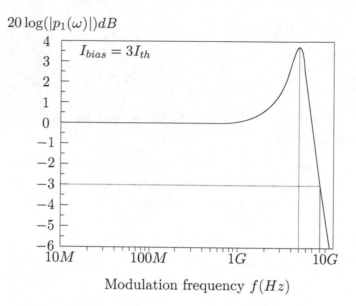

Modulation frequency $f(Hz)$

Figure 5-28   Modulation response of semiconductor laser[[79]]

As shown in Figure 5-28,

† The direct modulation response of the semiconductor laser displays a flat response in low frequency region.

† Then, it goes up to maximum at the "relaxation resonance frequency" ω_R.

† After then, it drops down near $f \approx 10 GHz$.

Where, from (5.193), the "relaxation frequency " ω_R may be obtained by

$$\left[\omega^2 - i\omega\left(\frac{1}{\tau} + \Gamma A p_0\right) - \frac{A p_0 \Gamma}{\tau_p}\right]\Bigg|_{\omega=\omega_R} = 0 \qquad (5.194)$$

The amplitude of (5.194) give us

$$\left[\left(\omega^2 - \frac{A p_0 \Gamma}{\tau_p}\right)^2 + \omega^2\left(\frac{1}{\tau} + \Gamma A p_0\right)^2\right]\Bigg|_{\omega=\omega_R} = 0 \qquad (5.195)$$

For a typical semiconductor laser, $L = 300\mu m$, $\tau_p \sim 10^{-12} s$, $\tau \sim 4 \times 10^{-9} s$, and $A p_0 \sim 2 \times 10^9 s^{-1}$. Thus, $\frac{A p_0 \Gamma}{\tau_p} >> \left(\frac{1}{\tau} + \Gamma A p_0\right)$. From which, we have

$$\omega_R \approx \sqrt{\frac{A p_0 \Gamma}{\tau_p}} \qquad (5.196)$$

Equation (5.196) tell us that to make wider bandwidth for the direct modulation of semiconductor laser,

† We have to decrease the carrier life τ_p of the semiconductor laser.

† Then, increase the DC photon density p_0.

† Then, increase the gain coefficient A of the semiconductor laser.

Important information

† Equation (5.196) give us a very important information for the 10Gbps optical fiber transponder design. The bandwidth of the direct modulation of the semiconductor laser can reach to 10GHz modulation bandwidth by decreasing the carrier life τ_p. In deed, the performance of the 10Gbps optical fiber transponder designed and manufactured by the Multiplexer Inc. New Jesse, USA has met the requirement of IEEE and then has been installed in the long-haul optical fiber communications link in USA, Canada, China and India from 2002 till now.

† The frequency response of the direct modulation of the semiconductor laser in Figure 5-28 is based on the injection bias current $I_{bias} = 3I_{th}$. The frequency response of this laser will be different at various bias current. And then the ω_R will be different at different bias current I_{bias}.

5.5 Fabry-Perot semiconductor laser amplifier

Recall the knowledge of RF and microwave, the RF or microwave oscillator is consist of a resonance circuit (or resonator) and a device. The function of the resonance circuit is to store the oscillation energy in the resonance circuit after the device amplified the oscillation signal until the device saturated. Where, the quality factor Q is used to measure the quality of the resonance circuit (or resonator). Which is defined as

$$Q = \omega \frac{Field\ enger\ stored\ by\ resonantor}{Power\ dissipaped\ by\ resonator} \tag{5.197}$$

The value of Q is high means that a small part of the stored energy in the resonator is enough to pay for the dissipation energy in each period of the oscillation.

Now, in a semiconductor laser, a device called "Fabry-Perot Etalon" or "interferometer" can be used as the optical resonator, named after its inventor [80]

5.5.1 Original Fabry-Perot Etalon

The Original Fabry-Perot Etalon is constructed of three layer dielectrics $n'/n/n'$ as shown in Figure 5-29. In which

$$n \geq n' \tag{5.198}$$

For a Fabry-Perot as shown in Figure 5-29,

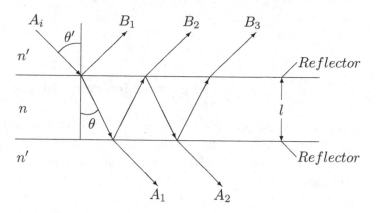

Figure 5-29   Fabry-Perot Etalon

$$The\ incident\ wave = A_i \qquad\qquad (5.199)$$

† When the wave A_i is incident into n-layer of the Fabry-Perot, the wave will be reflected again and again inside the n-layer. Which is because that the n-layer is a planar waveguide with cladding layer n' outside the n-layer, so as the wave inside the n-layer will be reflected at each boundary of n/n'.

† In other words, each boundary n/n' in the n-layer is a reflector since the refractive index n is different from the refract index n'. So as the incident wave will be reflected at each boundary of the n-layer.

† The output wave from the Fabry-Perot is the interference of all of the output wave A_1, A_2, ... Namely, the total output wave A_t is

$$A_t = A_1 + A_2 + \ldots \qquad\qquad (5.200)$$

† When the length l of n layer is fixed,

– The total output $A_t = Maximum$ for some wavelength since the constructive interference.

– The total output $A_t = Minimum$ for some other wavelength since the destructive interference.

Those characteristic is the same as a resonator in microwave.

5.5.2   Fabry-Perot semiconductor laser amplifier

The way to discuss the Fabry-Perot semiconductor laser is

† To discuss a Fabry-Perot semiconductor laser amplifier first.

† Then to form the Fabry-Perot semiconductor laser based on the discussion above.

Now we will discuss the Fabry-Perot semiconductor laser amplifier as shown in Figure 5-30 based on the discussion above. In which, a laser medium, which is the one with inverse population and a complex propagation constant $k'(\omega)$, is plated between two reflected mirrors.

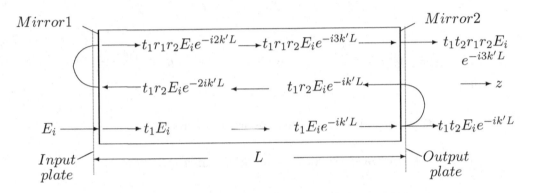

Figure 5-30   Mode used to analyze a laser amplifier

In fact,

† The active layer is a waveguide along the z-direction. So as the light wave is amplified along the z-direction.

† Then two mirrors are placed at the end points of the waveguide. Where two mirrors are formed by the difference between n' and n_0. Where n' is the refractive index of the active layer and $n_0 = 1$ is the refractive index of air outside the active layer. Thus, inside the active layer, the reflection coefficient from two mirrors is

$$r = \frac{n' - 1}{n' + 1} \tag{5.201}$$

Usually, the refractive index of the active layer $n' = 3.5$ (GaAlAs), and then $r = 30\%$.

Now, when the wave E_i is incident into the active waveguide from the left side, the wave propagation along the waveguide in z-direction. Then it is reflected from Mirror 2, propagation along $-z$ direction. Then it is reflected from Mirror 1, propagation along z-direction. This process repeat again and again.

In this process,

† The active layer is in inverse population, the wave will be amplified in the propagation along the z-direction and $-z$-direction inside the active layer. The gain obtained in each transmission of L is g_s (measured for the electric field E).

† The part wave will be reflected at the mirror 1 and the mirror 2, respectively. And the part wave will be transmitted into the air at the end of the right side.

Set $e^{-i\theta} = e^{-ik'L}$, we have the total transmission wave at the end outside the right side as follows,

$$E_t = t_1 t_2 g_s E_i e^{-i\theta} + t_1 t_2 r_1 r_2 g_s{}^3 E_i e^{-i3\theta} + t_1 t_2 r_1{}^3 r_2{}^3 g_s{}^5 E_i e^{-i5\theta} + \cdots \qquad (5.202)$$

or

$$E_t = t_1 t_2 g_s E_i e^{-i\theta}[1 + r_1 r_2 g_s{}^2 e^{-i2\theta} + r_1{}^2 r_2{}^2 g_s{}^4 e^{-i4\theta} + \cdots]E_i \qquad (5.203)$$

Where

† t_1 is the transmission coefficient for the mirror 1.

† t_2 is the transmission coefficient for the mirror 2.

† r_1 is the reflectivity for the mirror 1.

† r_2 is the reflectivity for the mirror 1.

† g_s is the gain in each transmission of L.

† $e^{-i\theta}$ is the phase delay in each transmission of L.

Now (5.203) is a series formed as

$$\sum_{k=0}^{\infty} aq^k = \frac{a}{1-q} \qquad (5.204)$$

If we set $q = r_1 r_2 g_s{}^2 e^{-i2\theta}$, we have,

† $q^0 = 1$,

† $q = r_1 r_2 g_s{}^2 e^{-i2\theta}$,

† $q^2 = r_1{}^2 r_2{}^2 g_s{}^4 e^{-i4\theta}$,

† $\cdots\cdots$,

Thus, equation (5.203) is the same as the series (5.204), and then

$$E_t = t_1 t_2 g_s e^{-i\theta} \frac{1}{1-q} E_i = \frac{t_1 t_2 g_s e^{-i\theta}}{1 - r_1 r_2 g_s^{\,2} e^{-i2\theta}} E_i \qquad (5.205)$$

Set

$$a = r_1 r_2 g_s^{\,2} \qquad (5.206)$$

$$\delta = 2\theta \qquad (5.207)$$

we have, from (5.205),

$$
\begin{aligned}
E_t &= \frac{t_1 t_2 g_s e^{-i\delta/2}}{1 - a(\cos\delta - i\sin\delta)} E_i \\
&= \frac{t_1 t_2 g_s e^{-i\delta/2}}{(1 - a\cos\delta) + ia\sin\delta} E_i \\
&= \frac{t_1 t_2 g_s e^{-i\delta/2}}{\sqrt{(1 - a\cos\delta)^2 + (a\sin\delta)^2}} E_i e^{-i\tan^{-1}\frac{a\sin\delta}{1 - a\cos\delta}} \qquad (5.208)
\end{aligned}
$$

Define: The gain of the light power

$$G = \frac{E_t \cdot E_t^{\,*}}{E_i \cdot E_i^{\,*}} \qquad (5.209)$$

We have, from (5.208),

$$G = \frac{(t_1 t_2 g_s)^2}{(1 - a\cos\delta)^2 + (a\sin\delta)^2} \qquad (5.210)$$

Where [8]

$$(1 - a\cos\delta)^2 + (a\sin\delta)^2 = (1-a)^2 + 2a(1 - \cos\delta) \qquad (5.211)$$

Considering that $\sin\delta/2 = \sqrt{\dfrac{1 - \cos\delta}{2}}$, we have $1 - \cos\delta = 2\sin^2\delta/2$. Thus, from (5.210),

$$G = \frac{(t_1 t_2 g_s)^2}{(1-a)^2 + 4a\sin^2\delta/2} \qquad (5.212)$$

---

[8]

$$
\begin{aligned}
(1 - a\cos\delta)^2 + (a\sin\delta)^2 &= 1 - 2a\cos\delta + a^2\cos^2\delta + a^2\sin^2\delta \\
&= 1 + a^2 - 2a\cos\delta \\
&= (1 - 2a + a^2) + 2a - 2a\cos\delta \\
&= (1-a)^2 + 2a(1 - \cos\delta)
\end{aligned}
$$

Considering, from (5.206) and (5.207), $a = r_1 r_2 g_s{}^2$, $\delta = 2\theta$, we have, from (5.212),

$$G = \frac{(t_1 t_2 g_s)^2}{(1 - r_1 r_2 g_s{}^2)^2 + 4 r_1 r_2 g_s{}^2 \sin^2 \theta} \tag{5.213}$$

Where, from light power point of view, $R_1 = r_1{}^2$, $R_2 = r_2{}^2$, $t_1{}^2 = 1 - R_1$, $t_2{}^2 = 1 - R_2$, $G_s = g_s{}^2$. Thus, equation (5.213) becomes

$$G = \frac{(1 - R_1)(1 - R_2)G_s}{(1 - \sqrt{R_1 R_2}G_s)^2 + 4\sqrt{R_1}\sqrt{R_2}G_s \sin^2 \theta} \tag{5.214}$$

$$\theta = k'L = \frac{2\pi}{\lambda}n'L \tag{5.215}$$

The Gain Spectrum of the Fabry-Perot semiconductor laser amplifier

Now, considering the definition of the light power gain $G = \dfrac{E_t \cdot E_t{}^*}{E_i \cdot E_i{}^*}$, the equation (5.215) can be rewritten as

$$G = \frac{E_t \cdot E_t{}^*}{E_i \cdot E_i{}^*} = \frac{(1 - R_1)(1 - R_2)G_s}{(1 - \sqrt{R_1 R_2}G_s)^2 + 4\sqrt{R_1}\sqrt{R_2}G_s \sin^2 \theta} \tag{5.216}$$

Where

$$\theta = k'L = \frac{2\pi}{\lambda}n'L, \quad \lambda = \frac{c}{\nu}, \quad \theta = k'L = \frac{2\pi\nu}{c}n'L \tag{5.217}$$

† R_1, R_2: is the power reflected coefficients from the left facet and the right facet of the active layer, respectively.

† G_s: is the light power gain obtained by the light wave traveling one L in the active layer.

Obviously, from (5.216) and (5.217), The gain G is variation associated with the variation of the frequency $\omega = 2\pi\nu$.

The maximum gain is

$$G_{max} = \frac{(1 - R_1)(1 - R_2)G_s}{(1 - \sqrt{R_1 R_2}G_s)^2} \tag{5.218}$$

at

$$\theta = \theta_{max} = \frac{2\pi\nu_m}{c}n'L = m\pi \tag{5.219}$$

The minimum gain is

$$G_{min} = \frac{(1 - R_1)(1 - R_2)G_s}{(1 + \sqrt{R_1 R_2}G_s)^2} \tag{5.220}$$

at

$$\theta = \theta_{min} = \frac{2\pi\nu_{min}}{c}n'L = \left(m + \frac{1}{2}\right)\pi \tag{5.221}$$

Thus,

$$\frac{G_{max}}{G_{min}} = \frac{(1 + \sqrt{R_1 R_2}G_s)^2}{(1 - \sqrt{R_1 R_2}G_s)^2} \tag{5.222}$$

The free spectral range $\Delta\nu$ is defined as the range between two adjacent points of the maximum gain, namely, from (5.219),

$$\Delta\nu = \nu_{m+1} - \nu_m = \frac{c}{2\pi L} \tag{5.223}$$

The discussion above results in the gain spectrum of the Fabry-Perot semiconductor laser as shown in Figure 5-31.

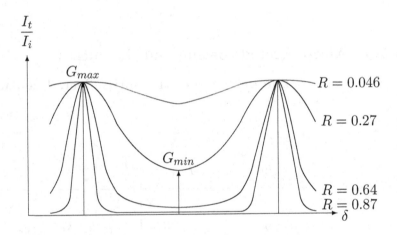

Figure 5-31   Transmission characteristics (theoretical) of a Fabry-Perot etalon
(After Reference [4])

The spectrum of the Fabry-Perot etalon is getting narrower as the reflectivity R is getting larger. Which is because that the stored energy in the etalon is getting lager as the reflectivity R is getting larger. And then the quality of the etalon Q is getting larger.

The theoretical results of a real Fabry-Perot etalon is shown in Figure 5-32. In which, $R = 0.90$ and $A = 0.98$.

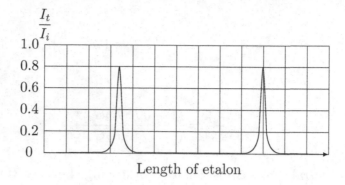

Figure 5-32   Transmission characteristics (theoretical)
of a Fabry-Perot etalon (After Reference [4])

Where,

† The peak corresponds to the constructive interference.

† Two adjacent peaks are in the optical length

$$\Delta(n'l) = \lambda/2 \tag{5.224}$$

5.5.3   Anti-reflected coating and the output

The principle of the anti-reflected coating

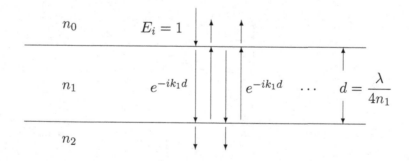

Figure 5-33   The principle of the anti-reflected coating

The anti-reflected coating is placed on the output of the Fabry-Perot etalon, which is used to make the output of the Fabry-Perot semiconductor laser amplifier being maximum at the carrier frequency.

The anti-reflected coating is consist of a three layer dielectrics. To understand the anti-reflected coating, we need to analyze the three layer dielectrics. In fact, the function of the anti-reflected coating is the interference of the total output light waves, namely, the interference is constructive interference at the carrier frequency. Which is the same as the Fabry-Perot etalon.

Now, we will analyze the three layer dielectrics with the perpendicular incident wave $E_i = 1$ as shown in Figure 5-33.

† When a light wave $E_i = 1$ is incident onto the interface of n_0/n_1, the wave will be reflected with a reflected coefficient

$$r_{01} = \frac{n_0 - n_1}{n_0 + n_1} \tag{5.225}$$

† After the wave travels inside the dielectric n_1, the wave will be reflected at the interface n_1/n_2 with the reflected coefficient

$$r_{12} = \frac{n_1 - n_2}{n_1 + n_2} \tag{5.226}$$

† After the wave travels inside the dielectric n_1, the wave will be reflected at the interface n_1/n_0 with the reflected coefficient

$$r_{10} = \frac{n_1 - n_0}{n_1 + n_0} \tag{5.227}$$

Meanwhile, the light wave will be transmitted through the interface n_1/n_0 into the dielectric n_0.

In those process of

$$E_i = 1 \rightarrow n_0/n_1 \overset{e^{-ik_1 d}}{\twoheadrightarrow} n_1/n_2 \overset{e^{-ik_1 d}}{\twoheadrightarrow} n_1/n_0 \overset{e^{-ik_1 d}}{\twoheadrightarrow} \cdots$$

The travel in each d inside the dielectric n_1 will results in a phase delay $e^{-ik_1 d}$. Those process will repeat again and again. Thus, we obtain the sum of the total reflected waves in the dielectric n_0 as follows

$$E_r = [r_{01} + t_{01}t_{10}r_{12}e^{-ik_1 2d} + t_{01}t_{10}r_{12}r_{10}r_{12}e^{-ik_1 4d} + \cdots]E_i \tag{5.228}$$

Where

† t_{01}:   is the transmission coefficient from n_0 into n_1.

† t_{10}:   is the transmission coefficient from n_1 into n_0.

† e^{-ik_1d}:   is the phase delay when the light wave travel one "d" inside the dielectric n_1.

† r_{01}:   is the reflected coefficient from the interface n_0/n_1.

† r_{12}:   is the reflected coefficient from the interface n_1/n_2.

† r_{10}:   is the reflected coefficient from the interface n_1/n_0.

Thus, from (5.228), the total reflected coefficient from the interface n_0/n_1 is

$$r = \frac{E_r}{E_i} = r_{01} + t_{01}t_{10}r_{12}e^{-ik_12d}[1 + r_{10}r_{12}e^{-ik_12d} + (r_{10}r_{12}e^{-ik_12d})^2 + \cdots] \quad (5.229)$$

Now, equation (5.229) is a series of

$$\sum_{k=0}^{\infty} aq^k = \frac{a}{1-q} \quad (5.230)$$

as long as we set

$$q = r_{10}r_{12}e^{-ik_1d} \quad (5.231)$$

and then $q^0 = 1$. Considering that

$$e^{-ik_12d} = e^{-i\pi} = -1, \quad when \quad d = \frac{\lambda}{4n_1}, \quad (5.232)$$

equation (5.229) can be rewritten as

$$r = r_{01} + \frac{t_{01}t_{10}r_{12}e^{-ik_12d}}{1 - r_{10}r_{12}e^{-ik_12d}} \quad (5.233)$$

or

$$r = r_{01} - \frac{t_{01}t_{10}r_{12}}{1 + r_{10}r_{12}} \quad (5.234)$$

It is well known, that

$$T + R = 1 \quad (5.235)$$

Which means that the sum of the transmission power T plus the reflective power R is equal to the input power "1". Where $T_{01} = t_{01}^2$, $R_{01} = r_{01}^2$.

Thus,

$$t_{01} = \sqrt{1 - R_{01}} = \sqrt{1 - \left(\frac{n_0 - n_1}{n_0 + n_1}\right)^2} = \sqrt{\frac{(n_0 + n_1)^2 - (n_0 - n_1)^2}{(n_0 + n_1)^2}} = \frac{\sqrt{4n_0n_1}}{n_0 + n_1}$$
$$(5.236)$$

In the same way, we have

$$t_{10} = \sqrt{1 - R_{10}} = \frac{\sqrt{4n_0n_1}}{n_0 + n_1} \quad (5.237)$$

Thus, we have

$$t_{01}t_{10} = \frac{4n_0n_1}{(n_0 + n_1)^2} \tag{5.238}$$

Substituting (5.238) into (5.234), we may have [9]

$$r = -\frac{(n_1^2 - n_0n_2)}{(n_1^2 + n_0n_2)} \tag{5.244}$$

The power reflectivity is

$$R = r^2 = \left(\frac{n_1^2 - n_0n_2}{n_1^2 + n_0n_2}\right)^2 \tag{5.245}$$

Summery

When a light wave is perpendicular incident onto a three layer dielectrics $n_0/n_1/n_2$, the light power reflectivity R is

$$R = 0, \quad when \quad d = \frac{\lambda}{4n_1}, \quad and \quad n_1 = \sqrt{n_0n_2} \tag{5.246}$$

$$R = \left(\frac{n_1^2 - n_0n_2}{n_1^2 + n_0n_2}\right)^2, \quad when \quad d = \frac{\lambda}{4n_1}, \quad and \quad n_1 \neq \sqrt{n_0n_2} \tag{5.247}$$

Where d is the thickness of the n_1 layer. The calculation curve of (5.247) is shown in Figure 5-34.

---

[9]   Substituting (5.238) into (5.234), we have

$$r = r_{01} - \frac{\dfrac{4n_0n_1}{(n_0 + n_1)^2} \cdot \dfrac{n_1 - n_2}{n_1 + n_2}}{1 + \dfrac{n_1 - n_0}{n_1 + n_0} \cdot \dfrac{n_1 - n_2}{n_1 + n_2}}$$

or

$$r = r_{01} - \frac{4n_0n_1(n_1 - n_2)}{[(n_1 + n_0)(n_1 + n_2) + (n_1 - n_0)(n_1 - n_2)](n_0 + n_1)} \tag{5.239}$$

Where

$$[(n_1 + n_0)(n_1 + n_2) + (n_1 - n_0)(n_1 - n_2)] = 2n_1^2 + 2n_0n_2$$

Thus (5.239) can be rewritten as

$$r = \frac{n_0 - n_1}{n_0 + n_1} - \frac{4n_0n_1^2 - 4n_0n_1n_2}{2(n_1^2 + n_0n_2)(n_0 + n_1)} \tag{5.240}$$

or

$$r = \frac{(n_0 - n_1)(n_1^2 + n_0n_2) - 2n_0n_1^2 + 2n_0n_2}{(n_0 + n_1)(n_1^2 + n_0n_2)} \tag{5.241}$$

Where

$$(n_0 - n_1)(n_1^2 + n_0n_2) - 2n_0n_1^2 + 2n_0n_2 = (n_0 + n_1)(n_0n_2 - n_1^2) \tag{5.242}$$

Therefore, (5.241) can be rewritten as

$$r = -\frac{(n_1^2 - n_0n_2)}{(n_1^2 + n_0n_2)} \tag{5.243}$$

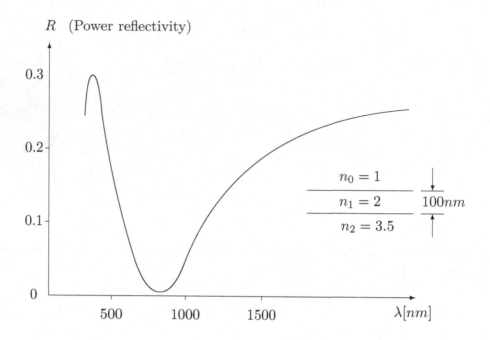

Figure 5-34   The calculation curve of the anti-reflected coating

Indeed, the light power reflectivity R is minimum near one certain wave length λ at which $d = \dfrac{\lambda}{4n_1}$. Which is because the constructive interference occurs at this point.

The construction of the anti-reflected coating in semiconductor laser amplifier

Now the real anti-reflected coating is connected at the output end of the semiconductor laser amplifier as shown in Figure 5-35. In which,

† The left side is the active layer (GaAlAs) $n_2 = 3.5$ with two coating $n_1 = 3.2$ to form a sandwich, which is a waveguide (with $n_2 > n_1$) to limit the light wave propagation along the z direction.

† The right side is the output end of the waveguide with an anti-reflected coating.

† The three layers are as follows:

- The layer 1 is the waveguide, which is consist of a sandwich $n_1/n_2/n_1$.
- The layer 2 is the coating n_3.
- The layer 3 is the air $n_0 = 1$.

In this case, recall the anti-reflected coating with three layer $n_0/n_1/n_2$, the requirement of an anti-reflected coating is indicated in (5.246). Now the layer 1 is a sandwich but not a dielectric. How should we do? A smart answer is to meet the requirements as follows:

$$\sqrt{n_1 \cdot n_0} < n_3 = n_{opt} < \sqrt{n_2 \cdot n_0} \qquad (5.248)$$

and

$$d = d_{opt} > \frac{\lambda}{4 n_{opt}} \qquad (5.249)$$

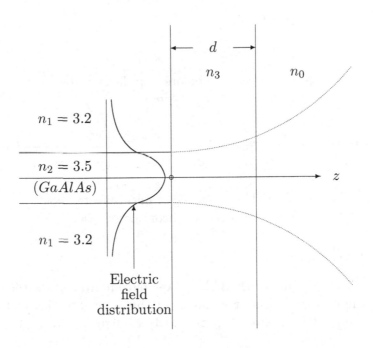

Figure 5-35   The construction of the anti-reflected coating
in semiconductor laser amplifier

- Under these conditions, on the output end of the laser,

    - The power reflectivity R is minimum near the wavelength λ at (5.249) as long as the requirement of (5.248) is satisfied.

    - Thus, the light output power of the semiconductor laser amplifier will be maximum near the wavelength λ at (5.249) as long as the requirement of (5.248) is satisfied.

The influence of anti-reflected coating on a real semiconductor laser

Now we will show you the influence of anti-reflected coating on a real semiconductor laser as indicated in Figure 5-36.

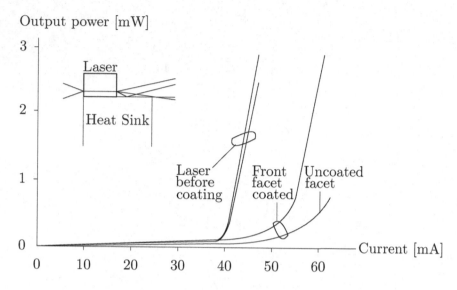

Figure 5-36   The influence of anti-reflected coating
on a real semiconductor laser

- In which

    - Before coating, the light power output from both the front facet and the end facet almost equally. In this case, the threshold current I_{th} is round $40mA$. After $I_{bias} > 40mA$, the light power is going up rapidly.

    - If the coating is that the front facet has coated only, the light power mainly output from the coated front facet. Which is just what we want. In this case, the threshold current T_{th} is round $60mA$. After $I_{bias} > 60mA$, the light power is going up not so rapidly as above.

5.6   The performance of the Fabry-Perot semiconductor laser

A mode to be used to analyze the *semiconductor laser* is shown in Figure 5-37. At first glance, seem like it is the same as the mode in Figure 5-30. However, the mode in Figure 5-30 is a semiconductor amplifier, the mode in Figure 5-37 is a semiconductor laser. Thus, for the semiconductor laser mode in Figure 5-37,

† There is no input E_i but the reflective waves from both of mirror 1 and 2.

† We hope that the bandwidth of the light wave output is as narrow as possible. So that the reflectivity r_1 and r_2 from the mirror 1 and mirror 2, respectively, should be both large enough.

† In fact, in semiconductor laser mode in Figure 5-37, the refractive index of the active layer (GaAlAs) is $n = 3.5$ and the refractive index outside the front facet and the end facet are both $n_0 = 1$ (the air). So that the reflectivity of r_1 and r_2 are both 30%. Thus, the active layer in Figure 5-37 forms a resonator automatically.

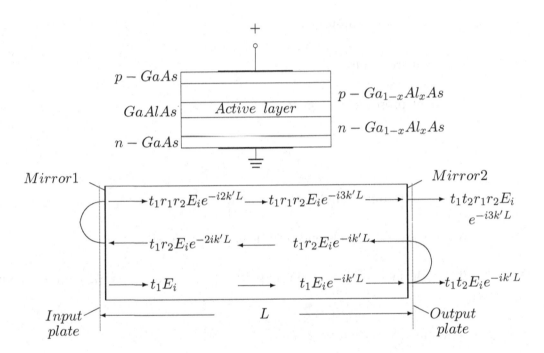

Figure 5-37   Mode used to analyze a semiconductor laser

5.6.1   Threshold condition

Population inverse

As shown in Figure 5-37, when the bias current I_{bias} is applied to a point $I_{bias} \geq I_{th}$, a "population inverse" ($N_2 > N_1$) occurs inside the active layer.

In this case, there is enough stimulated emission associated with some spontaneous emission in the active layer. In other words, the active layer becomes a gain medium for the light wave propagation.

As shown in Figure 5-37, when the gain is larger than the loss caused by the front facet and the end facet as the light travels one turn inside the Fabry-Perot cavity, the light energy will be stored inside the Fabry-Perot.

Threshold condition:

When the light wave travels one turn inside the Fabry-Perot cavity, the light oscillation will be set up if the gain is larger than the loss. Now we will discuss the threshold condition for the light oscillation. For a light mode inside the Fabry-Perot cavity,

† The initial light amplitude is: A_0.

† The light frequency is: ν.

† The light propagation constant is:

$$\beta = \overline{n}\frac{2\pi\nu}{c} \tag{5.250}$$

Where $\overline{n}$ is the refractive index for the light mode.

† When the light travels one turn inside the Fabry-Perot:

— The light mode earns a gain: $\exp[2(G/2)L]$. Where G is the gain for the light power.

— The phase variation is: $2\beta L$

— The light mode earns a loss: $\sqrt{R_1 R_2}\exp[-\alpha_{int}L]$. R_1, R_2 are the reflectivity from the front facet and the end facet, respectively. α_{int} is the interior loss caused by the free-carrier absorbtion and the the scattering from the interface between the active layer and two coating layer.

In the steady-state, at least, the amplitude of the light mode should be no changed after it travels one turn inside the Fabry-Perot cavity. It follows,

$$A_0 \exp[GL]\sqrt{R_1 R_2}\exp[-\alpha_{int}L]\exp(-i2\beta L) = A_0 \tag{5.251}$$

The real part and the imaginary part of (5.251) should be equal to each other in both side of the equal-sign, respectively. Which gives[10]

$$G = \alpha_{int} + \frac{1}{2L} \ln \left(\frac{1}{R_1 R_2} \right) \tag{5.254}$$

$$2\beta L = 2m\pi \tag{5.255}$$

Which means that the threshold condition is that, when the light mode travels one turn inside the Fabry-Perot cavity,

† The gain G should be enough to pay for the loss (from α_{int} and R_1, R_2) at least. — "Amplitude matching" (5.254).

† The phase change should be equal to $2\pi m$ to form the constructive interference inside the cavity. — "Phase matching" (5.255).

5.6.2   The longitudinal modes

Now, from the phase matching (5.255), each m corresponds to one longitudinal mode along L. Considering that $\beta = \overline{n}(2\pi\nu)/c$ in (5.255), we have the frequency of the longitudinal mode

$$\nu = \nu_m = \frac{mc}{2\overline{n}L} \tag{5.256}$$

Which means that there are a lot of the longitudinal modes in a Fabry-Perot cavity as shown in Figure 5-38. Each frequency ν_m corresponds one longitudinal mode.

† All of those longitudinal modes forms a wideband spectrum (~ 10THz) as shown in Figure 5-38. The central point of the spectrum corresponds to the maximum point, which is the dominant mode in the Fabry-Perot cavity. Here, the dominant mode is the lasing mode.

† An idea situation is that the amplitude of all of the longitudinal modes can not reach to the threshold point ($A \cdots A$) [11] except the dominant mode with two adjacent mode in both side of it, since the gain difference between the dominant mode and the two adjacent modes is very small ($\sim 0.1 cm^{-1}$) and then the phase velocity of them are very close.

† The multi-modes behavior of the Fabry-Perot cavity always limits the transmission rate for the long-haul optical fiber transmission link, in which we use the $1.5 \mu m$ laser.

---

[10]   Since

$$\ln \sqrt{R_1 R_2} = -\frac{1}{2} \ln \frac{1}{R_1 R_2} \tag{5.252}$$

we have

$$\sqrt{R_1 R_2} = \exp \left[-\frac{1}{2} \ln \left(\frac{1}{R_1 R_2} \right) \right] \tag{5.253}$$

[11]   At which, the gain is equal to the loss.

† The solution to deal with this issue is to design a single longitudinal mode laser at $1.5\mu m$.

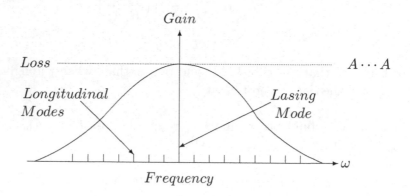

Figure 5-38   Schematic of gain and loss profiles in semiconductor laser [81]

5.6.3   The saturation problem

The equations (5.254) and (5.255) are not only the threshold condition but also the oscillation condition beyond the threshold point in the Fabry-Perot cavity.

As shown in Figure 5-36, when the bias current $I_{bias} \geq I_{th}$, the oscillation condition (5.254) and (5.255) holds until the bias current I_{bias} reaches to **the saturation point** I_s. At this point the gain G is dropping down $3dB$. Thus the amplitude matching (5.254) fails.

The operation point I_{bias}

Therefore, the operation point I_{bias} should be between the saturation point I_s and the threshold point I_{th}. Namely, $I_{th} < I_{bias} < I_s$. Usually,

$$I_{bias} < (0.3 \sim 0.4)(I_s - I_{th}) \tag{5.257}$$

5.7   Single Longitudinal Mode Semiconductor Laser

As shown in Figure 5-39, the *grating* can be used to make the DFB semiconductor laser and the DBR semiconductor laser. Where

† The distributed feedback semiconductor laser is called the DFB semiconductor laser shown in Figure 5-39(a).

† The distributed Bragg reflector semiconductor laser is called the DBR semi-conductor laser shown in Figure 5-39(b).

The DFR laser (a) and the DBR laser (b) are both of the single longitudinal mode semiconductor lasers working at $1.55\mu m$. This is **a magnificent step** to make the $1.55\mu m$ long-haul optical fiber transmission system.

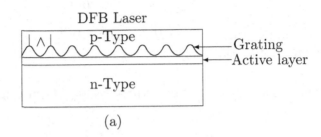

(a)

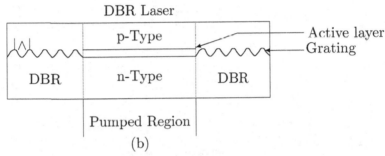

(b)

Figure 5-39   DFB and DBR laser structure

Where,

† The grating is manufactured above the active layer along the whole length L of the cavity in Figure 5-39(a).

† The gratings are manufactured by the ends of the active layer in Figure 5-39(b).

A Bragg Grating is a transparent device with a periodic refractive index (n), so that a large reflectivity may be reached in some wavelength. As an example, in a DFB semiconductor laser, a corrugated construction of Bragg Grating will caused the coupling between the forward light wave and the reflected light wave. This coupling occurs at the light wavelength λ_g, which satisfies

$$\wedge = m\frac{\lambda_g}{2\overline{n}} \qquad (5.258)$$

Where

† $\wedge$ shown in Figure 5-39(a) and (b) is the period of the Bragg Grating.

† $\overline{n}$ is the average of the mode index.

† m is the order number of the Bragg Grating, in which the strongest one corresponds to $m = 1$.

[Example] A DFB semiconductor laser working at the wavelength $\lambda = 1.55\mu m$. If we use $m = 1$, $\overline{n} = 3.3$, we have, from (5.258),

$$\wedge = m\frac{\lambda_g}{2\overline{n}} \simeq 255nm \qquad (5.259)$$

Which can be manufactured by *the Holographic Technique*.

Now, we will discuss the Bragg Grating step by step as follows.

5.7.1   Passive periodic waveguide - Bragg Grating

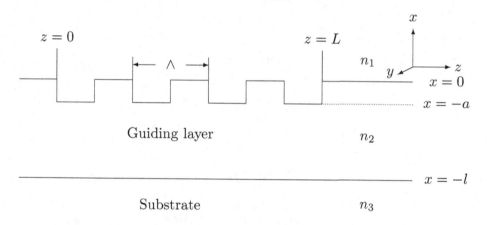

Figure 5-40   Bragg Grating and the periodic waveguide

The principle of the DFB semiconductor laser in based on the analysis of the Bragg Grating as shown in Figure 5-40. General speaking, which is a corrugated periodic waveguide.

The waveguide is a planar waveguide with $\dfrac{\partial}{\partial y} = 0$. In which, there are two modes, which can match the boundary condition:

† TE modes with E_y, H_x and H_z.
† TM modes with H_y, E_x and E_z.
† $\wedge$ is the period of the Bragg Grating along the z direction.
† L is the Bragg Grating length along the z direction.

Now, considering that the corruption is along x direction only, there is no coupling between TE mode and TM mode but the coupling between different TE mode (for instance between TE_m mode and TE_{m+1} mode) and the coupling between different TM mode (for instance between TM_m mode and TM_{m+1} mode).

The coupled equations

Assume that,

† When the waveguide is a normal planar waveguide without the corrugation, the dielectric constance distribution of the waveguide is $\varepsilon(\vec{r})$.

† After corrugated, the dielectric constant distribution is

$$\varepsilon'(\vec{r}) = \varepsilon(\vec{r}) + \Delta\varepsilon(\vec{r}) \tag{5.260}$$

In this case, the wave equation of the normal planar waveguide [$\varepsilon'(\vec{r}) = \varepsilon(\vec{r})$] is

$$\nabla^2 E_y - \mu\varepsilon(\vec{r})\frac{\partial^2 E_y}{\partial t^2} = 0 \tag{5.261}$$

Now, a perturbation polarization $\vec{P}_{pert}$ caused by the $\Delta\varepsilon(\vec{r})$ is

$$\vec{P}_{pert}(\vec{r}, t) = \Delta\varepsilon(\vec{r})\vec{E}(\vec{r}, t) = \Delta n^2(\vec{r})\vec{E}(\vec{r}, t) \tag{5.262}$$

So as the wave equation of the corrugated periodic waveguide becomes

$$\nabla^2 E_y - \mu\varepsilon(\vec{r})\frac{\partial^2 E_y}{\partial t^2} = \mu\frac{\partial^2}{\partial t^2}[\vec{P}_{pert}(\vec{r}, t)]_y \tag{5.263}$$

In fact, equation (5.263) can be obtained by substitution of (5.260) into (5.261).

The solution of (5.263) is the superposition of all of the normal modes in the normal waveguide($\Delta\varepsilon(\vec{r}) = 0$). In which the normal modes in the normal waveguide is consist of the discrete modes and the radiation modes. For the sake of simplification, assume that the coupling between the discrete modes and the radiation mode is ignored. And then the field in the corrugation waveguide, $E_y(\vec{r}, t)$, is the superposition of the discrete modes and the radiation mode in the normal waveguide. Namely,

$$E_y = \frac{1}{2}\sum_m A_m(z)\mathcal{E}_y^{(m)}(x)e^{i(\omega t - \beta_m z)} + c.c. \tag{5.264}$$

Where $c.c.$ are the radiation modes,

† The discrete mode $\mathcal{E}_y^{(m)}$ satisfies the wave equation in normal waveguide as follows,

$$\left(\frac{\partial^2}{\partial x^2} - \beta_m^2\right)\mathcal{E}_y^{(m)}(\vec{r}) + \omega^2\mu\varepsilon(\vec{r})\mathcal{E}_y^{(m)} = 0 \tag{5.265}$$

† Meanwhile, the discrete mode $\mathcal{E}_y^{(m)}$ satisfies the normalization condition in the normal waveguide. Namely,

$$\int_{-\infty}^{\infty} \mathcal{E}_y^{(l)}\mathcal{E}_y^{(m)}dx = \frac{2\omega\mu}{\beta_m}\delta_{l,m} \tag{5.266}$$

Now, substitution of (5.264) into (5.262) gives

$$\left[\vec{P}_{pert}(\vec{r},t)\right]_y = \frac{\Delta n^2(\vec{r})\varepsilon_0}{2} \sum_m [A_m \mathcal{E}_y^{(m)}(x)e^{i(\omega t - \beta_m z)} + c.c] \qquad (5.267)$$

Using (5.267) in (5.263), considering that $\dfrac{\partial}{\partial y} = 0$, $\dfrac{\partial}{\partial z} = -i\beta_m$ and $\dfrac{\partial}{\partial t} = i\omega$, we have [12]

$$\sum_m \left[\frac{A_m}{2}\left(\frac{\partial^2 \mathcal{E}_y^{(m)}}{\partial x^2} - \beta_m^2 \mathcal{E}_y^{(m)} + \omega^2 \mu \varepsilon(\vec{r})\mathcal{E}_y^{(m)}\right)e^{-i\beta_m z}\right.$$

$$+ \frac{1}{2}\left(-2i\beta_m \frac{dA_m}{dz} + \frac{d^2 A_m}{dz^2}\right)\mathcal{E}_y^{(m)}e^{-i\beta_m z}\right]e^{i\omega t}$$

$$+ c.c = \mu \frac{\partial^2}{\partial t^2}\left[P_{pert}(\vec{r},t)\right]_y \qquad (5.270)$$

Where, from (5.265),

$$\dagger \left(\frac{\partial^2 \mathcal{E}_y^{(m)}}{\partial x^2} - \beta_m^2 \mathcal{E}_y^{(m)} + \omega^2 \mu \varepsilon(\vec{r})\mathcal{E}_y^{(m)}\right) = 0.$$

$\dagger$ Considering that the corrugated periodic waveguide is a slow variation in geometric along z, we have

$$\left|\frac{d^2 A_m}{dz^2}\right| << \beta_m \left|\frac{dA_m}{dz}\right| \qquad (5.271)$$

$\dagger$ c.c are the radiation modes.

Thus (5.270) may be rewritten as

$$\sum_m -i\beta_m \frac{dA_m}{dz}\mathcal{E}_y^{(m)}e^{i(\omega t - \beta_m z)} + c.c = \mu \frac{\partial^2}{\partial t^2}[P_{pert}(\vec{r},t)] \qquad (5.272)$$

Now, multiplying (5.272) by $\mathcal{E}_y^{(s)}$, integral of which from $-\infty$ to $+\infty$ along x gives

$$\frac{dA_s^{(-)}}{dz}e^{i(\omega t + \beta_s z)} - \frac{dA_s^{(+)}}{dz}e^{i(\omega t - \beta_s z)} - c.c = \frac{i}{2\omega}\frac{\partial^2}{\partial t^2}\int_{-\infty}^{\infty}[P_{pert}(\vec{r},t)]_y \mathcal{E}_y^{(s)}(x)dx \quad (5.273)$$

Where,

---

[12]   Where

$$\frac{\partial}{\partial z}[A_m(z)e^{-i\beta_m z}] = -i\beta_m A_m e^{-i\beta_m z} + \frac{\partial A_m}{\partial z}e^{-i\beta_m z} \qquad (5.268)$$

$$\frac{\partial^2}{\partial z^2}[A_m(z)e^{-i\beta_m z}] = (-i\beta_m)^2 A_m e^{-i\beta_m z} - i2\beta_m \frac{\partial A_m}{\partial z}e^{-i\beta_m z} + \frac{\partial^2 A_m}{\partial z^2}e^{-i\beta_m z} \quad (5.269)$$

† The normalization condition (5.266) has been considered, and then only $m = s$ terms exit.

† $A_s^{(-)}e^{i(\omega t + \beta_s z)}$:   is the light wave along the $-z$ direction.

† $A_s^{(+)}e^{i(\omega t - \beta_s z)}$:   is the light wave along the z direction.

† Thus, (5.273) is just the coupled equations between the incident wave $A_s^{(+)}e^{i(\omega t - \beta_s z)}$ and the reflected wave $A_s^{(-)}e^{i(\omega t + \beta_s z)}$.

† Those coupling is caused by $\Delta\varepsilon(\vec{r})$ and then by the perturbation polarization $[P_{pert}(\vec{r}, t)]_y$.

The solutions of the coupled equations

Now, substitution of the perturbation polarization $[P_{pert}(\vec{r}, t)]_y$ in (5.262) into the coupled equations (5.273) gives

$$\frac{dA_s^{(-)}}{dz}e^{i(\omega t + \beta_s z)} - \frac{dA_s^{(+)}}{dz}e^{i(\omega t - \beta_s z)} - c.c$$
$$= -\frac{i\varepsilon_0}{4\omega}\frac{\partial^2}{\partial t^2}\sum_m \left[A_m \int_{-\infty}^{\infty} \Delta n^2(x,z)\mathcal{E}_y^{(m)}(x)\mathcal{E}_y^{(s)}(x)dx \, e^{i(\omega t - \beta_m z)} + c.c\right]$$

$$(5.274)$$

Where, $\Delta n^2(x, z)$ is a rectangular function along z direction and the period is $\wedge$. Thus, $\Delta n^2(x, z)$ may be expressed as a Fourier series along z direction as follows:
13

$$\Delta n^2(x, z) = \Delta n^2(x)\sum_{n=0}^{\infty} a_n \cos\frac{n\pi}{\frac{\wedge}{2}}z \qquad (5.276)$$

or

$$\Delta n^2(x, z) = \Delta n^2(x)\sum_{-\infty}^{\infty} a_q e^{i\frac{2q\pi}{\wedge}z} \qquad (5.277)$$

Substituting (5.277) into (5.274), we find that only one term (q=l, m=s) in right side of (5.274), $A_s^{(+)}\exp[i(\frac{2l\pi}{\wedge} - \beta_s)z]$, matches one term $\frac{dA_s^{(-)}}{dz}\exp[i\beta_s z]$ in left side of (5.274) as long as

$$\frac{2l\pi}{\wedge} - \beta_s \approx \beta_s \qquad (5.278)$$

---

13   The definition of the Fourier series: If $f(x)$ is a period function with the period $2l$, and the function is integrable in $[-l, l]$, then

$$f(x) = \frac{a_0}{2} + \sum_{n=1}^{\infty}\left(a_n \cos\frac{n\pi}{l}x + b_n \sin\frac{n\pi}{l}x\right) \qquad (5.275)$$

is called the Fourier series of $f(x)$.

This result can be written as, from (5.274),

$$\frac{dA_s^{(-)}}{dz} = \kappa \; A_s^{(+)} \; e^{-2i(\Delta\beta)z} \tag{5.279}$$

Similarly, we have [14]

$$\frac{dA_s^{(+)}}{dz} = \kappa^* \; A_s^{(-)} \; e^{i2(\Delta\beta)z} \tag{5.281}$$

Where[34]

$$\kappa = \frac{i\omega\varepsilon_0 a_l}{4} \int_{-\infty}^{\infty} \Delta n^2(x) [\varepsilon_y^{(s)}(x)]^2 dx \tag{5.282}$$

with

$$a_l = \begin{cases} \dfrac{-i}{\pi l} & : \quad l = odd \\ 0 & : \quad l = even \\ \dfrac{1}{2} & : \quad l = 0 \end{cases} \tag{5.283}$$

For $l = odd$,

$$\kappa = \frac{-\omega\varepsilon_0}{4\pi l} \int_{-\infty}^{\infty} \Delta n^2(x) [\varepsilon_y^{(s)}(x)]^2 dx \tag{5.284}$$

with

$$\Delta\beta = \beta_s - \frac{l\pi}{\wedge} = \beta_s - \beta_0, \quad \beta_0 = \frac{l\pi}{\wedge} \tag{5.285}$$

In practise, the period $\wedge$ is so chosen, so as it satisfies the phase matching

$$\Delta\beta = 0 \tag{5.286}$$

for some l. In this case, from (5.285), $\beta_s = \dfrac{2\pi}{\lambda_g^{(s)}} = \dfrac{l\pi}{\wedge}$, and then

$$\wedge = l\frac{\lambda_g^{(s)}}{2} \tag{5.287}$$

Where $\lambda_g^{(s)}$ is the wavelength of the normal mode in the normal waveguide, called the waveguide wavelength.

Which means that the phase matching (5.286) can be reached if the period $\wedge$ is the integer of $\dfrac{\lambda_g^{(s)}}{2}$.

---

[14] Substituting (5.277) into (5.274), we find that only one term (q=-l, m=s) in right side of (5.274), $A_s^{(-)} \exp[i(\beta_s - \frac{2l\pi}{\wedge})z]$, corresponds to $\dfrac{dA_s^{(+)}}{dz} \exp[-i\beta_s z]$ in left side of (5.274) as long as

$$\beta_s - \frac{2l\pi}{\wedge} \approx -\beta_s \tag{5.280}$$

Equations (5.279) and (5.281) are the coupled equations between the incident wave $A_s^{(+)}$ and the reflected wave $A_s^{(-)}$. Which describes the process of the energy exchanging between the incident wave $A_s^{(+)}$ and the reflected wave $A_s^{(-)}$. After we get the solutions of the coupled equations we may seen the process clearly.

To find the solutions of the coupled mode equations (5.279) and (5.281), we need to set up the boundary condition as shown in Figure 5-41. In which

† The incident light wave is $A^{(+)}(0)$ at $z = 0$.

† The reflected light wave is $A^{(-)}(L)$ at $z = L$.

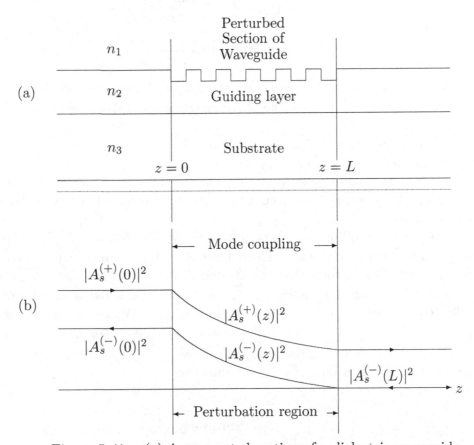

Figure 5-41   (a) A corrugated section of a dielectric waveguide

(b) The incident and the reflected intensities

inside the corrugated section

In this case, the solutions of the coupled mode equations are

$$A_s^{(-)}(z)e^{i\beta z} = A^{(+)}(0)\frac{iK\,e^{i\beta_0 z}\,\sinh[K(z-L)]}{-\Delta\beta\sinh KL + iK\cosh KL} \tag{5.288}$$

$$A_s^{(+)}(z)e^{-i\beta z} = A^{(+)}(0)\frac{ie^{-i\beta_0 z}}{-\Delta\beta\sinh KL + iK\cosh KL}$$
$$\cdot\left\{\Delta\beta\sinh[K(z-L)] + iK\cosh[K(z-L)]\right\} \tag{5.289}$$

Where

$$K = \sqrt{\kappa^2 - (\Delta\beta)^2} \tag{5.290}$$

Under the phase matching condition $\Delta\beta = 0$, the solutions of the coupled equations (5.279) are

$$A_s^{(-)}(z) = A_s^{(+)}(0)\frac{\sinh[\kappa(z-L)]}{\cosh\kappa L} \tag{5.291}$$

$$A_s^{(+)}(z) = A_s^{(+)}(0)\frac{\cosh[\kappa(z-L)]}{\cosh\kappa L} \tag{5.292}$$

Figure 5-41 shows the power converted from the incident wave $A_s^{(+)}(z)$ to the reflected wave $A_s^{(-)}(z)$ under the phase matching condition. In which,

† The corrugation region is in the range from $z = 0$ to $z = L$.

† The incident intensity is $A_s^{(+)}(0)$ at $z = 0$.

† When the incident wave is incident into the corrugation region along the z direction, the reflected wave occurs along the $-z$ direction because of the every corrugation.

† The sum of all of the reflected waves is an interference of all of the reflected wave. And the solution of the reflected intensity $A_s^{(-)}(z)$ in (5.291) is a constructive interference of all of the reflected waves because of the phase matching condition.

† Thus, the incident intensity $A_s^{(+)}(0)$ at $z = 0$ converts its energy into the reflected intensity in all of the corrugation region. Which results in the incident intensity $A_s^{(+)}(z)$ in the corrugation region decreases in exponential law along z direction. And results in the reflected intensity $A_s^{(-)}(z)$ in the corrugation region increases in exponential law along $-z$ direction.

Summery

The discussion above comes to an important conclusion:

1. The incident intensity $A_s^{(+)}(z)$ converts its energy into the reflected intensity $A_s^{(-)}(z)$ in all of the corrugation region. And both of them are in an opposite

direction propagation.

2. In the design, if we choice

$$\wedge = l\frac{\lambda_g}{2} \tag{5.293}$$

for the central frequency, the transformation efficiency from the incident intensity $A_s^{(+)}(z)$ into the reflected intensity $A_s^{(-)}(z)$ in the corrugation region is in maximum, as it is in phase matching $\Delta\beta = 0$ at the central frequency. Where $\wedge$ is the period of the corrugation region. λ_g is the waveguide mode wavelength for the central frequency.

The approximation solutions of the coupled equations

Now, near the central frequency, $\Delta\beta \neq 0$. we have to use the solutions in (5.288) and (5.289). In which, $\sinh\theta = \frac{1}{2}(e^\theta - e^{-\theta})$, $\cosh\theta = \frac{1}{2}(e^\theta + e^{-\theta})$, $\theta = s(z - L)$.

$$s = \sqrt{\kappa^2 - \Delta\beta^2} \tag{5.294}$$

For a large $\theta = s(z - L)$, $\sinh\theta \approx \frac{e^\theta}{2}$, $\cosh\theta \approx \frac{e^\theta}{2}$. Thus, the solutions in (5.288) and (5.289) may be approximately written as

$$A_s^{(-)}(z)e^{i\beta z} = A^{(+)}(0)\frac{i\,K\,e^{i\beta_0 z}\,\frac{1}{2}\,e^{s(z-L)}}{(-\Delta\beta + is)\,\frac{1}{2}\,e^{sL}} \sim e^{i(\beta_0 - is)z} \tag{5.295}$$

$$A_s^{(+)}(z)e^{-i\beta z} = A^{(+)}(0)\frac{ie^{-i\beta_0 z}}{(-\Delta\beta + is)\,\frac{1}{2}\,e^{sL}}\left\{(\Delta\beta + is)\,\frac{1}{2}\,e^{s(z-L)}\right\}$$
$$\sim e^{-i(\beta_0 + is)z} \tag{5.296}$$

Which means that the propagation constants of the incident light wave $A_s^{(+)}(z)e^{-i\beta z}$ and the reflected light wave $A_s^{(-)}(z)e^{i\beta z}$ in the corrugation region in Figure 5-41, are, respectively,

$$\beta'^{(+)} = \beta_0 + is = \frac{l\pi}{\wedge} + i\sqrt{\kappa^2 - [B(\omega) - \beta_0]^2} \tag{5.297}$$

$$\beta'^{(-)} = \beta_0 - is = \frac{l\pi}{\wedge} - i\sqrt{\kappa^2 - [B(\omega) - \beta_0]^2} \tag{5.298}$$

Where, again

$$s = \sqrt{\kappa^2 - \Delta\beta^2} \tag{5.299}$$

Which means that in the range of $\Delta\beta < \kappa$, the propagation constant in the corrugation region $\beta'^{(\pm)}$ contain an *imaginary part (is)*, which will result in *the coupling* between the incident wave $A_s^{(+)}(z)$ and the reflected wave $A_s^{(-)}(z)$ and the incident wave power $|A_s^{(+)}(z)|^2$ drops off exponentially along the perturbation region and reflection wave of power into the backward traveling mode power $|A_s^{(-)}(z)|^2$.

In the central frequency, from (5.293), we choice

$$\wedge = l\frac{\lambda_g}{2} \tag{5.300}$$

Thus, from (5.285), $\Delta\beta = 0$. In this case as shown in Figure 5-41, the transformation efficiency from the incident intensity $A_s^{(+)}(z)e^{-i\beta z}$ into the reflected intensity $A_s^{(-)}(z)e^{i\beta z}$ is in maximum.

In a frequency range near the central frequency, $\Delta\beta(\omega) < \kappa$, from (5.295) and (5.296), the energy transformation from the incident wave $A_s^{(+)}(z)e^{-i\beta z}$ into the reflected intensity $A_s^{(-)}(z)e^{i\beta z}$ is the same as that in the central frequency, however, the transformation efficiency is smaller than that in the central frequency.

5.7.2 Active periodic waveguide

Based on the discussion of the corrugation waveguide, we may discuss DFB (distributed Feedback) semiconductor laser as shown in Figure 5-42. In which,

† The active layer of the DBF semiconductor laser may be considered as an active period waveguide.

† Compared with the passive period waveguide in Figure 5-41, the n_2 layer in Figure 5-41 is now a pumped layer or the active layer in Figure 5-42. Thus, this is an active period waveguide with gain γ inside the guiding layer.

Thus, the coupling equations of the passive period waveguide in Figure 5-41, (5.279) and (5.281), may be used as the coupling equations of the active period waveguide in Figure 5-42, except we need to add the gain term in the coupling equations as follows:

$$\frac{dA_s^{(-)}}{dz} = \kappa \ A_s^{(+)} \ e^{-2i(\Delta\beta)z} - \gamma A_s^{(-)} \tag{5.301}$$

$$\frac{dA_s^{(+)}}{dz} = \kappa^* \ A_s^{(-)} \ e^{i2(\Delta\beta)z} + \gamma A_s^{(+)} \tag{5.302}$$

Where, from (5.284),

$$\kappa = \frac{-\omega\varepsilon_0 a_l}{4\pi l} \int_{-\infty}^{\infty} \Delta n^2(x)[\varepsilon_y^{(s)}(x)]^2 dx \tag{5.303}$$

$$\gamma : \quad \text{is a gain constant in } e^{\gamma z} \tag{5.304}$$

And then,

† $+\gamma A_s^{(+)}$ is the positive gain of the incident wave $A_s^{(+)}$ propagation along z direction.

† $-\gamma A_s^{(-)}$ is the negative gain of the reflected wave $A_s^{(-)}$ propagation along z direction.

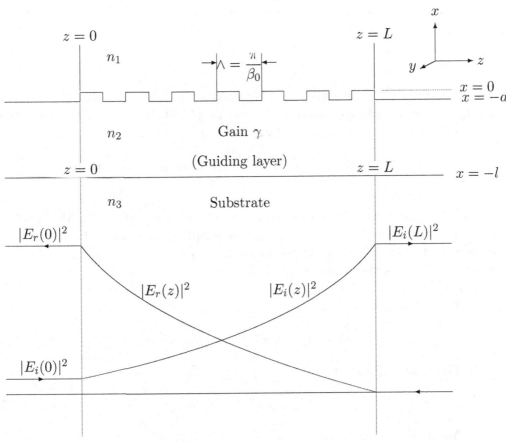

Figure 5-42   Incident and reflected waves
inside an amplifying periodic waveguide

Thus, we may make a definition of the incident wave $A_s^{(+)}(z)$ and the reflected wave $A_s^{(-)}(z)$ as follows:

$$A_s^{(-)}(z) = A_s'^{(-)}(z)e^{-\gamma z} \tag{5.305}$$

$$A_s^{(+)}(z) = A_s'^{(+)}(z)e^{\gamma z} \tag{5.306}$$

And then, the passive coupling equations (5.301) and (5.302) become active coupling equations as follows: [15]

$$\frac{dA_s'^{(-)}}{dz} = \kappa \ A_s'^{(+)} \ e^{-2i(\Delta\beta + i\gamma)z} \tag{5.310}$$

$$\frac{dA_s'^{(+)}}{dz} = \kappa^* \ A_s'^{(-)} \ e^{i2(\Delta\beta + i\gamma)z} \tag{5.311}$$

Now, recall the the passive coupling equations, from (5.279) and (5.281),

$$\frac{dA_s^{(-)}}{dz} = \kappa \ A_s^{(+)} \ e^{-2i(\Delta\beta)z} \tag{5.312}$$

$$\frac{dA_s^{(+)}}{dz} = \kappa^* \ A_s^{(-)} \ e^{i2(\Delta\beta)z} \tag{5.313}$$

we understand that the active coupling equations (5.310) and (5.311) are the same as the passive coupling equations (5.279) and (5.281), respectively, except that $\Delta\beta$ in passive one has to be replaced by $\Delta\beta + i\gamma$ in active one.

The solution of the active coupling equations

In this case, the solution of the active coupling equations (5.310) and (5.311) are the same as the solution of the passive coupling equations, (5.288) and (5.289), except that that $\Delta\beta$ in passive solution has to be replaced by $\Delta\beta + i\gamma$ in active solution. Now, as shown in Figure 5-42,

† The incident wave is

$$E_i = A_s'^{(+)}(z)e^{-i\beta z} = A_s^{(+)}(z)e^{(-i\beta + \gamma)z} \tag{5.314}$$

† The reflected wave is

$$E_r = A_s'^{(+)}(z)e^{i\beta z} = A_s^{(+)}(z)e^{(i\beta - \gamma)z} \tag{5.315}$$

† Assume that:

---

[15]   From (5.305) and (5.306), we have

$$A_s'^{(-)}(z) = A_s^{(-)}(z)e^{\gamma z} \tag{5.307}$$

$$A_s'^{(+)}(z) = A_s^{(+)}(z)e^{-\gamma z} \tag{5.308}$$

Thus,

$$\begin{aligned} \frac{dA_s'^{(-)}(z)}{dz} &= \frac{dA_s^{(-)}(z)}{dz}e^{\gamma z} + \gamma A_s^{(-)}(z) = \kappa A_s^{(+)}(z)e^{i2\Delta\beta z}e^{\gamma z} - \gamma A_s^{(-)}(z)e^{\gamma z} + -\gamma A_s^{(-)}(z)e^{\gamma z} \\ &= \kappa \ A_s'^{(+)}(z) \ e^{-2i(\Delta\beta + i\gamma)z} \end{aligned} \tag{5.309}$$

— the incident wave at the incident point $z = 0$ of the period waveguide is $A_s^{(+)}(0)$,

— the reflected wave at the output of the period waveguide is $A_s^{(-)}(L)$.

Under those boundary conditions, the solutions of the active coupling equations (5.310) and (5.311) are

$$E_i(z) = A^{(+)}(0) \frac{e^{-i\beta_0 z} \left\{ (\gamma - i\Delta\beta) \sinh[S(L-z)] - S \cosh[S(L-z)] \right\}}{(\gamma - i\Delta\beta) \sinh SL - S \cosh SL} \quad (5.316)$$

$$E_r(z) = A^{(+)}(0) \frac{S e^{i\beta_0 z} \sinh[S(L-z)]}{(\gamma - i\Delta\beta) \sinh SL - S \cosh SL} \quad (5.317)$$

Where

$$S^2 = \kappa^2 + (\gamma - i\Delta\beta)^2 \quad (5.318)$$

Discussion

(5.316) and (5.317) are the solutions of the active period waveguide. And (5.288) and (5.289) are the solution of the passive period waveguide. The difference between those two solutions are

† $S = \sqrt{\kappa^2 + (\gamma - i\Delta\beta)^2}$ is a complex number in the active period waveguide,

† $s = \sqrt{\kappa^2 - \Delta\beta^2}$ is a real number in the passive period waveguide as the discussion in the range of $\kappa > \Delta\beta$.

† Thus, from (5.316) and (5.317), if the following condition is satisfied,

$$(\gamma - i\Delta\beta) \sinh SL = S \cosh SL \quad (5.319)$$

we will have

$$\frac{E_r(0)}{E_i(0)} = \frac{S \sinh SL}{(\gamma - i\Delta\beta) \sinh SL - S \cosh SL} \to \infty \quad (5.320)$$

$$\frac{E_i(L)}{E_i(0)} = \frac{-S e^{-i\beta_0 L}}{(\gamma - i\Delta\beta) \sinh SL - S \cosh SL} \to \infty \quad (5.321)$$

Which means that the device is under an oscillation state. Where $\dfrac{E_r(0)}{E_i(0)}$ is the reflective gain at the point $z = 0$. $\dfrac{E_i(L)}{E_i(0)}$ is the transmission gain from $z = 0$ to $z = L$.

† However, if $\gamma = 0$, we have $S = s = real$. Thus,

$$\frac{E_r(0)}{E_i(0)} < 1 \quad (5.322)$$

$$\frac{E_i(L)}{E_i(0)} < 1 \tag{5.323}$$

Which is the situation of the passive period waveguide.

When the frequency is near the Bragg frequency ω_0, i.e. $\Delta\beta \approx 0$, and the gain γ is large enough, so as the condition (5.319) is satisfied, the function of the active period waveguide is just an amplifier. And the outputs of the amplifier may be expressed as the reflective gain

$$\frac{E_r(0)}{E_i(0)} = \frac{S \sinh SL}{(\gamma - i\Delta\beta)\sinh SL - S\cosh SL} \tag{5.324}$$

or the transmission gain

$$\frac{E_i(L)}{E_i(0)} = \frac{-Se^{-i\beta_0 L}}{(\gamma - i\Delta\beta)\sinh SL - S\cosh SL} \tag{5.325}$$

Where the power reflective gain $\left[\frac{E_r(0)}{E_i(0)}\right]^2$ and the power transmission gain $\left[\frac{E_i(L)}{E_i(0)}\right]^2$ are shown in Figure 5-42. In which,

† The transmission gain can be used to analyze a DFB semiconductor laser.

† The reflective gain can be used to analyze a DBR semiconductor laser.

Which means, that

† When the central operation frequency ω is equal to the Bragg grating frequency ω_0, we have $\Delta\beta = 0$ and the condition (5.319) is satisfied. Thus, the output of the DFB semiconductor laser is very high at central frequency ω_0.

† Which can be realized by the following design formula

$$\wedge = l(\lambda_g/2\pi) \tag{5.326}$$

Where $\wedge$ is the period of the Bragg grating and λ_g is the waveguide wavelength of the normal waveguide without the Bgagg grating.

† Which means that the spectrum of the output of theDFB semiconductor laser is very narrow since the output exits only if the frequency is near the central frequency ω_0 ($\Delta\beta \approx 0$).

5.7.3   Introduce to the single longitudinal mode laser(1)

Now we may discuss the conception of the single longitudinal mode semiconductor laser. Which is involved with two conception:

† The mode suppression ratio — MSR.

† The mode selectivity.

The mode suppression ratio — MSR

First, we will introduce the conception of the mode suppression ratio (MSR). Recall the two kinds of the single longitudinal mode semiconductor laser as shown in Figure 5-43. In which,

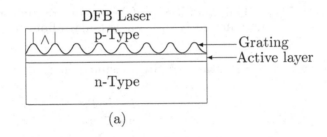

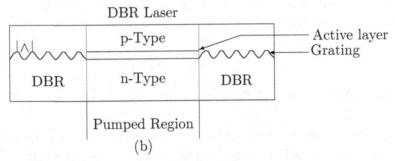

Figure 5-43   DFB and DBR laser structure

† the Bragg grating is made in/on an active layer in the DFB (distributed feedback) semiconductor laser in Figure 5-43(a),

† the Bragg gratings are made as the frequency selective reflectors at two ends of the active layer in the DBR (distributed Bragg reflector) semiconductor laser in Figure 5-43(b),. If the condition (5.326) is satisfied, the reflective of the reflectors are maximum at the waveguide wavelength λ_g satisfied (5.326). Thus, the loss in the cavity are minimum at the waveguide wavelength λ_g satisfied (5.326) as shown in Figure 5-44.

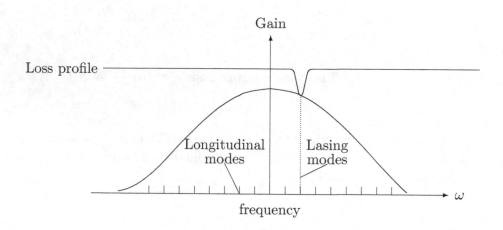

Figure 5-44   Gain and loss profile for semiconductor lasers oscillating
predominantly in a single longitudinal mode. [82]

The mode suppression ratio — MSR is defined as

$$MSR = \frac{I_m}{I_s} = \frac{P_m}{P_s} \qquad (5.327)$$

Where I_m is the light intensity of the main longitudinal mode, whose waveguide
wavelength λ_g satisfies

$$\wedge = \frac{\lambda_g}{2n_{eff}} \qquad (5.328)$$

I_s is the light intensity of the side-mode, which is the one nearest to the domain
longitudinal mode.

The MSR may be obtained by the carrier density rate equation. For a steady
state, means the continuous operation state, the rate equation is, from (5.182),

$$\frac{dP}{dt} = 0 = \Gamma A(N - N)P - \frac{P}{\tau_p} \qquad (5.329)$$

Where P is the number of the photons and is $P = P_0$ for the steady state.

In this case, as shown in Figure 5-45, in all of the Fabry-Perot cavity modes,
there is a domain longitudinal mode (λ_N) with two side-modes λ_{N+1} and λ_{N-1}.
Where Δg_c is the gain difference between the domain longitudinal mode (λ_N) and
the side-mode (λ_{N+1} and λ_{N-1}). Which is because that the domain longitudinal
mode is formed by the complete constructive interference inside the Fabry-Perot
cavity, all of other modes can not be complete constructive interference.

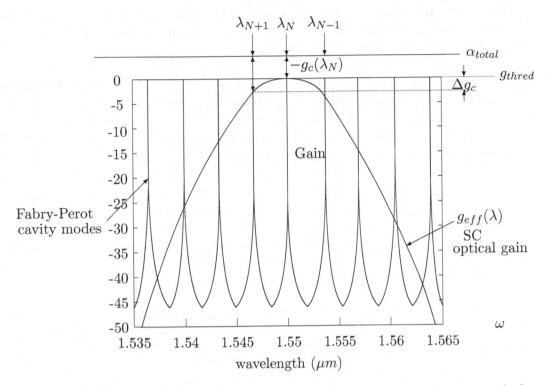

Figure 5-45   Gain and loss profile for semiconductor lasers oscillating [83]

For the steady state operation (or continuous wave operation), the domain longitudinal mode and the side-modes satisfy the steady state rate equations, respectively, as follows

$$\frac{dP_N}{dt} = 0 = \Gamma A(N_N - N_{tr})P_N - \frac{P_N}{\tau_p} \tag{5.330}$$

$$\frac{dP_{N+1}}{dt} = 0 = \Gamma A(N_{N+1} - N_{tr})P_N - \frac{P_{N+1}}{\tau_p} \tag{5.331}$$

From which we may obtain the MSR as follows

$$MSR = \frac{P_N}{P_{N+1}} = \frac{\Gamma A(N_{N+1} - N_{tr}) - \frac{1}{\tau_p}}{\Gamma A(N_N - N_{tr}) - \frac{1}{\tau_p}} \tag{5.332}$$

The mode selectivity

As we mentioned before, the domain longitudinal mode is formed by the complete constructive interference inside the Fabry-Perot cavity and all other mode inside the

Fabry-Perot cavity can not reach the complete constructive interference. Therefore, the gain of the domain longitudinal mode is the maximum gain over the gain of all other modes inside the Fabry-Perot cavity. Thus, we may obtain a single longitudinal mode in the Fabry-Perot cavity by the mode selectivity.

For a single longitudinal mode operation in a digital modulated laser, numerical simulation of multi-mode rate equations show that the excess of the gain of the domain longitudinal mode over the gain of all other modes must be $3 \sim 5 \ cm^{-1}$.

Here, we use the "gain margin" Δg_c to express the excess of the gain of the domain mode over the gain of all of the other modes.

The gain margin Δg_c is defined as the excess of the domain mode gain g over the threshold gain g_{thred} which the nearest side-mode can reach. Namely,

$$\Delta g_c = g - g_{thred} \tag{5.333}$$

The formula of Δg_c is [16]

$$\Delta g_c = MSR \frac{n_{sp}}{2} \, h\nu \, v_g \, \alpha_m(\alpha_i + \alpha_m) \frac{1}{P_{off}} \tag{5.334}$$

Where

† n_{sp} is the spontaneous emission factor.
† v_g is the group velocity of the domain mode.
† P_{off} is the power at "zero" bit (or at "off" state).

Now, we will take two examples to express that how to obtain the single longitudinal mode operation in DFB semiconductor laser.

[Example 1] Considering that $n_{sp} = 3$; $h\nu = 0.8eV$; $v_g = c/n_{eff} = 3 \times 10^8/4$; $\alpha_m = \alpha_i = 30cm^{-1}$; $P_{off} = 0.025mW$, when $MSR = 100$, $\Delta g_c = 10cm^{-1}$.

[Example 2] When a DFB semiconductor laser works in a continuous operation state: $3 \sim 5cm^{-1}$ of gain margin can reach $MSR > 30dB$. [85] When it works in a digital directed modulated operation state: we need more gain margin (for instance, $> 10cm^{-1}$) to reach $MSR > 30dB$. [85]

5.7.4   The performance of the single longitudinal mode laser

Now, we will discuss some basic knowledge for the design of the single longitudinal mode laser.

For a passive Bragg grating in Figure 5-41

----

[16]   Prof. D. J. Blumenthal, ECE228B, Lecture 5.

We start from the coupling equations for the passive Bragg grating in Figure 5-41, from (5.312) and (5.313),

$$\frac{dA_s^{(-)}}{dz} = -\kappa\, A_s^{(+)}\, e^{-i2\Delta\beta z}$$

$$\frac{dA_s^{(+)}}{dz} = \kappa^*\, A_s^{(-)}\, e^{+i2\Delta\beta z} \tag{5.335}$$

Set the reflective wave $A_s^{(-)}$ and the incident wave $A_s^{(+)}$, respectively, as follows,

$$A_s^{(-)}(z) = S(z)\, e^{-i\Delta\beta z} \tag{5.336}$$

$$A_s^{(+)}(z) = R(z)\, e^{+i\Delta\beta z} \tag{5.337}$$

Take a derivative of (5.336) and (5.337) with respected to z, respectively, we have
17

$$R'(z) + i\Delta\beta R(z) = \kappa^* S(z) \tag{5.342}$$

$$S'(z) - i\Delta\beta S(z) = \kappa R(z) \tag{5.343}$$

Where, from (5.284),

$$\kappa = \frac{i\omega\varepsilon_0 a_l}{4} \int_{-\infty}^{\infty} \Delta n^2(x)[\varepsilon_y^{(s)}(x)]^2 dx \tag{5.344}$$

Now, if the κ in (5.344) is replaced by a new κ_{new} as following

$$\kappa_{new} = -i\kappa = \frac{\omega\varepsilon_0 a_l}{4} \int_{-\infty}^{\infty} \Delta n^2(x)[\varepsilon_y^{(s)}(x)]^2 dx \tag{5.345}$$

17

- Proof of (5.342) and (5.343): First, from (5.336),
$$A_s^{\prime(-)}(z) = S'(z)e^{-i\Delta\beta z} - i\Delta\beta S(z)e^{-i\Delta\beta z} = \kappa R(z)e^{i\Delta\beta z}e^{-i2\Delta\beta z} \tag{5.338}$$

or
$$S'(z) - i\Delta\beta S(z) = \kappa R(z) \tag{5.339}$$

Here, from (5.335) and (5.337), $A_s^{\prime(-)}(z) = \kappa A_s^{(+)}(z)e^{-i2\Delta\beta z}$ and $A_s^{(+)}(z) = R(z)e^{i\Delta\beta z}$ have been used in driving (5.339).

Now, from (5.337),
$$A_s^{\prime(+)}(z) = R'(z)e^{i\Delta\beta z} + i\Delta\beta R(z)e^{i\Delta\beta z} = \kappa^* S(z)e^{-i\Delta\beta z}e^{i2\Delta\beta z} \tag{5.340}$$

or
$$R'(z) + i\Delta\beta R(z) = \kappa^* S(z) \tag{5.341}$$

Here, from (5.335) and (5.336), $A_s^{\prime(+)}(z) = \kappa^* A_s^{(-)}(z)e^{i2\Delta\beta z}$ and $A_s^{(-)}(z) = S(z)e^{-i\Delta\beta z}$ have been used in driving (5.341).

the equations (5.342) and (5.343) may be rewritten as a new coupling equations as follows,

$$R'(z) + i\Delta\beta R(z) = -i\kappa_{new}S(z) \qquad (5.346)$$

$$S'(z) - i\Delta\beta S(z) = i\kappa_{new}R(z) \qquad (5.347)$$

Where, κ is a complex number; κ_{new} is a real number. The solutions of the new coupling equations (5.346) and (5.347) are, (see Appendix A),

$$R(z) = \left[\cosh\gamma z - \frac{i\Delta\beta}{\gamma}\sinh\gamma z\right]R(0) - \frac{i\kappa_{new}}{\gamma}\sinh\gamma z\, S(0) \qquad (5.348)$$

$$S(z) = \left[\cosh\gamma z - \frac{i\Delta\beta}{\gamma}\sinh\gamma z\right]S(0) - \frac{i\kappa_{new}}{\gamma}\sinh\gamma z\, R(0) \qquad (5.349)$$

How to find the reflectivity at $z = 0$ of a passive period waveguide in Figure 5-41?

For the sake of the simplification, from now on, we'd better to use κ to replace κ_{new}. Thus, (5.348) and (5.349) may be rewritten as

$$R(z) = \left[\cosh\gamma z - \frac{i\Delta\beta}{\gamma}\sinh\gamma z\right]R(0) - \frac{i\kappa}{\gamma}\sinh\gamma z\, S(0) \qquad (5.350)$$

$$S(z) = \left[\cosh\gamma z - \frac{i\Delta\beta}{\gamma}\sinh\gamma z\right]S(0) - \frac{i\kappa}{\gamma}\sinh\gamma z\, R(0) \qquad (5.351)$$

Now, we try to find the reflectivity at $z = 0$ defined as

$$r_{per} = \frac{S(0)}{R(0)} \qquad (5.352)$$

First, (5.350) and (5.351) at $z = L$ are

$$R(L) = \left[\cosh\gamma L - \frac{i\Delta\beta}{\gamma}\sinh\gamma L\right]R(0) - \frac{i\kappa}{\gamma}\sinh\gamma L\, S(0) \qquad (5.353)$$

$$S(L) = \left[\frac{i\kappa}{\gamma}\sinh\gamma L\right]R(0) + \left[\cosh\gamma L + \frac{i\Delta\beta}{\gamma}\sinh\gamma L\right]S(0) \qquad (5.354)$$

Which can be written as a matrix form as

$$\begin{bmatrix} R(L) \\ S(L) \end{bmatrix} = F_{per}(L)\begin{bmatrix} R(0) \\ S(0) \end{bmatrix} \qquad (5.355)$$

Where the transmission matrix $F_{per}(L)$ is

$$F_{per}(L) = \begin{bmatrix} \cosh\gamma L - \dfrac{i\Delta\beta}{\gamma}\sinh\gamma L, & -\dfrac{i\kappa}{\gamma}\sinh\gamma L \\[2ex] \dfrac{i\kappa}{\gamma}\sinh\gamma L, & \cosh\gamma L + \dfrac{i\Delta\beta}{\gamma}\sinh\gamma L \end{bmatrix} \qquad (5.356)$$

In which,

$$\gamma^2 = \kappa^2 - \Delta\beta^2 \tag{5.357}$$

And $\kappa = \kappa_{new}$ in (5.345) as we mentioned before.

When $S(L) = 0$, (that means that no any incident from the right side of the period waveguide), we may defined the reflectivity at $z = 0$ as

$$r_{per} = \frac{S(0)}{R(0)} = \frac{-\dfrac{i\kappa}{\gamma}\sinh\gamma L}{\left[\cosh\gamma L + \dfrac{i\Delta\beta}{\gamma}\sinh\gamma L\right]} \approx \frac{-i\kappa L}{[1 + i\Delta\beta L]} \tag{5.358}$$

In which, two conditions have been considered.

† $\gamma L \to 0$, and then $\sinh\gamma \to \gamma L$.
† $\kappa \to \Delta\beta$.

Thus, the power reflectivity R_{per} is, from (5.358),

$$R_{per} = |r_{per}|^2 = \frac{\left(-\dfrac{i\kappa}{\gamma}\right)\left(-\dfrac{i\kappa}{\gamma}\right)\sinh^2\gamma L}{\cosh^2\gamma L + \left(\dfrac{\Delta\beta}{\gamma}\right)^2\sinh^2\gamma L}, \quad For\ \kappa \approx \Delta\beta,\ \gamma L \to 0. \tag{5.359}$$

For $\kappa < \Delta\beta$ and $\gamma L \to 0$, (and then $cosh\gamma L \to 1$, $\sinh\gamma \to \gamma L$), we have

$$R_{per} = |r_{per}|^2 \approx \frac{\left(\dfrac{\kappa}{\gamma}\right)^2\sinh^2\gamma L}{\left[1 + \left(\dfrac{\Delta\beta}{\gamma}\right)^2(\gamma L)^2\right]} = \left(\dfrac{\kappa}{\gamma}\right)^2\frac{\sinh^2\gamma L}{1 + (\Delta\beta L)^2} \tag{5.360}$$

Or [18]

$$R_{per} = |r_{per}|^2 \approx \frac{(\kappa L)^2\sin^2\sqrt{(\Delta\beta L)^2 - (\kappa L)^2}}{(\Delta\beta L)^2 - (\kappa L)^2}, \quad For\ \kappa < \Delta\beta, \gamma L \to 0. \tag{5.363}$$

---

[18]   Proof of (5.363)

$$
\begin{aligned}
R_{per} &\approx \left(\frac{\kappa}{\gamma}\right)^2\frac{\sinh^2\gamma L}{1 + (\Delta\beta L)^2} \\[2mm]
&\approx \left(\frac{\kappa}{\gamma}\right)^2\sinh^2[\sqrt{(\kappa L)^2 - (\Delta\beta L)^2}], \quad Since\ \ 1 >> (\Delta\beta L)^2 \\[2mm]
&= \frac{\kappa^2\{\sinh[i\sqrt{(\Delta\beta L)^2 - (\kappa L)^2}]\}^2}{\kappa^2 - \Delta\beta^2}, \quad Since\ \ \gamma^2 = \kappa^2 - \Delta\beta^2 \\[2mm]
&= \frac{(\kappa L)^2\{i\sin\sqrt{(\Delta\beta L)^2 - (\kappa L)^2}\}^2}{(\kappa L)^2 - (\Delta\beta L)^2}, \quad Since\ \ \sinh i\theta = i\sin\theta \quad (5.361)
\end{aligned}
$$

Thus

$$R_{per} = |r_{per}|^2 \approx \frac{(\kappa L)^2\sin^2\sqrt{(\Delta\beta L)^2 - (\kappa L)^2}}{(\Delta\beta L)^2 - (\kappa L)^2} \tag{5.362}$$

For $\kappa > \Delta\beta$ and $\gamma L \to 0$, (and then $\cosh\gamma L \to 1$, $\sinh\gamma L \to \gamma L$), we have

$$R_{per} = |r_{per}|^2 = \frac{\kappa^2 \sinh^2 \gamma L}{\gamma^2 \cosh^2 \gamma L + (\Delta\beta)^2 \sinh^2 \gamma L} \tag{5.364}$$

In which, the denominator is

$$\begin{aligned}
\gamma^2 \cosh^2 \gamma L + (\Delta\beta)^2 \sinh^2 \gamma L &= (\kappa^2 - \Delta\beta^2) \cosh^2 \gamma L + \Delta\beta^2 \sinh \gamma L \\
&= \kappa^2 \cosh^2 \gamma L - \Delta\beta^2 (\cosh^2 \gamma L - \sinh^2 \gamma L) \\
&= \kappa^2 \cosh^2 \gamma L - \Delta\beta^2 \\
&\approx \kappa^2 \cosh^2 \gamma L, \quad Since \; \kappa > \Delta\beta.
\end{aligned} \tag{5.365}$$

Thus

$$R_{per} = |r_{per}|^2 \approx \tanh^2 \gamma L = \tanh^2 \sqrt{(\kappa L)^2 - (\Delta\beta L)^2}, \quad For \; \kappa > \Delta\beta, \gamma L \to 0 \tag{5.366}$$

In summery, from (5.363) and (5.366),

$$R_{per} = |r_{per}|^2 = \begin{cases} \dfrac{(\kappa L)^2 \sin^2 \sqrt{(\Delta\beta L)^2 - (\kappa L)^2}}{(\Delta\beta L)^2 - (\kappa L)^2}, & For \; \Delta\beta > \kappa, \gamma L \to 0. \\[4mm] \tanh^2 \sqrt{(\kappa L)^2 - (\Delta\beta L)^2}, & For \; \Delta\beta < \kappa, \gamma L \to 0. \end{cases} \tag{5.367}$$

Where, from (5.358),

$$r_{per} = \frac{S(0)}{R(0)} = \frac{-\dfrac{i\kappa}{\gamma} \sinh \gamma L}{\cosh \gamma L + \dfrac{i\Delta\beta}{\gamma} \sinh \gamma L} \approx \frac{-i\kappa L}{[1 + i\Delta\beta L]}, \quad For \; \kappa \to \Delta\beta, \gamma L \to 0. \tag{5.368}$$

Summery

Note, the results (5.367) and (5.368) are obtained under the conditions of $\kappa \approx \Delta\beta$, $\gamma L \to 0$. In this case,

† The reflectivity $|r_{per}|$ is increasing with the increasing of κL. Which means that the larger the coupling coefficient κL the higher the reflectivity $|r_{per}|$.

† The reflectivity $|r_{per}|$ is decreasing with the increasing of $\Delta\beta$. Which means that the larger the wavelength deviation from the central wavelength of the Bragg grating the smaller the reflectivity.

† When $\Delta\beta = 0$, the reflectivity $|r_{per}|$ suffers from a phase shift "$\pi/2$".

5.7.5 Distributed Bragg Reflector (DBR) semiconductor laser

In DBR principle Figure 5-46(a)

Now we will take an example to explain the principle of the DBR semiconductor laser. Where, The DBR principle is shown in Figure 5-46(a) and Figure 5-46(b) is the real laser structure - the Butt-Joint DBR semiconductor laser.

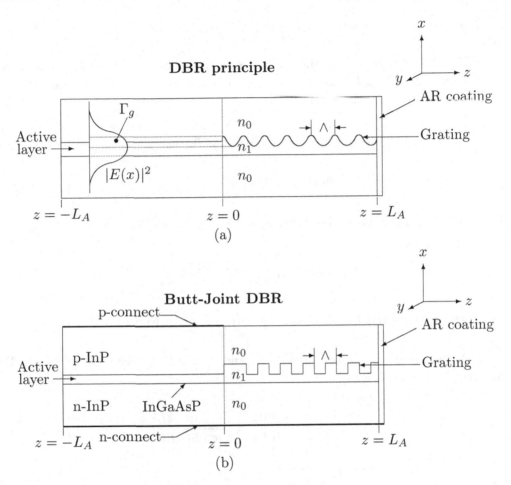

Figure 5-46   DBR laser structure

In DBR principle Figure 5-46(a),

† The end of the left-side is a mirror with the reflectivity of R_1 at $z = -L_A$.

† The end of the right-side is a Bragg Grating, which is a wavelength dependent mirror with reflectivity of R_{per} at $z = L_A$.

† When the light wave travels one turn ($2L_A$) inside the laser, the light wave earns a gain

$$Gain = \exp(2g_{net}L_A) \tag{5.369}$$

and suffer two reflections R_1 and R_{per}.

† And then a new amplitude balance is set up according to

$$R_1 R_{per} \exp(2g_{net}L_A) = 1 \qquad (5.370)$$

Where

† g_{net} is the net gain in the unit length inside the laser caused by the active layer.

† "1" on the right side of the quality is the incident amplitude of the light wave.

† $R_1 R_{per} \exp(2g_{net}L_A)$ on the left side is the amplitude of the light wave when it travels one turn inside the laser.

Now, taking the ln operation of (5.370), we have

$$\ln R_1 + \ln R_{net} + 2g_{net}L_A = 0 \qquad (5.371)$$

From which, for the normal operation of the laser, it is required that [19]

$$g_{net} \geq \frac{1}{2L_A}\left[\ln\frac{1}{R_1} + \ln\frac{1}{R_{per}}\right] \qquad (5.372)$$

In the Distributed Bragg Reflector (DBR) laser in Figure 5-46(b),

† At he Bragg frequency, $\Delta\beta = 0$, and $\gamma = \kappa$, the reflectivity of the Bragg reflector is, from (5.368),

$$|r_{per}| = \left|\frac{S(0)}{R(0)}\right| = \left|\frac{-i\sinh(\kappa L)}{\cosh(\kappa L)}\right| \qquad (5.373)$$

$$R_{per} = |r_{per}|^2 = tanh^2(\kappa L) \qquad (5.374)$$

† Figure 5-47 shown the calculation results from (5.373) and (5.374). Which tell us that $|r_{per}|$ is increasing with the increasing of κL.

† Where, we will prove that $\kappa \sim \dfrac{n_2^2 - n_1^2}{n_2}$. Which means that the larger the Δn the larger the κ and then the higher the reflectivity $|r_{per}|$.

---

[19] $\ln\dfrac{1}{R_1} = -\ln R_1$.

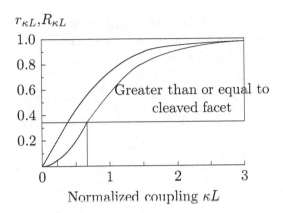

Figure 5-47   Reflectivity r_{per} vs coupling coefficient κL [86]

Near the Bragg frequency ($\Delta\beta \approx 0$), from (5.368) and (5.367),

$$r_{per} = \frac{S(0)}{R(0)} = \frac{-\dfrac{i\kappa}{\gamma}\sinh\gamma L}{\cosh\gamma L + \dfrac{i\Delta\beta}{\gamma}\sinh\gamma L} \approx \frac{-i\kappa L}{[1 + i\Delta\beta L]}, \quad For\ \kappa \to \Delta\beta, \gamma L \to 0.$$

(5.375)

$$R_{per} = |r_{per}|^2 = \begin{cases} (\kappa L)^2 \dfrac{\sin^2 \sqrt{(\Delta\beta L)^2 - (\kappa L)^2}}{(\Delta\beta L)^2 - (\kappa L)^2}, & For\ \Delta\beta > \kappa, \gamma L \to 0. \\[4mm] \tanh^2 \sqrt{(\kappa L)^2 - (\Delta\beta L)^2}, & For\ \Delta\beta < \kappa, \gamma L \to 0. \end{cases}$$

(5.376)

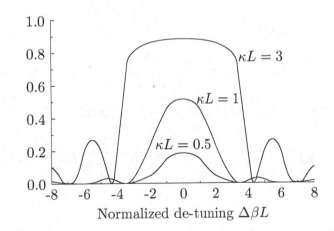

Figure 5-48   Reflectivity r_{per} vs normalized de-tuning $\Delta\beta L$ [87]

Figure 5-48 shows the calculation results from (5.375) and (5.376). In which,

† $|r_{per}|$ is increasing with the increasing of κL.

† $|r_{per}|$ is decreasing with the increasing of $\Delta\beta$.

† The zero point of $|r_{per}|$ is at, from (5.376),

$$\gamma L = N\pi \tag{5.377}$$

Considering that $\gamma = \sqrt{\kappa^2 - \Delta\beta^2}$, we have, from (5.377), the zero point of r_{per} is at

$$\Delta\beta L = \sqrt{(\kappa L)^2 + (N\pi)^2}, \quad N = 1, 2, 3, \cdots \tag{5.378}$$

as shown in Figure 5-48.

The reflection bandwidth of the Bragg mirror $\Delta\lambda_r$

From Figure 5-48, the peak point of reflectivity r_{per} in the Bragg mirror is at the central frequency ω_0 (corresponding to $\Delta\beta = 0$) of the Bragg mirror.

Now, the bandwidth of the reflectivity r_{per} corresponds to the point, that r_{per} drops down to 0.707 of the peak point (or drops down 3dB). This point, from (5.375), can be obtained with

$$\cosh^2 \gamma L + \left(\frac{\Delta\beta}{\gamma}\right)^2 \sinh^2 \gamma L = 2 \tag{5.379}$$

since at the central frequency ω_0, $\Delta\beta = 0$ and $\gamma L \to 0$, we have $\cosh^2 \gamma L + \left(\frac{\Delta\beta}{\gamma}\right)^2 \sinh^2 \gamma L = 1$ at the peak point.

From (5.379), we may obtain the reflection bandwidth of Bragg mirror $\Delta\lambda_r$ as follows,

$$\Delta\lambda_r = \frac{\lambda_B^2 \kappa}{\pi n_{g,eff}} \tag{5.380}$$

Where, from (5.303)

$$\kappa = \frac{-\omega\varepsilon_0}{4\pi l} \int_{-\infty}^{\infty} \Delta n^2(x)[\varepsilon_y^{(s)}(x)]^2 dx \tag{5.381}$$

This integral has been obtained in [34] and then

$$\kappa \approx -\frac{2\pi s^2}{3L\lambda} \frac{(n_2^2 - n_1^2)}{n_2} \left(\frac{a}{l}\right)^3 \left[1 + \frac{3}{2\pi} \frac{\lambda/a}{(n_2^2 - n_1^2)^{1/2}} + \frac{3}{4\pi^2} \frac{(\lambda/a)^2}{(n_2^2 - n_1^2)^2}\right] \tag{5.382}$$

The design consideration of a single longitudinal mode DBR semiconductor laser

Now, it is the time to discuss the design consideration of a single longitudinal mode DBR semiconductor laser as shown in Figure 5-49.

The idea of the design is

† Using the active layer as the cavity.
† Using two Bragg mirrors as the ends of the active layer cavity.

In this case,

† The active layer cavity will lunch a lot of modes as shown in Figure 5-45, all of the modes are in constructive interference inside the active layer cavity if two ends are normal mirrors.
† Now, the two ends are Bragg mirrors, which are the selective reflectors as shown in Figure 5-48. Thus, only a single longitudinal mode is in constructive interference inside the active layer cavity. Where, the central frequency of the single longitudinal mode is just the central frequency of the Bragg mirrors.

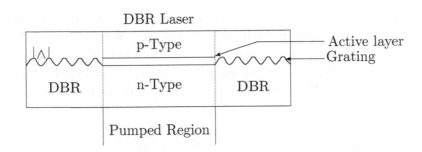

Figure 5-49   Single longitudinal mode DBR semiconductor laser

As shown in Figure 5-49,

† The active layer is InGaAsP layer with the p-InGaAsP layer and the n-InGaAsP layer to make a sandwich structure.
† Two ends of the active layer are the Bragg mirrors - the selective reflectors.
† For an active layer cavity without the Bragg mirrors on two ends, the active layer cavity will lunch Fabry-Perot modes as shown in Figure 5-45. In which, the frequency interval $\Delta\nu$ between two adjacent Fabry-Perot modes is, from [88], (see Appendix C),

$$\Delta\nu = \nu_{m+1} - \nu_m = \frac{c}{2n_{eff}L} \tag{5.383}$$

Where

$$\nu_{m+1} = \frac{1}{\lambda_{m+1}}, \qquad \nu_m = \frac{1}{\lambda_m} \tag{5.384}$$

Substituting (5.384) into (5.383) we have

$$\frac{1}{\lambda_{m+1}} - \frac{1}{\lambda_m} = \frac{c}{2n_{eff}L} \tag{5.385}$$

or

$$\frac{\lambda_m - \lambda_{m+1}}{\lambda_{m+1}\lambda_m} = \frac{1}{2n_{eff}L} \tag{5.386}$$

Where λ_{m+1} and λ_m are the wavelengths of two adjacent modes. Thus, we may consider that $\lambda_{m+1} \approx \lambda_m \approx \lambda$. Therefore, from (5.386), we have the wavelength interval $\Delta\lambda_m$ between two adjacent modes

$$\Delta\lambda_m \approx \frac{\lambda^2}{2n_{eff}L} \tag{5.387}$$

† Now, in Figure 5-49, the two ends of the active layer are connected to two Bragg mirrors. The bandwidth of the Bragg mirror is $\Delta\lambda_r$ in (5.380). In this case, if the the bandwidth of the Bragg mirror $\Delta\lambda_r$ is less than the interval between two adjacent Fabry-Perot modes, only single longitudinal mode exits in the active cavity.

† *Thus, the single longitudinal mode operation condition is*

$$\Delta\lambda_r < \Delta\lambda_m \tag{5.388}$$

or

$$\frac{\lambda_B^2 \kappa}{\pi n_{g,eff}} < \frac{\lambda^2}{2n_{eff}L} \tag{5.389}$$

or

$$\kappa L < \frac{\pi}{2} \tag{5.390}$$

Where, $\lambda_B \approx \lambda$ has been considered in driven (5.390). Meanwhile $n_{g,eff}$ is the same as n_{eff} since both are the active layer cavity.

† *And then, (5.390) is the single longitudinal mode operation condition in a DBR semiconductor laser. Where L is the length of the active layer cavity and κ can be obtained from (5.382).*

5.7.6 Distributed Feedback (DFB) semiconductor laser

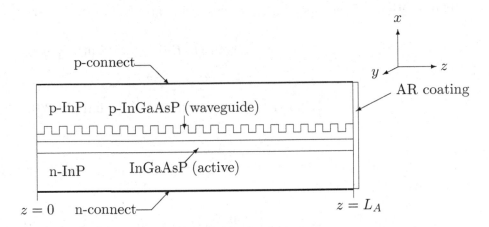

Figure 5-50 DBR laser structure

The distributed feedback (DFB) semiconductor laser is shown in Figure 5-50. In which, the Bragg grating $p - InGaAsP$ is made on the active layer, and then the period waveguide become a active period waveguide. The ends of the active period waveguide are both the AR-coatings - the anti-reflected coatings.

The coupling equations in an active period waveguide

In subsection 5.7.2, we have discussed the coupling equations of the active period waveguide, in which, we have mentioned that from the coupling equations of the passive period waveguide to the coupling equations of the active period waveguide, we only need to do is to replace $\Delta\beta$ by $\Delta\beta + ig_0$. Namely

$$\Delta\beta \to \Delta\beta + ig_0 \tag{5.391}$$

Where g_0 is the gain in the active layer cavity.

Now, the coupling equations in a passive period waveguide are, from (5.279) and (5.281),

$$R'(z) + i\Delta\beta R(z) = -i\kappa S(z) \tag{5.392}$$
$$S'(z) - i\Delta\beta S(z) = i\kappa R(z) \tag{5.393}$$

Replacing $\Delta\beta$ by $\Delta\beta + ig_0$, we may obtain the coupling equations of the active period waveguide as follows,

$$R'(z) + i(\Delta\beta + ig_0)R(z) = -i\kappa S(z) \tag{5.394}$$
$$S'(z) - i(\Delta\beta + ig_0)S(z) = i\kappa R(z) \tag{5.395}$$

The solutions in the active period waveguide

The solutions of the passive period waveguide are, from (5.350) and (5.351)

$$R(z) = \left[\cosh\gamma z - \frac{i\Delta\beta}{\gamma}\sinh\gamma L\right]R(0) - \left[\frac{i\kappa}{\gamma}\sinh\gamma L\right]S(0)$$

$$(5.396)$$

$$S(z) = \left[\frac{i\kappa}{\gamma}\sinh\gamma L\right]R(0) + \left[\cosh\gamma z + \frac{i\Delta\beta}{\gamma}\sinh\gamma L\right]R(0)$$

$$(5.397)$$

Replacing $\Delta\beta$ by $\Delta\beta + ig_0$, we may obtain the coupling equations of the active period waveguide as follows,

$$R(z) = \left[\cosh\gamma z - \frac{i(\Delta\beta + ig_0)}{\gamma}\sinh\gamma L\right]R(0) - \left[\frac{i\kappa}{\gamma}\sinh\gamma L\right]S(0)$$

$$(5.398)$$

$$S(z) = \left[\frac{i\kappa}{\gamma}\sinh\gamma L\right]R(0) + \left[\cosh\gamma z + \frac{i(\Delta\beta + ig_0)}{\gamma}\sinh\gamma L\right]R(0)$$

$$(5.399)$$

Where two boundary conditions

$$R(z)|_{z=0} = R(0) \tag{5.400}$$
$$S(z)|_{z=0} = S(0) \tag{5.401}$$

have been used in driven (5.398) and (5.399). And

$$\gamma^2 = \kappa^2 - \Delta\beta^2 \tag{5.402}$$

At $z = L$ point, we have, from (5.398) and (5.399),

$$R(L) = \left[\cosh\gamma L - \frac{i(\Delta\beta + ig_0)}{\gamma}\sinh\gamma L\right]R(0) - \left[\frac{i\kappa}{\gamma}\sinh\gamma L\right]S(0)$$

$$(5.403)$$

$$S(L) = \left[\frac{i\kappa}{\gamma}\sinh\gamma L\right]R(0) + \left[\cosh\gamma L + \frac{i(\Delta\beta + ig_0)}{\gamma}\sinh\gamma L\right]S(0)$$

$$(5.404)$$

Which can be written as the matrix form as

$$\begin{bmatrix} R(L) \\ S(L) \end{bmatrix} = F_{per}(L)\begin{bmatrix} R(0) \\ S(0) \end{bmatrix} \tag{5.405}$$

Resonant condition in DFB semiconductor laser

To find the resonant condition, we may considered that only one incident $R(0)$ from $z = 0$ point exits. In this case, the reflected wave $S(L)$ at $z = L$ is, from (5.404)

$$S(L) = \frac{\dfrac{i\kappa}{\gamma} \sinh \gamma L}{\cosh \gamma L + \dfrac{i(\Delta\beta + ig_0)}{\gamma} \sinh \gamma L} R(0) \qquad (5.406)$$

Now, if

$$\cosh \gamma L + \frac{i(\Delta\beta + ig_0)}{\gamma} \sinh \gamma L = 0 \qquad (5.407)$$

the reflected wave at $z = L$, $S(L)$ will be infinity for any value of the incident $R(0)$ at $z = 0$. Thus, (5.407) is the resonant condition inside the DFB semiconductor laser obviously.

The resonant condition (5.407) may be rewritten as

$$\gamma L \coth \gamma L = -i(\Delta\beta + ig_0)L \qquad (5.408)$$

with

$$\gamma^2 = \kappa^2 - (\Delta\beta + ig_0)^2 \qquad (5.409)$$

Where

† κ can be a known value, from (5.382).
† $\Delta\beta$ is an unknown phase.
† g_0 is an unknown amplitude.

Equations (5.408) and (5.409) are the complex coupling equations, which determinate both of the amplitude balance and the phase balance inside the laser.

† The phase value $\Delta\beta$ and the amplitude value g_0 are determined by the coupling coefficient κ and the coupling length L.
† From the resonant condition (5.408), we may obtain a set of solutions. Each of them is consist of $\Delta\beta L$ and $g_0 L$.

— The central frequency is determined by the $\Delta\beta L$.
— The amplitude of the oscillation is determined by the $g_0 L$.

Now we need to deal with the problem that γ is a complex number since

$$\gamma = \sqrt{\kappa^2 - \left(\Delta\beta + ig_0\right)^2} \qquad (5.410)$$

or

$$\gamma = \sqrt{\gamma_0^2 - i2\Delta\beta g_0} \qquad (5.411)$$

Where

$$\gamma_0^2 = \kappa^2 - \Delta\beta^2 + g_0^2 \tag{5.412}$$

Set

$$\gamma_0^2 = R\cos\alpha, \qquad -2\Delta\beta g_0 = R\sin\alpha \tag{5.413}$$

we have

$$\gamma = \sqrt{R\left(\cos\alpha + i\sin\alpha\right)} = \sqrt{R\left(\cos\alpha + i\sin\alpha\right)} = \gamma_e + i\gamma_m \tag{5.414}$$

or

$$\gamma = Re[\gamma] + iIm[\gamma] \tag{5.415}$$

with

$$\gamma_e = Re[\gamma] = \sqrt{R}\cos\alpha/2 = \sqrt{R\frac{1+\cos\alpha}{2}} = \sqrt{\frac{R+\gamma_0^2}{2}} \tag{5.416}$$

$$\gamma_m = Im[\gamma] = \sqrt{R}\sin\alpha/2 = \sqrt{R\frac{1-\cos\alpha}{2}} = \sqrt{\frac{R-\gamma_0^2}{2}} \tag{5.417}$$

Thus, the resonant condition (5.408) becomes

$$(\gamma_e + i\gamma_m)L\coth(\gamma_e + i\gamma_m)L = g_0L + i\Delta\beta L \tag{5.418}$$

Now we will proof that [20]

$$\coth\left(\gamma_e + i\gamma_m\right)L = Re^{-i(\theta_1+\theta_2)} \tag{5.421}$$

---

[20]   Proof of (5.421):

$$
\begin{aligned}
\coth\left(\gamma_e L + i\gamma_m L\right) &= \frac{1 - e^{-2\gamma_e L}e^{-i2\gamma_m L}}{1 + e^{-2\gamma_e L}e^{-i\gamma_m L}} \\[2mm]
&= \frac{1 - e^{-2\gamma_e L}(\cos 2\gamma_m L - i\sin 2\gamma_m L)}{1 + e^{-2\gamma_e L}(\cos 2\gamma_m L - i\sin 2\gamma_m L)} \\[2mm]
&= \frac{(1 - e^{-2\gamma_e L}\cos 2\gamma_m L) + ie^{-2\gamma_e L}\sin 2\gamma_m L}{(1 + e^{-2\gamma_e L}\cos 2\gamma_m L) - ie^{-2\gamma_e L}\sin 2\gamma_m L} \\[2mm]
&= \frac{\sqrt{\left(1 - e^{-2\gamma_e L}\cos 2\gamma_m L\right)^2 + \left(e^{-2\gamma_e L}\sin 2\gamma_m L\right)^2}}{\sqrt{\left(1 + e^{-2\gamma_e L}\cos 2\gamma_m L\right)^2 + \left(e^{-2\gamma_e L}\sin 2\gamma_m L\right)^2}} \\[2mm]
&\times \frac{\exp i\left[-\tan^{-1}\left(\dfrac{e^{-2\gamma_e L}\sin 2\gamma_m L}{1 - e^{-2\gamma_e L}\cos 2\gamma_m L}\right)\right]}{\exp i\left[\tan^{-1}\left(\dfrac{e^{-2\gamma_e L}\sin 2\gamma_m L}{1 + e^{-2\gamma_e L}\cos 2\gamma_m L}\right)\right]}
\end{aligned} \tag{5.419}
$$

or

$$\coth\left(\gamma_e + i\gamma_m\right)L = Re^{-i(\theta_1+\theta_2)} \tag{5.420}$$

with

$$R = \frac{\sqrt{\left(1 - e^{-2\gamma_e L} \cos 2\gamma_m L\right)^2 + \left(e^{-2\gamma_e L} \sin 2\gamma_m L\right)^2}}{\sqrt{\left(1 + e^{-2\gamma_e L} \cos 2\gamma_m L\right)^2 + \left(e^{-2\gamma_e L} \sin 2\gamma_m L\right)^2}} \qquad (5.422)$$

$$\theta_1 = \tan^{-1}\left(\frac{e^{-2\gamma_e L} \sin 2\gamma_m L}{1 - e^{-2\gamma_e L} \cos 2\gamma_m L}\right) \qquad (5.423)$$

$$\theta_2 = \tan^{-1}\left(\frac{e^{-2\gamma_e L} \sin 2\gamma_m L}{1 + e^{-2\gamma_e L} \cos 2\gamma_m L}\right) \qquad (5.424)$$

Rewritten (5.421) as

$$\coth\left(\gamma_e L + i\gamma_m L\right) = R\cos(\theta_1 + \theta_2) - i\sin(\theta_1 + \theta_2) \qquad (5.425)$$

Substitution of (5.425) into the resonant condition (5.418), we have

$$\left(\gamma_e + i\gamma_m\right)L\left[R\cos(\theta_1 + \theta_2) - i\sin(\theta_1 + \theta_2)\right] = g_0 L - i\Delta\beta L$$

or the resonant condition is

$$\left[\gamma_e R\cos(\theta_1+\theta_2)+\gamma_m R\sin(\theta_1+\theta_2)\right]+i\left[\gamma_m R\cos(\theta_1+\theta_2)-\gamma_e R\sin(\theta_1+\theta_2)\right] = g_0 - i\Delta\beta$$
$$(5.426)$$

Thus, the resonant condition (5.426) contains

† The amplitude balance condition

$$g_0 = \gamma_e R\cos(\theta_1 + \theta_2) + \gamma_m \sin(\theta_1 + \theta_2) \qquad (5.427)$$

† The phase balance condition

$$\Delta\beta = \gamma_e R\sin(\theta_1 + \theta_2) - \gamma_m R\sin(\theta_1 + \theta_2) \qquad (5.428)$$

(5.427) and (5.428) can be used to design the DFB semiconductor laser. Where R, θ_1 and θ_2 have given in (5.422), (5.423) and (5.424), respectively.

5.7.7   Appendix

5.7.7.1   Appendix A

Now we will give the Proof of (5.348) and (5.349). To this end rewrite the coupling equation (5.346) and (5.347) as follows,

$$R'(z) + i\Delta\beta R(z) = -i\kappa_{new}S(z) \tag{5.429}$$

$$S'(z) - i\Delta\beta S(z) = i\kappa_{new}R(z) \tag{5.430}$$

† Take a derivative of (5.429) with respect to z, we have

$$R''(z) + i\Delta\beta R'(z) = -i\kappa_{new} S'(z) \tag{5.431}$$

Where, from (5.429) and (5.430), $R'(z) = -i\kappa_{new} S(z) - i\Delta\beta R(z)$ and $S'(z) = i\kappa_{new}R(z) + i\Delta\beta S(z)$. Thus, (5.431) may be rewritten as

$$R''(z) + (\Delta\beta^2 - \kappa_{new}^2)R(z) = 0 \tag{5.432}$$

Set

$$\gamma^2 = \kappa_{new}^2 - \Delta\beta^2 \tag{5.433}$$

The equation (5.432) may be rewritten as

$$R''(z) - \gamma^2 R(z) = 0 \tag{5.434}$$

And the solution of (5.434) is

$$R(z) = A\cosh\gamma z + B\sinh\gamma z \tag{5.435}$$

† Take a derivative of (5.430) with respect to z, we have

$$S''(z) - i\Delta\beta S'(z) = i\kappa_{new}R'(z) \tag{5.436}$$

Where, from (5.429) and (5.430), (5.436) may be rewritten as

$$S''(z) + (\Delta\beta^2 - \kappa_{new}^2)S(z) = 0 \tag{5.437}$$

Considering that $\gamma^2 = \kappa_{new}^2 - \Delta\beta^2$, the equation (5.437) may be rewritten as

$$S''(z) - \gamma^2 S(z) = 0 \tag{5.438}$$

And the solution of (5.434) is

$$S(z) = C\cosh\gamma z + D\sinh\gamma z \tag{5.439}$$

† The boundary conditions may be obtained from (5.435) and (5.439), i.e.

$$S(0) = C, \quad R(0) = A \tag{5.440}$$

Thus, from (5.439) and (5.435), we have

$$R(z) = R(0)\cosh\gamma z + B\sinh\gamma z \tag{5.441}$$
$$S(z) = S(0)\cosh\gamma z + D\sinh\gamma z \tag{5.442}$$

Substitution of (5.441) and (5.442) into (5.346), we have

$$\gamma[R(0)\sinh\gamma z + B\cosh\gamma z] + i\Delta\beta[R(0)\cosh\gamma z + B\sinh\gamma z]$$

$$= -i\kappa_{new}[S(0)\cosh\gamma z + D\sinh\gamma z] \tag{5.443}$$

The equation (5.443) at $z = 0$ give us

$$\gamma B + i\Delta\beta R(0) = -i\kappa_{new}S(0) \tag{5.444}$$

Thus, we have

$$B = -\frac{i\kappa_{new}}{\gamma}S(0) - \frac{i\Delta\beta}{\gamma}R(0) \tag{5.445}$$

Substitution of (5.441) and (5.442) into (5.347), we have

$$\gamma[S(0)\sinh\gamma z + D\cosh\gamma z] - i\Delta\beta[S(0)\cosh\gamma z + D\sinh\gamma z]$$

$$= i\kappa_{new}[R(0)\cosh\gamma z + B\sinh\gamma z] \tag{5.446}$$

From equation (5.443) at $z = 0$, we have

$$\gamma D - i\Delta\beta S(0) = i\kappa_{new}R(0) \tag{5.447}$$

From which we have

$$D = \frac{i\kappa_{new}}{\gamma}R(0) + \frac{i\Delta\beta}{\gamma}S(0) \tag{5.448}$$

Substitution of A, B, C, D into (5.441) and (5.442), we have

$$R(z) = R(0)\cosh\gamma z + \left[-\frac{i\kappa_{new}}{\gamma}S(0)\sinh\gamma z - \frac{i\Delta\beta}{\gamma}R(0)\sinh\gamma z\right] \tag{5.449}$$

$$S(z) = S(0)\cosh\gamma z + \left[\frac{i\kappa_{new}}{\gamma}R(0)\sinh\gamma z + \frac{i\Delta\beta}{\gamma}S(0)\sinh\gamma z\right] \tag{5.450}$$

or

$$R(z) = \left[\cosh\gamma z - \frac{i\Delta\beta}{\gamma}\sinh\gamma z\right]R(0) - \frac{i\kappa_{new}}{\gamma}\sinh\gamma z S(0) \tag{5.451}$$

$$S(z) = \left[\cosh\gamma z + \frac{i\Delta\beta}{\gamma}\sinh\gamma z\right]S(0) + \frac{i\kappa_{new}}{\gamma}\sinh\gamma z R(0) \tag{5.452}$$

Which are agreement with (5.348) and (5.349), respectively.

5.7.7.2   Appendix B

- Proof of (5.380):

We will start from the bandwidth of the reflectivity r_{per} corresponds to the point, that r_{per} is dropping down 0.707 of it-self (or dropping down 3dB). This point, from (5.376), can be obtained with

$$\cosh^2 \gamma L + \left(\frac{\Delta\beta}{\gamma}\right)^2 \sinh^2 \gamma L = 2 \tag{5.453}$$

From which, we have

$$\gamma^2 \cosh^2 \gamma L + (\Delta\beta)^2 \sinh^2 \gamma L = 2\gamma^2 \tag{5.454}$$

or

$$(\kappa^2 - \Delta\beta^2)\cosh^2 \gamma L + \Delta\beta^2 \sinh^2 \gamma L = 2(\kappa^2 - \Delta\beta^2) \tag{5.455}$$

Considering that $\cosh^2 x - \sinh^2 x = 1$, we have, from (5.455),

$$\kappa^2 \cosh^2 \gamma L - \Delta\beta^2 = 2\kappa^2 - 2\Delta\beta^2 \tag{5.456}$$

From which, we may obtain the point that r_{per} dropping down 3dB as

$$\Delta\beta^2 = \kappa^2 (2 - \cosh^2 \gamma L) \tag{5.457}$$

Considering that inside the reflection bandwidth, we may consider that $\cosh \gamma L \approx 1$ and $\Delta\beta^2 \approx \kappa^2$. we have

$$\Delta\beta \approx \pm\kappa \tag{5.458}$$

Where [21]

$$\Delta\beta = \beta - \beta_0 = \frac{2\pi n'_{eff}\lambda}{\lambda} - \frac{2\pi n'_{eff}\lambda_B}{\lambda_B} \approx \frac{2\pi n_{g,eff}}{\lambda_B^2}\Delta\lambda \tag{5.460}$$

Substituting (5.460) into (5.458), we have

$$2\pi n_{g,eff}\left(\frac{\lambda_B - \lambda}{\lambda\lambda_B}\right) \approx 2\pi n_{g,eff}\left(\frac{\Delta\lambda}{\lambda_B^2}\right) = \kappa \tag{5.461}$$

Thus, we have

$$\Delta\lambda = \frac{\lambda_B^2 \kappa}{2\pi n_{g.eff}} \tag{5.462}$$

And then the reflection bandwidth of Bragg mirror is

$$\Delta\lambda_r = 2\Delta\lambda = \frac{\lambda_B^2 \kappa}{\pi n_{g,eff}} \tag{5.463}$$

Which is the equation (5.380).

---

[21]

$$\frac{1}{\lambda} - \frac{1}{\lambda_B} = \frac{\lambda_B - \lambda}{\lambda\lambda_B} \approx \frac{\Delta\lambda}{\lambda_B^2} \tag{5.459}$$

5.7.7.3 Appendix C

- Proof of (5.383):

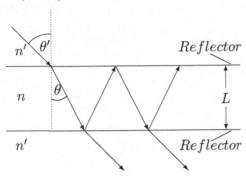

Figure 5-51   Fabry-Perot etalon

When a light wave is incident obliquely onto Fabry-Perot cavity as shown in Figure 5-51, the interval of two adjacent Fabry-Perot modes is

$$\Delta \nu = \nu_{m+1} - \nu_m = \frac{c}{2nL\cos\theta} \qquad (5.464)$$

Now, in active layer cavity, the light wave is perpendicularly incident into the active layer along L. Therefore $\theta = 0$ and $\cos\theta = 1$. $n = n_{eff}$, which is the effective refractive index of the active layer cavity. In this case,

$$\Delta \nu = \nu_{m+1} - \nu_m = \frac{c}{2n_{eff}L} \qquad (5.465)$$

Chapter 6

Optical Photo-detector

6.1 Introduction

6.2 PIN photo-detector diode

6.2.1 p-n photodiode

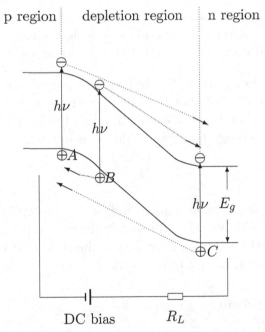

Figure 6-1   Electron-hole pairs created by absorbed photons
in p-n junction

p-n junction has been widely used as the photo-detector diode. In which, a bias is applied to the p-n junction as shown in Figure 6-1. Now,

- If a photon is absorbed in the depletion region near the p region "A", a pair of electron-hole is created, if this take places in a diffusion length [1] (means before recombining with a carrier of the opposite type), the free electron will drift along the dash line then cross the layer boundary to give a contribution of charge e to the external circuit under the influence of the field in the depletion region.

- If a photon is absorbed in the depletion region "B", a pair of electron-hole is created, if this take places in a diffusion length the free electron will drift along the dash line and reaches to the layer boundary and then cross the boundary to give a contribution of charge e to the external circuit. Meanwhile a "hole" will diffuse along the dash line and cross the layer boundary to give a contribution of *hole* to the external circuit under the influence of the field in the depletion region.

- If a photon is absorbed in the depletion region near the n region "C", a pair of electron-hole is created, if this take places in a diffusion length the "hole" will diffuse along the dash line then cross the layer boundary to give a contribution of *hole* to the external circuit under the influence of the field in the depletion region.

As well known, a p-n junction with a bias will result in a DC current in the external circuit. And now, the photons absorbed in the depletion region will give additional current in the external circuit. Which includes two processes:

- The photons absorption in the depletion region will create the electron-hole pairs or a charge e. The photon can give up its energy to excite the electron from the valance band to the conduction band only if it has energy equal or greater than the energy bandgap of the semiconductor material that is used. namely, only if

$$h\nu \geq E_g \qquad (6.1)$$

- The carriers (electrons and holes) diffuse in the depletion region under the influence of the field. Which will result in the delayed current response caused by the diffusion time. This diffuse time is the time that each carrier traverses a distance less than the full junction width.

6.2.2 PIN photodiode

Therefore, photodiode often use a p-i-n structure, in which, a lightly doped intrinsic high resistivity "I" layer is sandwiched between the p and n region as shown in Figure 6-2(a) and (b). The potential drop occurs mostly across this layer, which can be made long enough to insure that most of the incident photons are absorbed

---

[1] The diffuse length is defined as the distance in which an excess minority concentration is reduced to e^{-1} of its peak value, or in physical terms the average distance a minority carrier traverses before recombining with a carrier of the opposite type.

in this region. And then the carriers (electrons and holes) drift in a diffusion length to cross the boundary to give contribution of "e" or "hole" to the external circuit, respectively.

In depletion region, the photon can give up its energy to excite the electron from the valance band to the conduction band only if it has energy equal or greater than the energy bandgap of the semiconductor material that is used. Therefore, equation (6.1) is also hold for this depletion region "I".

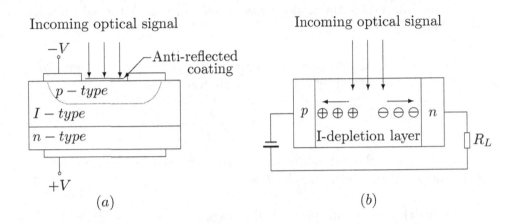

Figure 6-2   PIN photodiode

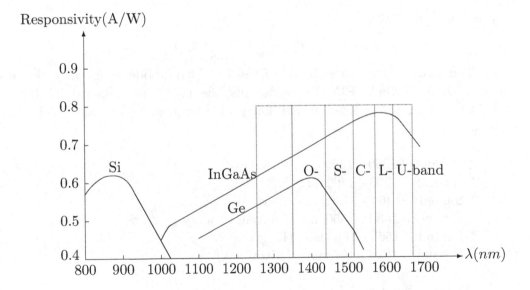

Figure 6-3   Responsivity of PIN photodiodes vs wavelength [90]

6.2.3   The frequency response of a PIN photodiode

As we mentioned before, the photon can be absorbed in the depletion region only if $h\nu \geq E_g$, where $h\nu$ is the energy of the incoming photon, E_g is the band-gap in the depletion region of the semiconductor. In this case, the absorbtion will be increased strongly with frequency. If $h\nu >> E_g$, the absorption in depletion region will take place entirely near the input face and the minority carriers generated by the absorption will be recombined with the majority carriers before diffusing to the external circuit. This is why the frequency response of the photodiode drops off when $h\nu > E_g$. The typical frequency response curve of the photodiode is illustrated in Figure 6-3. Where the responsivity R is a new parameter, which will be discussed as follows.

Definition: Quantum efficiency η is the number of carriers following in the external circuit caused by per incident photon. Namely,

$$\eta = \frac{I/e}{P_{in}/(h\nu)} \tag{6.2}$$

in which, I/e is the number of the electrons caused by the absorption, $P_{in}/(h\nu)$ is the number of the incoming photons.

Definition: Responsivity R is the output current I generated by the absorption in PIN photodiode caused by per unit of incident optical power P_{in}, namely,

$$R = \frac{I}{P_{in}}(A/W) \tag{6.3}$$

which is, from (6.2),

$$R = \frac{\eta e}{h\nu} \approx \frac{\eta \lambda}{1.24}(A/W) \tag{6.4}$$

The frequency responsivity R of the PIN photodiode is shown in Figure 6-3, in which the InGaAs PIN photodiode displays maximum responsivity R near the light wavelength $\lambda = 1550nm$ with fairly wide frequency response, which can cover following bands:

```
* O-band:   1260 - 1360nm   (Original)
* E-band:   1360 - 1460nm   (External)
* S-band:   1460 - 1530nm   (Short)
* C-band:   1530 - 1550nm   (Conventional)
* L-band:   1565 - 1605nm   (Long)
* U-band:   1625 - 1675nm   (Ultra-long)
```

Where C-band and L-band are the most important today. Which have been used in the long-haul optical fiber communications link over world wide.

For the commonly used PIN photodiode, the responsivity R are in the ranges

* $(0.6 \sim 0.85)$ A/W   for   InGaAs-based PIN photodiode.
* $(0.5 \sim 0.7)$ A/W   for   germanium-based PIN photodiode.
* $(0.4 \sim 0.6)$ A/W   for   silicon-based PIN photodiode.

For a specific photodiode, for instance a InGaAs-based photodiode, the frequency response of the PIN photodiode is limited by three factors:

* Finite diffusion time (τ_h for hole and τ_e for electron): which may be minimum by chosen the length of the depletion region properly.

* Junction capacitor of the photodiode (C_d): as shown in Figure 6-4, which will place an upper limit of

$$\omega_m \simeq \frac{1}{R_e C_d} \tag{6.5}$$

Where R_d is the AC resistor of the photodiode, R_s is the contact resistor, R_L is the load resistor. And R_e is the equivalent resistor in parallel with C_d.

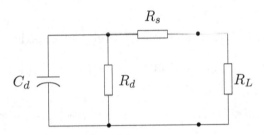

Figure 6-4   Equipment circuit of a PIN photodiode

* The finite transit time τ_{tr} of the carriers diffusing across the depletion layer. The three factors result in the cutoff frequency f_c as follows,

$$f_c = \frac{1}{2\pi(\tau_{tr} + \tau_{rc})} \tag{6.6}$$

Where τ_{tr} is caused by the finite transit time of the carriers diffusing across the depletion layer, τ_{rc} is caused by the junction capacitor of the photodiode C_d and the external load resistor R_L. And f_c also characterizes the bandwidth of the frequency response of the photodiode, therefore, it limits the response speed of the photodiode.

The cutoff frequency f_c is inversely proportional to the width of the I-layer and the capacitance of the reverse biased PIN structure. The width of the I-layer can be decreased in order to increase f_c, however, the responsivity of the photodiode will be decrease. Therefore, the responsivity R of the PIN photodiode used in high speed optical receiver is lower than the responsivity of the photodiode used in low speed optical receiver. Where, τ_{tr} is the transit time of the carrier in the I-layer and τ_{rc} is the RC constant of the circuit which consist of the photodiode reverse capacitance $"C_d"$ and the external load $"R_L"$ as shown in Figure 6-4.

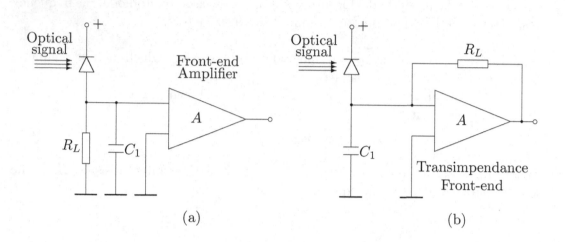

(a)   (b)

Figure 6-5   Optical receiver front-end schematic

To high the speed of the photodiode, both of τ_{tr} and τ_{rc} should be decreased, for instance, we need

$$(\tau_{tr} + \tau_{rc}) < 10ps \quad for \quad a\ system\ operating\ above\ 10Gb/s \tag{6.7}$$

$$(\tau_{tr} + \tau_{rc}) < 3ps \quad for \quad a\ system\ operating\ above\ 40Gb/s \tag{6.8}$$

* In this case, to high the speed of the photodiode, the load resistor R_L should be decreased.

* However, to minimize the thermal noise component, the load resistor R_L should be increased. Which will be given by (6.18).

Thus, the choice of the load resistor R_L is involved some trad-off mentioned above.

As shown in Figure 6-5(b), the transimpedance front-end amplifier design has been used to decrease the equivalent input load resistor $(R_L)_{in}$ by $(1+A)$, namely,

$$(R_L)_{in} = \frac{R_L}{1+A} \tag{6.9}$$

so as the cutoff frequency or the bandwidth of the frequency response of the photodiode f_c will be increased by $(1+A)$. However, from (6.18), the thermal noise component $\overline{i_{N2}^2}$ will be increased by $(1+A)$, since the thermal noise $\overline{i_{N2}^2}$ is now,

$$\overline{i_{N2}^2} = \frac{4kT_e\Delta\nu}{(R_L)_{in}} \tag{6.10}$$

6.2.4   Detection sensitivity of photodiode

The total noise generated during the detection process includes
* The shot noise: which is a noise associated with the random generation of carriers.
* The thermal noise: which is the white noise associated with a resistor.
* The dark current: which exits even though the incident optical power $P_{in} = 0$.
Now we will discuss the detection sensitivity of the photodiode based on the total noise above. To this end, the noise equivalent of a photodiode is shown in Figure 6-6.

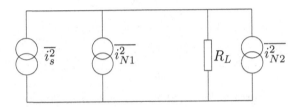

Figure 6-6   Noise equivalent circuit of a photodiode.

Where
* $\overline{i_s^2}$ is the signal current of the photodiode.

* $\overline{i_{N1}^2}$ is the shot noise of the photodiode.

* $\overline{i_{N2}^2}$ is the thermal noise from the load resister R_L.

Now the modulated signal current is given by [2]

$$i_s(t) = \frac{P_{in}e\eta}{h\nu}\left(1 + \frac{m^2}{2}\right) + \frac{P_{in}e\eta}{h\nu}2me^{i\omega_m t} \tag{6.12}$$

and ω_m is the modulation frequency, m is the modulation factor. In practices, the modulation frequency ω_m is low enough that

$$\omega_m << \frac{1}{C_d R_e} \tag{6.13}$$

---

[2]   Considering that, from (6.2), we have

$$I = \frac{P_{in}e\eta}{h\nu} \tag{6.11}$$

The modulation current associated with the modulation frequency ω_m can be expressed as (6.12). In which, the first term is the DC component and the second term is the modulation signal.

In this case, the shunt of the signal by the capacitor C_d can be neglected. Where R_e is the equivalent resistance in parallel with the capacitor C_d. Considering that the signal power is proportional to the mean-square value of the sinusoidal current i_s, we have, for $m = 1$,

$$\overline{i_s^2} = 2\left(\frac{P_{in}e\eta}{h\nu}\right)^2 \tag{6.14}$$

The first noise is the shot noise associated with the random generation of carriers, which is given by

$$\overline{i_{N1}^2} = 2e\overline{I}\Delta\nu \tag{6.15}$$

Where I is the average current of the photodiode, which is, from (6.12),

$$\overline{I} = \frac{P_{in}e\eta}{h\nu}\left(1 + \frac{m^2}{2}\right) \tag{6.16}$$

For $m = 1$, we have from (6.15) and (6.16),

$$\overline{i_{N1}^2} = \frac{3e^2(P_{in} + P_B)\eta}{h\nu}\Delta\nu + 2e\overline{i_d}\Delta\nu \tag{6.17}$$

Where P_B is the background optical power entering the photo-detector, i_d is the dark current that exits even though $P = P_B = 0$.

The second noise is the thermal noise generated by the load resistor R_L,

$$\overline{i_{N2}^2} = \frac{4kT_e\Delta\nu}{R_L} \tag{6.18}$$

Where T_e is used to include the input noise power of the preamplifier following the photodiode [3].

Thus, the signal-to-noise power ratio at the preamplifier output is

$$\frac{S}{N} = \frac{\overline{i_s^2}}{(\overline{i_{N1}^2} + \overline{i_{N2}^2})} = \frac{2(P_{in}e\eta/h\nu)^2}{3e^2(P_{in} + P_B)\eta\Delta\nu/h\nu + 2e\overline{i_d}\Delta\nu + 4kT_e\Delta\nu/R_L} \tag{6.20}$$

Considering, in practices, the condition $\omega_m << \dfrac{1}{R_eC_d}$ force that R_L should be small since $R_e \approx R_L$, and the optical power P_{in} is near the detectability limit (S/N=1), we have, from (6.17) and (6.18), $\overline{i_{N2}^2} >> \overline{i_{N1}^2}$, and then

$$\frac{S}{N} \simeq \frac{2(P_{in}e\eta/h\nu)^2}{4kT_e\Delta\nu/R_L} \tag{6.21}$$

---

[3]In practice, the signal-to-noise ratio of the photodiode always take account of the noise power contributed by the following preamplifier, this is down by characterizing the "noise" of the preamplifier by the effective input noise "temperature" T_A. The output noise of the preamplifier is $GkT_A\Delta\nu$, where G is the power gain. And the total effective thermal noise "temperature" T_e of the photodiode is

$$T_e = T + T_A \tag{6.19}$$

The sensibility of the photodiode is defined as the "minimum detectable optical power" $(P_{in})_{min}$, which is by the definition that results in $S/N = 1$. Thus, from (6.21),

$$(P_{in})_{min} = \frac{h\nu}{e\eta}\sqrt{\frac{2kT_e\Delta\nu}{R_L}} \qquad (6.22)$$

6.2.5 Photodiode preamplifier

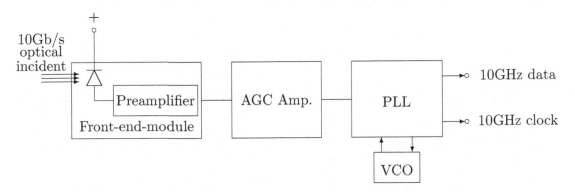

Figure 6-7   Optical receiver front-end schematic

Now we will introduce some preamplifiers following the photodiode. Figure 6-7 shows an optical receiver diagram. In which,

    * A preamplifier following the photodiode is manufactured as a front-end module to converts the photocurrent into voltage signal and then been amplified. In which, the photodiode and the front-end with the preamplifier can be integrated on the same substractrate.

    * An AGC amplifier is used to automatically control the amplitude of the detecting signal so as to provide an appropriated signal to input to the PLL (phase-locked-loop),

    * A PLL, which used to recover the 10Gb/s clock and the 10Gb/s data.

A preamplifier schematic is shown in Figure 6-8. Where the photodiode is equivalent to a current source I_{pd} and capacitor C_{pd} inside the dash line.

    * This is a silicon bipolar and bi-CMOS optical receiver.

    * The on chip inductive load L_2 improves the bandwidth of the TIA with the input bond wire L_1.

    * The bond wire L_1 also improves the signal-noise-ratio (SNR).

This optical receiver can provide the following specifications:

- The sensitivity: -18.1dBm   at   10Gb/s.

- The bit error rate   (BER): $2^{23} - 1$.

- The dynamic range: $\simeq 20dB$.

Those specification is suitable for the optical receiver of the 10Gb/s optical fiber transponder. Which will be discussed in Chapter 7.

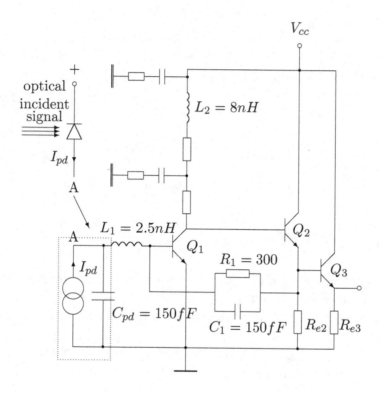

Figure 6-8   10GHz Si bipolar TIA with bandwidth enhancing (L_1 and L_2).

6.3   Avalanche photodiode

As shown in Figure 6-9(a), the structure of APD photodiode is slightly different from the structure of the PIN photodiode since an additional p-type layer is added to enhance the electric field inside the I-layer. This structure is optimized to provide an amplification of the generated electron-hole pairs before they reach the electrodes of the photodiode.

The mechanism of the avalanche photodiode is to achieve a gain inside the I-layer by the process of impact ionization, which is caused by the high reverse bias voltage in the I-layer as shown in Figure 6-9(b). In this case,

    * A electron-hole pair is generated by a photon incident on the I-layer, which is the same as the PIN photodiode.

    * However, when the reverse bias voltage in pn junction reaches to a point at which the carriers (electron or hole) is accelerated across the I-layer to gain enough kinetic energy to excite the electrons from the valence to the conductive band, while still travailing the layer. Which results in a dramatic increasing of pn junction current as shown in Figure 6-10.

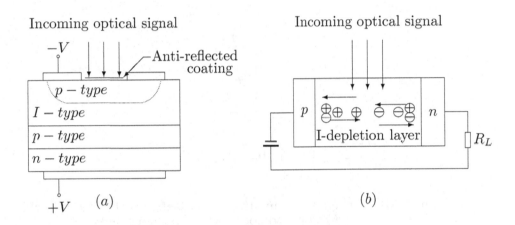

Figure 6-9   Avalanche photodiode

This process shown in Figure 6-9(b) is so called the process of impact ionization or avalanche multiplication. The result is a dramatic increase (avalanche) in pn junction current as shown in Figure 6-10.

The process of avalanche multiplication can be expressed by a parameter M as

$$M = (1 + p + p^2 + p^3 + \cdots) = \frac{1}{1-p} \qquad (6.23)$$

Where

    * p is defined as the probability that a photon-excited electron-hole pair will create another electron-hole pair during its drift in I-layer.

    * And then M is the current gain in avalanche photodiode.

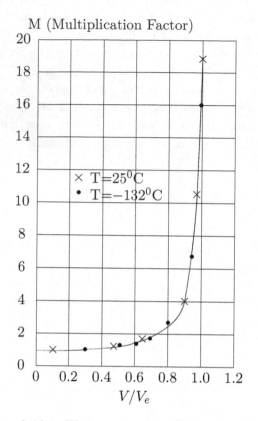

Figure 6-10   The current multiplication factor vs the electric field
in a APD photodiode (After Reference [89]

The characters of the APD photodiode

The avalanche photodiodes are similar to the PIN photodiode in construction except the current gain M caused by the process of the avalanche multiplication. Therefore,

* The equivalent circuit of the avalanche photodiode is similar to that of the PIN photodiode as shown in Figure 6-4.

* The frequency response of the avalanche photodiode is similar limited by the finite transit time of the carriers diffusing across the depletion layer τ_{tr} and the junction capacitor of the photodiode C_d and the external load resistor R_L as given in equation (6.6).

The sensitivity of the avalanche photodiode

Based on the discussion above, we understand that

* The current gain M leads to the signal power gain should be M^2.
* However, the shot noise is observed to increase as M^n, where $2 < n < 3.17$,

but not M^2 as we respect.

Thus, the signal-to-noise-ratio (S/N) is, from (6.20),

$$\frac{S}{N} = \frac{2M^2(P_{in}e\eta/h\nu)^2}{[3e^2(P_{in} + P_B)\eta\Delta\nu/h/nu]M^n + [2ei_d\Delta\nu]M^n + 4kT_e\Delta\nu/R_L} \quad (6.24)$$

Now,

* When $M = 1$, the S/N is no different for both of the avalanche photodiode and the PIN photodiode. In this case, the thermal noise term $4kT_e\Delta\nu/R_L$ is much larger than the shot noise terms. This will cause S/N increase with M. Until one value of M, the shot noise terms is compatible with the thermal term $4kT_e\Delta\nu/R_L$. After then the S/N will decrease with M since $n > 2$.

* Thus we may find an optimum value M' so as to obtain a minimum detectable power,

$$(P_{in})_{min} = \frac{2h\nu}{M'e\eta}\sqrt{\frac{2kT_e\Delta\nu}{R_L}} \quad (6.25)$$

that is the input power for which $S/N = 1$. Typically, the value of M' is between 30 and 100. Therefore, the improvement of the sensitivity is considerable compared with the PIN photodiode. The compensation for this improvement is that the width of the I-layer will be lager than that in PIN photodiode, which will narrow the bandwidth of the frequency response. Table 5.1 gives the the typical values for both of the avalanche photodiode and the PIN photodiode.

Table 6-1

Parameter	PIN	APD
responsivity in A/W	$0.7 \sim 0.95$	$0.7 \sim 0.9$
Cutoff frequency in GHz	up to 75	up to 15
internal Gain	1	up to 100

Chapter 7

10 Gbps Optical Fiber Transponder

7.1 Introduction to 10 Gbps optical fiber transponder

10Gbps optical fiber transponder, manufactured by Multiplexer Inc, New Jersey USA, has been installed first in USA, Canada, and then in China, India, $\cdots$ world wide area.

The long-haul conventional TDM (Time-Division-Multiplexing) transmission, SONET OC-192 (10Gbps optical transponder), must be periodically regenerated in core networks since optical signals become attenuated as they travel through the optical fiber. In SONET/SDH optical networks , each single mode fiber carrying a single optical data-stream, typically at 10Gbps (OC-192), required an electric regenerator every 40km (or beyond) as shown in Figure 7-1.

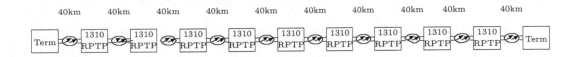

Figure 7-1   Long-haul conventional TDM transmission links - 10Gbps

Where "RPTP" working at $\lambda = 1310nm$ is a 10Gbps optical fiber transponder, in which, "RP" is the receiver part and "TP" is the transmitter part.

10Gbps optical fiber communications means that the 10Gbps optical data-stream transmission in a long haul optical fiber communication network, and the 10Gbps optical data-steam must be amplified after 40km (or beyond) transmission link, so as to overcome the bit error rate issue due to the dissipation and the dispersion of the single mode optical fiber. The function of each 10Gbps optical transponder are

,

329

- As optical amplifier: to amplify the 10Gbps optical data-stream to a properly level, with the recovery of the optical data clock and the data-stream inside the transponder.

- As digital add/drop: to drop down some electrical data stream into the local station and pick up some electrical data-steam from the local station into the long haul optical fiber network.

7.1.1   The block diagram of the 10Gbps optical fiber transponder

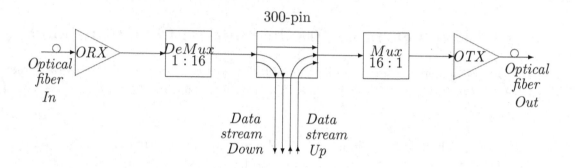

Figure 7-2   A 10Gbps optical fiber transponder

As shown in Figure 7-2, a 10Gbps optical fiber transponder is consist of

- A 10Gbps photo-receiver with an integrated limiting amplifier (ORX):

- An optical transmitter (OTX).

- A 300-pin with MSA (Multi Source Agreement) compliant as the data stream inter-cooperated with the local station:

Where

- The 10Gbps photo-receiver with an integrated limiting amplifier (ORX) is to convert the input 10Gbps optical series data-stream into 10Gbps electric series data-stream. The 10Gbps series electric data steam is then de-multiplexed by DeMux into 16 channels of parallel electric data stream at 622Mbps.

- At the optical transmitter (OTX), 16 channels of the parallel 622Mbps electric data stream are multiplexed by Mux into 10Gbps series data stream and then into the laser to convert the series 10Gbps electric data stream into 10Gbps series optical data stream.

- The 300-pin with MSA (Multi Source Agreement) compliant is used as the data stream inter-cooperated with the local station:

  - The 300-pin MSA is used to down load part of the electric data stream from the receiver side into the local station.

  - Meanwhile, the 300-pin MSA is used to pick up part of the electric data stream from the local station into the transmitter side. Which combine with part of 16 channels of electric data stream from the receiver side to send into the Mux to be multiplexed into 10Gbps series electric data stream.

7.1.2 The 10Gbps optical transponder functions

Thus, the 10Gbps optical transponder (manufactured by Multiplexer Inc. New Jersey, USA) functions are:

- As an optical amplifier: The sensitivity of a 10Gbps photo-receiver (ORX) is, for instance, -17.5dBm, which is far better than the MAS's requirements. The output of the optical transmitter is , for instance, +2dBm, which meets the MAS's requirement as well. So as the 10Gbps optical fiber transponder supports transmission link distance up to 40km and beyond, with exceptionally good eye diagram and jitter performance.

- As an electric data stream inter-cooperation with the local station through 300-pin with MSA compliant.

- To recover the data clock and the data-stream by using a CDR (clock and data recovery circuit) within the de-multiplexer DeMux.

De-multiplexer

The DeMux in Figure 7-3 is a fully integrated receiver, functions for SONET/FEC 10Gbps equipment. Figure 7-3 is the diagram of the DeMux. This receiver functions are

- To receiver the 9.95328Gbps CML [1] serial data input, then the De-Multiplexer is to convert the input 9.95328Gbps CML serial data into 16 channels of 622.08bps parallel data LVDS [2] output according to the FIFO (First-IN-First-OUT) in the output registers (See Table 7-1).

- Before the converting in De-Multiplexer, the CDR (clock and data recovery circuit) generates a clock that is at the same frequency (9.95328GHz) as incoming data rate (9.95328Gbps) at the CML series input.

---

[1]   CML - Current Mode Logic
[2]LVDS means low voltage differential signal

- Then the clock is divided by 16 (RXPOCLK- 622.06MHz) to make judgement and regeneration the 16 channels of 622.02Mbps parallel data output according the the FIFO (First-IN-First-OUT) in the output registers (See Table 7-2).

Table 7-1 Data Rate Selection

Series Data Rate	Parallel Date Rate	Rate Selection	Rate Selection
RDIN(Gbps)	TXDOUT(Mbps)	SELFECB(pin M10)	SEL10GEB(pin C13)
9.953	622.08	1	1

Table 7-2 Reference Clock Frequency Selection

Series Data Rate	Reference Clock Frequency	Reference Clock Frequency Selection
RDIN(Gbps)	RXREFCLK(MHz)	REF155ENB(pin C7)
9.953	622.08	1

Where

- RDINP/RDINN is the 9.5328Gbps CML serial data input, and "P" means positive, "N" means negative. CML is the current mode logic.

- RXDOUT0P/RXDOUT0N to RXDOUT15P/RXDOUT15N are the 16 channels LVDS parallel data outputs, and "P" means positive, "N" means negative. LVDS means low voltage data differential signal.

- LCKDET is the 9.5328Gbps clock detector. When the RXREFCLK frequency drift more than $\pm 150 ppm$ from the RXMCLK, CDR PLL Lock Detect goes low.

- Los DETECT is the data loss detector, and LOSB is used to inform the maintain station. When LOSB is low, the signal at RDIN is below the preset threshold.

- RESETB is used to reset the DeMux if it is out of work.

- RB/LD is an option only for applications. Living RB/LD open to save power approximately 20% over the power specified.

- RXPOCLKP(N) is the clock 622.08Mbps output.

- The RXMCLKP(N) is another reference clock 155.52Mbps. Which is an option for the user. In 10Gbps optical fiber transponder, this option is unnecessary and is omitted. There, Divider/by/4 is omitted too.

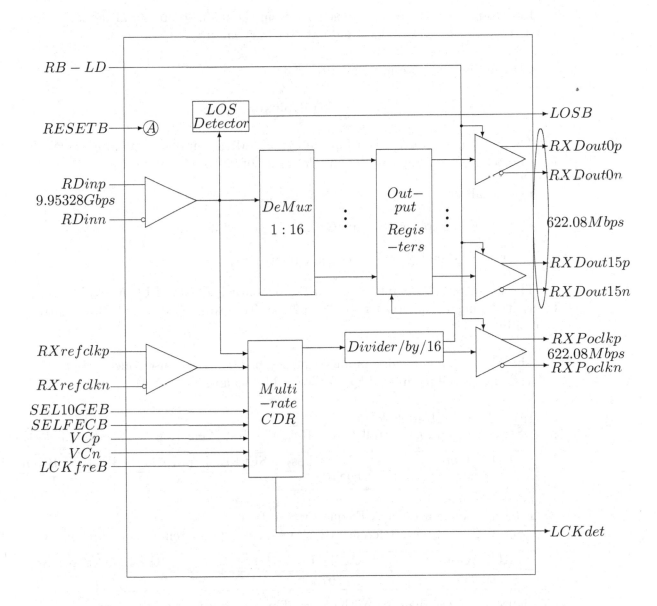

Figure 7-3   De-multiplexer-DeMux

Now

How to convert the 9.95328Gbps serial data input into 16 channels of 622.08Mbps parallel data output in De-multiplexer?

How to make judgement and generation the 16 channels of 622.08Mbps parallel data output according to the FIFO(First-IN-First-OUT) in the registers?

How to pick up the 9.5328Gbps clock from the 9.5328Gbps serial random data input by using the CDR (clock and data recovery circuit)?

We will discuss those in detail after then.

Multiplexer

The Mux in Figure 7-4 is a fully integrated MSA-compliant quadrate SONET transmitter operating at OC-192 and data rate of 9.95328Gbps with

- Serializer,

- Clock multiplication unit (CMU),

- Loss-of-signal (LOS) detection circuit.

The low-jitter LVDS interface and the onboard low-jitter PLL meets "Optical Internetworking Forum (OIF), IEEE 802.3ae, Telcordia, ANSI, and ITU-T jitter standards.

The Mux reference clock input frequency 622.08MHz is user-selectable to the line rate 9.95328Gbps divided by 16 (See Table 7-3 and Table 7-4).

Table 7-3 Data Rate Selection

Series Data Rate	Parallel Date Rate	Rate Selection	Rate Selection
RDIN(Gbps)	TXDOUT(Mbps)	SELFECB(pin A11)	SEL10GEB(pin C13)
9.953	622.08	1	1

Table 7-4 Reference Clock Frequency Selection

Series Data Rate	Reference Clock Frequency	Reference Clock Frequency Selection
RDIN(Gbps)	RXREFCLK(MHz)	REF155ENB(pin A8)
9.953	622.08	1

Figure 7-4 is the diagram of the Mux. This transmitter functions are

- To convert 16 channels of 622.08Mbps parallel data input into 9.95328Gbps serial data output in **16:1 MUX** based on the FIFO(First-IN-First-OUT) technology.

- Before the conversion, pick up the 9.95328Gbps clock from De-Multiplexer by means of PLL technology in **10GHz CMU**. This 9.95328Gbps clock functions are:

– To send to **Divider/by/16** to obtain 622.08Mbps clock, then send to **16X10 FIFO** to control FIFO(First-IN-First-OUT) in the **16:1 MUX** conversion

– To send to **Output Retime** to re-time the 9.985328Gbps serial data before output.

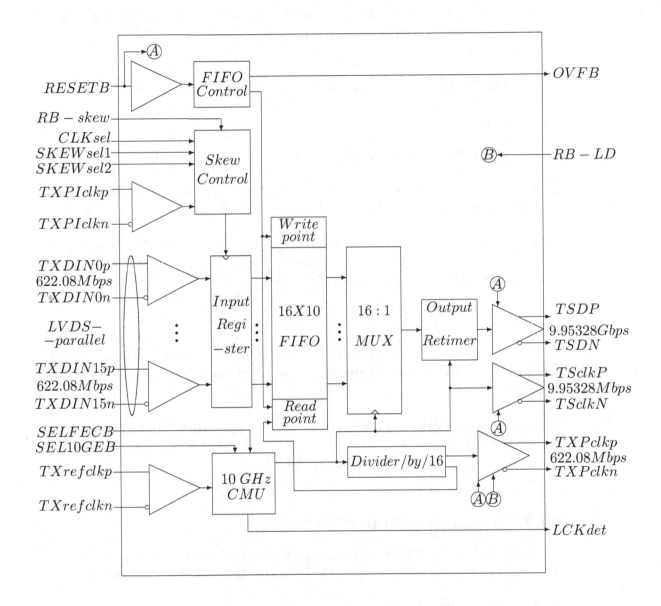

Figure 7-4   Multiplexer-Mux

Where

- RDINP/RDINN is the 9.5328Gbps CML serial data input, and "P" means positive, "N" means negative. CML is the current mode logic.

- RXDOUT0P/RXDOUT0N to RXDOUT15P/RXDOUT15N are the 16 channels LVDS parallel data output, and again "P" means positive, "N" means negative.

- LCKDET is the 9.5328Gbps clock detector. When the RXREFCLK frequency drift more than $\pm150ppm$ from the RXMCLK, CDR PLL Lock Detect goes low.

- Los DETECT is the data loss detector, and LOSB is used to inform the maintain station. When LOSB is low, the signal at RDIN is below the preset threshold.

- RESETB is used to reset the DeMux if it is out of work.

- RB/LD is an option only for applications. Living RB/LD open to save power approximately 20% over the power specified.

- RXMCLKP(N) is another reference clock 155.52Mbps. Which is an option for the user. In 10Gbps optical fiber transponder, this option is unnecessary and is omitted. There, Divider/by/4 is omitted too.

Now

How to convert the 16 channels of 622.08Mbps input parallel data into 9.95328Gbps serial data output in **16:1 Multiplexer** based on the FIFO(First-IN-First-OUT) technology?

How to re-time the 9.95328Gbps serial data from **16:1 Multiplexer** in **Out Retime**?

We will discuss those in detail after then.

7.1.3 Digital ADD/DROP Technology

In 10 Gbps optical fiber transponder, the digital ADD/DROP is realized in a 300-pin circuit as shown in Figure 7-5. In which there are three components,

- De-multiplexer,

- Multiplexer,

- 300-pin.

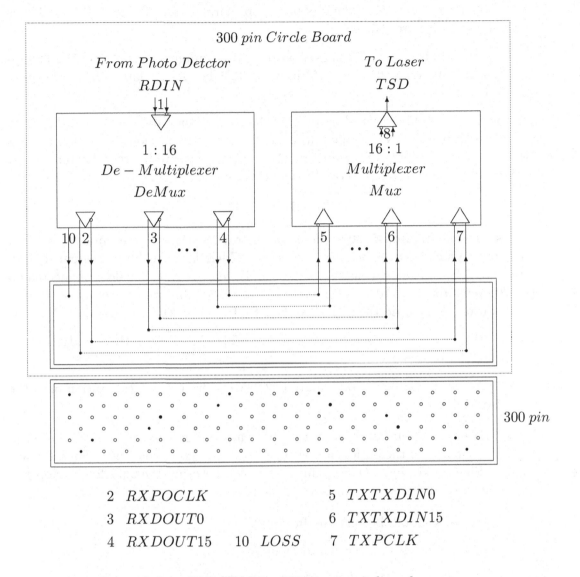

2 *RXPOCLK*		5 *TXTXDIN*0
3 *RXDOUT*0		6 *TXTXDIN*15
4 *RXDOUT*15	10 *LOSS*	7 *TXPCLK*

Figure 7-5   ADD/DROP—300-pin-circuit-board

As shown in Figure 7-5, De-multiplexer DeMux is a fully integrated MSA-compliant SONET/SDH/10GE receiver operation at OC-192 (9.053Gbps). Multiplexer Mux is a fully integrated MSA-compliant SONET/SDH/10GE transmitter operation at OC-192 (9.053Gbps). Both of them are connected on the 300-pin-circuit-board with ball-pin-connectors as shown in Figure 7-5. Where

- The input RDIN is 9.953Gbps electric series data stream.

- The output TSD is 9.953Gbps electric series data stream.

Where,

- The De-Multiplexer DeMux is used to convert the input 9.95328Gbps serial data RDIN into 16 channels of 622.08bps parallel data RXDOUT ($0 \rightarrow 15$) as shown in Figure 7-3.

- The Multiplexer Mux is used to convert the input 16 channels of 622.08bps parallel data TXDIN ($0 \rightarrow 15$) into 9.95328Gbps serial data TSD as shown in Figure 7-4.

Obviously, on 300-pin-circuit-board, all of the transmission data is 622.08Mbps data except the input RDIN and the output TSD. The transmission lines for the input RDIN and the output TSD are very short on the circuit board. Therefore, the design of the 300-pin-circuit-board is mainly based on the microwave technology.

ADD/DROP—300-pin

As shown in Figure 7-5, 300-pin is a Meg-Array 300 Position Receptacle from FCI company. Which is also a boll-pin-connectors soldered on the 300-pin-circuit-board. Just like a 300-pin cable, consist of a male and a female. The female is connected on the 300-pin-circuit-board and the male is a 300-pin cable can be connected to the local station so as to do the digital ADD/DROP data in the local station. Namely,

- Drop down any channel parallel data 622.08Mbps from the De-Multiplexer into the local station.

- Add any channel parallel data 622.08Mbps from the local station onto the Multiplexer.

- We need the input or output of the control signals onto the local station so as to maintain the optical fiber transponder in a good condition. For instance, in Figure 7-5, the "LOSS" point is to indicate wether the 9.95328Gbps optical data from the last station is loss or not?

Two process should be very careful,

- The soldering process: which includes

  - To solder the DE-Multiplexer DeMux onto the 300-pin-circuit-board.
  - To solder the Multiplexer Mux onto the 300-pin-circuit-board.
  - To solder the 300-pin onto the 300-pin-circuit-board.

- To connect the female and male together in 300-pin.

All of three soldering process are soldered by a special machine. So as to obtain a very high reliability.

Meanwhile, in 300-pin, to connect the female and male together is also a very careful process, which is also completed by a special machine.

Laser

On the 300-pin-circuit-board, the 9.9528Gbps electric series data TSD output from Multiplexer BCN2122 on the 300-pin-circuit-board is then sent to a laser via a mini-meter amplifier. The mini-meter amplifier is used to amplify the 9.95328Gbps data to a properly voltage. Which is applied to the bias circuit of the laser to do the amplitude modulation on the laser, so as to convert the 9.9528Gbps electric series data into a 9.953828Gbps optical series data by laser. The 9.95328Gbps optical series data is then sent to the long haul optical fiber link via single mode optical fiber.

Summery

The 10Gbps optical fiber transponder is involved in at least three high technologies:

- 1. Optical receiver (Photo-detector) and optical transmitter (Laser).

- 2. Electrical digital ADD/DROP technology: which includes digital receiver (i.e. digital de-multiplexer DeMux), in which, the 9.95328Gbps electric series data is converted into 16 channels of 622.08Mbps parallel data, and digital transmitter (i.e. digital Multiplexer Mux), in which the 16 channels of 622.08Mbps parallel data is converted into 9.95328Gbps electric series data.

- 3. microwave technology: i.e. 300-pin-circuit-board design, which includes

  - The transmission of 9.5328Gbps series data from photo-detector to the input of the de-multiplexer DeMux.
  - The transmission of 16 channels of 622.08Mbps parallel data from the output of de-multiplexer DeMux to 300-pin.
  - The transmission of 16 channels of 622.08Mbps parallel data from 300-pin to multiplexer Mux.
  - The transmission of 9.95328Gbps series data from the output of multiplexer Mux to mini-meter wave amplifier before laser.

  - The transmission of 622.08Mbps clock from De-multiplexer DeMux to 300-pin.
  - The transmission of 622.08Mbps clock from 300-pin to Multiplexer Mux.

The transmission of 16 channels of 622.08Mbps parallel data and the transmission of the 9.95328Gbps series data as well as the transmission of the 622.08Mbps clock, all of those are microwave signal on the 300-pin-circuit board. And then the circuit board design should be careful to follow the microwave design technology.

- 4.   Mini-meter wave technology: i.e. mini-meter wave amplifier before laser. The mini-meter wave amplifier is a independent circuit board made on a ceramic substrate. Which is placed inside the laser box and is connected to the laser by a bonding technology.

The size of the 10Gbps optical fiber transponder is about 60mm x 90mm x 15mm only. However, it contains a lot of the newest IT technologies.

7.2   Digital technology

Now we will discuss the digital technology in 10Gbps optical fiber transponder (in brief). Which includes

- 1.   1:16 De-multiplexer.

- 2.   16:1 Multiplexer.

- 3.   FIFO (First-in-first-out) technology.

- 4.   Resister.

- 5.   Synthesizer, counter and divider.

- 6.   Skew control.

7.2.1   16:1 Multiplexer

The function of the multiplexer 16:1 is to convert 16 channels of 622.08Mbps parallel data into 9.5328Gbps series data. The simplest multiplexer is the 2:1 multiplexer as shown in Figure 7-6.

2:1 Multiplexer

Figure 7-6   2:1 Multiplexer

A device as shown in Figure 7-6 functions as a switch,

- First, the output "Q" is connected to the first input "D_0",

- Then, the switch "S" forces the output "Q" to connect to the second input "D_1".

so as the device functions as a 2:1 multiplexer, namely, the inputs are a parallel data "D_0" and "D_1". Which has been converted into a series data output "Q".

Now, this 2:1 multiplexer can be realized by a digital circuit "AND" and "OR" as shown in Figure 7-7.

S
D1
D0
AND(1)
AND(2)
OR
Q

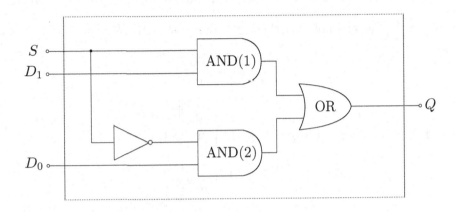

Figure 7-7 2:1 Multiplexer

It will be recalled that the functions of "AND", "OR" and "NO" are, respectively,

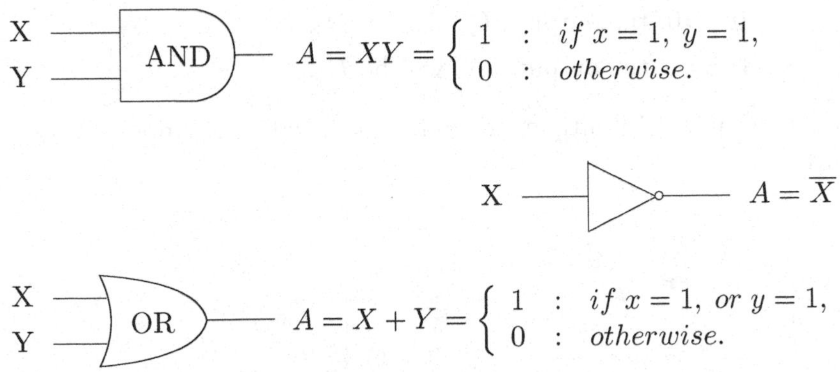

X
Y
AND
$A = XY = \begin{cases} 1 & : \ if\ x = 1,\ y = 1, \\ 0 & : \ otherwise. \end{cases}$
X
$A = \overline{X}$
X
Y
OR
$A = X + Y = \begin{cases} 1 & : \ if\ x = 1,\ or\ y = 1, \\ 0 & : \ otherwise. \end{cases}$

Figure 7-8 "AND", "OR" and "NO"

If you familia with the functions of "AND", "OR" and "NO", then it is easy to understand the function of 2:1 multiplexer as shown in Figure 7-7. Where

- S(0,1) is a select switch,

  – $D_1(0,1)$ can be transmitted into the output of "AND(1)" only if S=1.
  – $D_0(0,1)$ can be transmitted into the output of "AND(2)" only if S=0.

- That means that

  – When S=1, $D_1(0,1)$ can be transmitted into "Q", the output of "OR".
  – When S=0, $D_0(0,1)$ can be transmitted into "Q", the output of "OR".

Therefore, it is easy to obtain the truth table of 2:1 multiplexer as follows:

Truth Table of 2:1 multiplexer

S	D_1	D_0	Q
0	0	0	0
0	0	1	1
0	1	0	0
0	1	1	1
1	0	0	0
1	0	1	0
1	1	0	1
1	1	1	1

The Boolean expression of the 2:1 multiplexer is

$$Q = S'D_0 + SD_1 \tag{7.1}$$

Both of the Truth Table and the Boolean expression say that

- The multiplexer routers one of its data inputs (D_0 or D_1) to the output Q, based on the value of S:

  – If S=0, the output "Q" will be D_0.
  – If S=1, the output "Q" will be D_1.

The function of 2:1 multiplexer can be simplified to a symbol expression as shown in Figure 7-9.

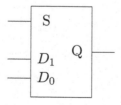

Figure 7-9   Symbol of 2:1 multiplexer

Enable Inputs

"Enable input" is used to "active" or "de-active" the multiplexer as the Truth Table shown:

Truth Table of Enable

EN	S	D_1	D_0	Q
0	0	0	0	0
0	0	0	1	0
0	0	1	0	0
0	0	1	1	0
0	1	0	0	0
0	1	0	1	0
0	1	1	0	0
0	1	1	1	0
1	0	0	0	0
1	0	0	1	1
1	0	1	0	0
1	0	1	1	1
1	1	0	0	0
1	1	0	1	0
1	1	1	0	1
1	1	1	1	1

Where

* EN=0 disables the multiplexer, which forces the output to be 0. (It does not turn off the multiplexer).

* EN=1 enable the multiplexer, and it works as specified earlier.

This Truth Table can be simplified to a Truth Table as follows:

Truth Table of Enable

EN	S	Q
0	x	0
1	0	D_0
1	1	D_1

Where S=x, can be S=1, or S=0.

4:1 Multiplexer

Following the same idea of 2:1 multiplexer, it is easy to set up the symbol of 4:1 multiplexer in Figure 7-10 and the Truth Table for 4:1 multiplexer as follows.

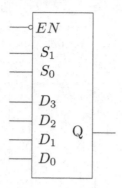

Figure 7-10   Symbol of 4:1 multiplexer

Truth Table of 4:1 multiplexer

EN'	S_1	S_0	Q
0	0	0	D_0
0	0	1	D_1
0	1	0	D_2
0	1	1	D_3
1	x	x	1

Where

- EN'=0 means EN=1. Which enables the 4:1 multiplexer, and then,

  - $Q = D_0$, if $S_1 = 0$, $S_0 = 0$.
  - $Q = D_1$, if $S_1 = 0$, $S_0 = 1$.
  - $Q = D_2$, if $S_1 = 1$, $S_0 = 0$.
  - $Q = D_3$, if $S_1 = 1$, $S_0 = 1$.

Which also can be expressed by the Boolean expression as

$$Q = S_1'S_0'D_0 + S_1'S_0D_1 + S_1S_0'D_2 + S_1S_0D_3 \qquad (7.2)$$

In which, $S_0 = 1$, $S_0' = 0$, $S_1 = 1$, $S_1' = 0$.

- EN'=1, means EN=0, which disables the multiplexer, and forces the output to be 0. (it does not turn off the multiplexer).

Compared with the 2:1 multiplexer, the 4:1 multiplexer needs 2 selects (S_0,S_1).

- Which means: for $2^n : 1$ multiplexer,

  - There are 2^n data inputs, thus, there must be "n" selects.

—   The output is a single bit "Q".

For example:

For $2^2 = 4$ data inputs (D_0, D_1, D_3, D_4),

We need $n = 2$ selects (S_0, S_1).

8:1 Multiplexer

- Figure 7-11 is a 8:1 multiplexer:

  —   There are $2^3 = 8$ data inputs. Thus there must be "3" selects (S_0, S_1, S_2).

  —   The output is a single bit "Q".

- The function of the 8:1 multiplexer should be

  — $Q = D_0$, if $S_2 = 0$, $S_1 = 0$, $S_0 = 0$

  — $Q = D_1$, if $S_2 = 0$, $S_1 = 0$, $S_0 = 1$

  — $Q = D_2$, if $S_2 = 0$, $S_1 = 1$, $S_0 = 0$

  — $Q = D_3$, if $S_2 = 0$, $S_1 = 1$, $S_0 = 1$

  — $Q = D_4$, if $S_2 = 1$, $S_1 = 0$, $S_0 = 0$

  — $Q = D_5$, if $S_2 = 1$, $S_1 = 0$, $S_0 = 1$

  — $Q = D_6$, if $S_2 = 1$, $S_1 = 1$, $S_0 = 0$

  — $Q = D_7$, if $S_2 = 1$, $S_1 = 1$, $S_0 = 1$

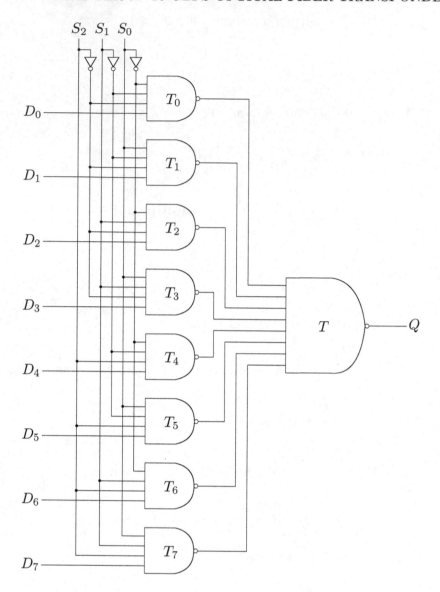

Figure 7-11   8:1 Multiplexer

- And the Truth Table of 8:1 multiplexer is

S_2	S_1	S_0	D_7	D_6	D_5	D_4	D_3	D_2	D_1	D_0	Q	
0	0	0								0	0	
0	0	0								1	1	D_0
0	0	1							0		0	
0	0	1							1		1	D_1
0	1	0						0			0	
0	1	0						1			1	D_2
0	1	1					0				0	
0	1	1					1				1	D_3
1	0	0				0					0	
1	0	0				1					1	D_4
1	0	1			0						0	
1	0	1			1						1	D_5
1	1	0		0							0	
1	1	0		1							1	D_6
1	1	1	0								0	
1	1	1	1								1	D_7

- The Boolean expression of 8:1 multiplexer is

$$\begin{aligned}
Q & = S_2' S_1' S_0' D_0 + S_2' S_1' S_0 D_1 + S_2' S_1 S_0' D_2 + S_2' S_1 S_0 D_3 \\
& + S_2 S_1' S_0' D_4 + S_2 S_1' S_0 D_5 + S_2 S_1 S_0' D_6 + S_2 S_1 S_0 D_7
\end{aligned} \tag{7.3}$$

- If we consider "ENABLE" input together, the abbreviated Truth Table of the 8:1 multiplexer is

The Truth Table of the 8:1 multiplexer

EN	S_2	S_1	D_0	Q
0		x		0
1	0	0	0	D_0
1	0	0	1	D_1
1	0	1	0	D_2
1	0	1	1	D_3
1	1	0	0	D_4
1	1	0	1	D_5
1	1	1	0	D_6
1	1	1	1	D_7

- The function of 8:1 multiplexer can be simplified to a symbol expression as shown in Figure 7-12.

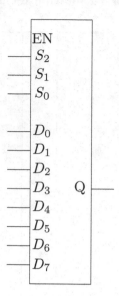

Figure 7-12   Symbol of 8:1 Multiplexer

The Truth Table of 8:1 multiplexer is very clear, which can be used to check the original circuit of the 8:1 multiplexer as shown in Figure 7-11.

For example, from Truth Table of 8:1 multiplexer, that

- $Q = D_0$, if $S_2 = 0$, $S_1 = 0$, ,$S_0 = 0$.

In this case, from Figure 7-11, T_0, T_1 are "NAND". The function of "NAND" is

$$A = \overline{XY} = \begin{cases} 0 & : \quad if\ x = 1,\ y = 1. \\ 1 & : \quad otherwish. \end{cases} \qquad (7.4)$$

And then,

- When $S_2 = 0$, $S_1 = 0$, $S_0 = 0$,

    - the inputs of T_1 are: $S_2' = 1$, $S_1' = 1$, $S_0' = 1$.

- In this case, when $D_0 = 0$,

    - $A_1 = 1$ and Q=0.

- When $D_0 = 1$,

    - $A_1 = 1$ and $Q = 1$.

Namely,

- $Q = D_0$, if $S_2 = 0$, $S_1 = 0$, $S_0 = 0$.

Where, all of $D_1 = 0$, $D_2 = 0$, $D_3 = 0$, $D_4 = 0$, $D_5 = 0$, $D_6 = 0$ and $D_7 = 0$ have been considered.

- Following the same process, it is not so difficult to prove that [as an exercise]
    - $Q = D_1$, if $S_2 = 0$, $S_1 = 0$, $S_0 = 1$.
    - $Q = D_2$, if $S_2 = 0$, $S_1 = 0$, $S_0 = 0$.
    - $Q = D_3$, if $S_2 = 0$, $S_1 = 1$, $S_0 = 1$.
    - $Q = D_4$, if $S_2 = 1$, $S_1 = 0$, $S_0 = 0$.
    - $Q = D_5$, if $S_2 = 1$, $S_1 = 0$, $S_0 = 1$.
    - $Q = D_6$, if $S_2 = 1$, $S_1 = 1$, $S_0 = 0$.
    - $Q = D_7$, if $S_2 = 1$, $S_1 = 1$, $S_0 = 1$.

Thus, the abbreviated Truth Table of 8:1 multiplexer can be used to design the 8:1 multiplexer.

- Figure 7-11 is a 8:1 multiplexer:
    - There are $2^3 = 8$ data inputs. Thus there must be "3" selects (S_0, S_1, S_2).
    - The output is a single bit "Q".

16:1 Multiplexer

Now we will set up a 16:1 Multiplexer based on the idea of the 8:1 multiplexer in Figure 7-11.

- Figure 7-13 is a 16:1 multiplexer:
    - There are $2^4 = 16$ data inputs. Thus there must be "4" selects (S_0, S_1, S_2, S_3).
    - The output is a single bit "Q".

Where, T is a completed unit. Here, it has been cut into two parts, one is on the left side (T) and another one is on the right side (T). However, we shall consider it to be a combination of the left side part and the right side part. In this case, the control lines S_3, S_2, S_1, S_0 and S_3', S_2', S_1', S_0' shall be continuously from the left side to the right side.

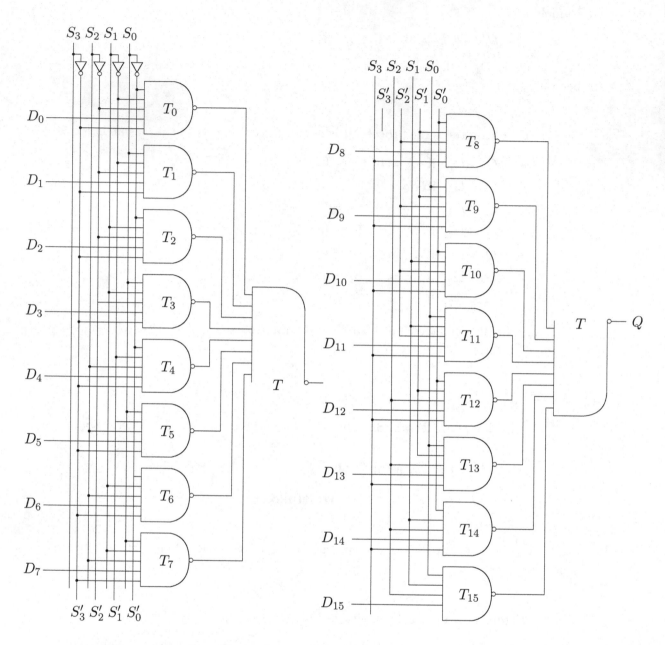

Figure 7-13   16:1 Multiplexer

- The function of the 16:1 multiplexer shall be

  - $Q = D_0$, if $S_3 = 0$, $S_2 = 0$, $S_1 = 0$, $S_0 = 0$,

  - $Q = D_1$, if $S_3 = 0$, $S_2 = 0$, $S_1 = 0$, $S_0 = 1$,
  - $Q = D_2$, if $S_3 = 0$, $S_2 = 0$, $S_1 = 1$, $S_0 = 0$,
  - $Q = D_3$, if $S_3 = 0$, $S_2 = 0$, $S_1 = 1$, $S_0 = 1$,

  - $Q = D_4$, if $S_3 = 0$, $S_2 = 1$, $S_1 = 0$, $S_0 = 0$,
  - $Q = D_5$, if $S_3 = 0$, $S_2 = 1$, $S_1 = 0$, $S_0 = 1$,
  - $Q = D_6$, if $S_3 = 0$, $S_2 = 1$, $S_1 = 1$, $S_0 = 0$,

  - $Q = D_7$, if $S_3 = 0$, $S_2 = 1$, $S_1 = 1$, $S_0 = 1$,
  - $Q = D_8$, if $S_3 = 1$, $S_2 = 0$, $S_1 = 0$, $S_0 = 0$,
  - $Q = D_9$, if $S_3 = 1$, $S_2 = 0$, $S_1 = 0$, $S_0 = 1$,

  - $Q = D_{10}$, if $S_3 = 1$, $S_2 = 0$, $S_1 = 1$, $S_0 = 0$,
  - $Q = D_{11}$, if $S_3 = 1$, $S_2 = 0$, $S_1 = 1$, $S_0 = 1$,
  - $Q = D_{12}$, if $S_3 = 1$, $S_2 = 1$, $S_1 = 0$, $S_0 = 0$,

  - $Q = D_{13}$, if $S_3 = 1$, $S_2 = 1$, $S_1 = 0$, $S_0 = 1$,
  - $Q = D_{14}$, if $S_3 = 1$, $S_2 = 1$, $S_1 = 1$, $S_0 = 0$,
  - $Q = D_{15}$, if $S_3 = 1$, $S_2 = 1$, $S_1 = 1$, $S_0 = 1$,

- Thus, the Truth Table of 16:1 multiplexer is

	S_3	S_2	S_1	S_0	D_{15}	14	13	12	11	10	9	D_8	Q	
	0	0	0	0										
0	0	0	0	0										
	0	0	0	1										
1	0	0	0	1										
	0	0	1	0										
2	0	0	1	0										
	0	0	1	1										
3	0	0	1	1										
	0	1	0	0										
4	0	1	0	0										
	0	1	0	1										
5	0	1	0	1										
	0	1	1	0										
6	0	1	1	0										
	0	1	1	1										
7	0	1	1	1										
	1	0	0	0								0	0	
8	1	0	0	0								1	1	D_8
	1	0	0	1							0		0	
9	1	0	0	1							1		1	D_9
	1	0	1	0						0			0	
10	1	0	1	0						1			1	D_{10}
	1	0	1	1					0				0	
11	1	0	1	1					1				1	D_{11}
	1	1	0	0				0					0	
12	1	1	0	0				1					1	D_{12}
	1	1	0	1			0						0	
13	1	1	0	1			1						1	D_{13}
	1	1	1	0		0							0	
14	1	1	1	0		1							1	D_{14}
	1	1	1	1	0								0	
15	1	1	1	1	1								1	D_{15}

- From the Truth Table of 16:1 multiplexer, we may write down the Boolean expression of 16:1 multiplexer as follows:

$$\begin{aligned} Q = {} & S_3'S_2'S_1'S_0'D_0 + S_3'S_2'S_1'S_0D_1 + S_3'S_2'S_1S_0'D_2 + S_3'S_2'S_1S_0D_3 + \\ & + S_3'S_2S_1'S_0'D_4 + S_3'S_2S_1'S_0D_5 + S_3'S_2S_1S_0'D_6 + S_3'S_2S_1S_0D_7 + \\ & + S_3S_2'S_1'S_0'D_8 + S_3S_2'S_1'S_0D_9 + S_3S_2'S_1S_0'D_{10} + S_3S_2'S_1S_0D_{11} + \\ & + S_3S_2S_1'S_0'D_{12} + S_3S_2S_1'S_0D_{13} + S_3S_2S_1S_0'D_{14} + S_3S_2S_1S_0D_{15} \qquad (7.5) \end{aligned}$$

	S_3	S_2	S_1	S_0	D_7	6	5	4	3	2	1	D_0	Q	
	0	0	0	0								0	0	
0	0	0	0	0								1	1	D_0
	0	0	0	1							0		0	
1	0	0	0	1							1		1	D_1
	0	0	1	0						0			0	
2	0	0	1	0						1			1	D_2
	0	0	1	1					0				0	
3	0	0	1	1					1				1	D_3
	0	1	0	0				0					0	
4	0	1	0	0				1					1	D_4
	0	1	0	1			0						0	
5	0	1	0	1			1						1	D_5
	0	1	1	0		0							0	
6	0	1	1	0		1							1	D_6
	0	1	1	1	0								0	
7	0	1	1	1	1								1	D_7
	1	0	0	0										
8	1	0	0	0										
	1	0	0	1										
9	1	0	0	1										
	1	0	1	0										
10	1	0	1	0										
	1	0	1	1										
11	0	0	1	1										
	0	1	0	0										
12	0	1	0	0										
	1	1	0	1										
13	1	1	0	1										
	1	1	1	0										
14	1	1	1	0										
	1	1	1	1										
15	1	1	1	1										

- From the Truth Table of 16:1 multiplexer, we may write down the Boolean expression of 16:1 multiplexer as follows:

$$
\begin{aligned}
Q = \ & S_3'S_2'S_1'S_0'D_0 + S_3'S_2'S_1'S_0D_1 + S_3'S_2'S_1S_0'D_2 + S_3'S_2'S_1S_0D_3 + \\
& + S_3'S_2S_1'S_0'D_4 + S_3'S_2S_1'S_0D_5 + S_3'S_2S_1S_0'D_6 + S_3'S_2S_1S_0D_7 + \\
& + S_3S_2'S_1'S_0'D_8 + S_3S_2'S_1'S_0D_9 + S_3S_2'S_1S_0'D_{10} + S_3S_2'S_1S_0D_{11} + \\
& + S_3S_2S_1'S_0'D_{12} + S_3S_2S_1'S_0D_{13} + S_3S_2S_1S_0'D_{14} + S_3S_2S_1S_0D_{15}\,(7.6)
\end{aligned}
$$

- The abbreviated Truth Table of the 16:1 multiplexer is

	EN	S_3	S_2	S_1	S_0	Q
	0			x		0
0	1	0	0	0	0	D_0
1	1	0	0	0	1	D_1
2	1	0	0	1	0	D_2
3	1	0	0	1	1	D_3
4	1	0	1	0	0	D_4
5	1	0	1	0	1	D_5
6	1	0	1	1	0	D_6
7	1	0	1	1	1	D_7
8	1	1	0	0	0	D_8
9	1	1	0	0	1	D_9
10	1	1	0	1	0	D_{10}
11	1	1	0	1	1	D_{11}
12	1	1	1	0	0	D_{12}
13	1	1	1	0	1	D_{13}
14	1	1	1	1	0	D_{14}
15	1	1	1	1	1	D_{15}

- The symbol of the 16:1 multiplexer is shown in Figure 7-14.

[Exercise 1] Check the schematic Figure 7-13 to see that it match the Boolean expression and the abbreviated Truth Table of 16:1 multiplexer.

[Exercise 2] Prove that $Q = S_3 S_2 S_1 S_0 D_{15}$ when $S_3 = 1$, $S_2 = 1$, $S_1 = 1$, $S_0 = 1$.

[Exercise 3] Prove that $Q = S_3 S_2' S_1 S_0 D_{11}$ when $S_3 = 1$, $S_2 = 0$, $S_1 = 1$, $S_0 = 1$.

[Exercise 4] Compared Figure 7-13 with Figure 7-8 to see if there is a "S" switch in each of

$$S_3{}^x, \ S_2{}^x, \ S_1{}^x, \ S_0{}^x \ D_n.$$

Where "x" can be "1" or "0". And "n" can be "0, 1, 2, 3, 4, 5, 6, 7, 8, 9, 10, 11, 12, 13, 14" or "15".

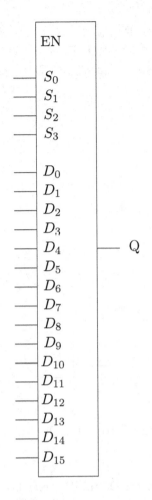

Figure 7-14   Symbol of 16:1 multiplexer

7.2.2   1:16 De-multiplexer

1:2 De-multiplexer

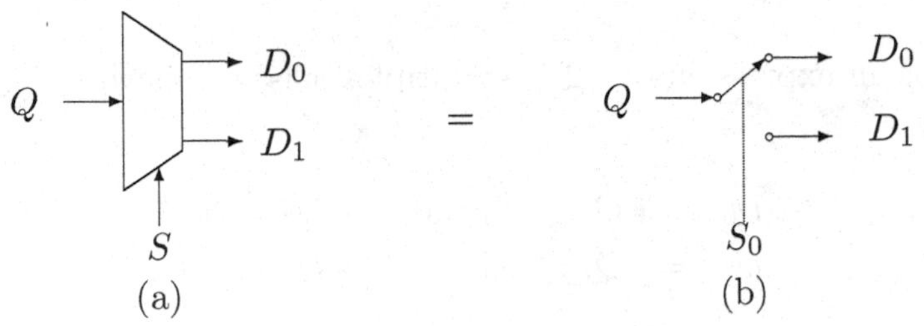

Figure 7-15   1:2 De-multiplexer

A 1:2 De-multiplexer in Figure 7-15(a) is a device to converter a single input "Q" into two outputs "D_0" and "D_1 by a switch "S". Which can be equated to a controlled switch "S_0" as shown in Figure 7-15(b).

The 1:2 De-multiplexer can be realized by a digital circuit as shown in Figure 7-16.

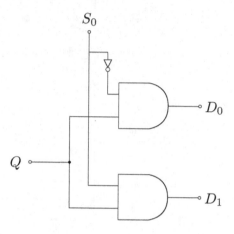

Figure 7-16   1:2 De-multiplexer

The Truth Table of the 1:2 De-multiplexer is

Q	S_0	D_1	D_0
0	0		0
1	0		1
0	1	0	
1	1	1	

- Which means that S_0 is a selected switch to select one of two outputs D_0 then D_1 to connect to the single input Q. Which is easy to be understood by the function of "AND".

- The Boolean expression of 1:2 De-multiplexer is as follows:

$$D_0 = Q, \; if \; S_0 = 0, \; \rightarrow D_0 = S_0'Q.$$
$$D_1 = Q, \; if \; S_0 = 1, \; \rightarrow D_0 = S_0Q. \tag{7.7}$$

1:4 De-multiplexer

Following the same process, it is easy to figure out the 1:4 De-multiplexer as shown in Figure 7-17.

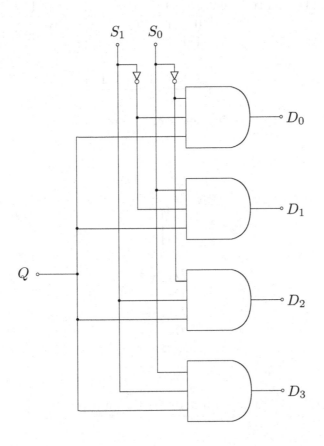

Figure 7-17   1:4 De-multiplexer

- Where the 1:4 De-multiplexer is used to convert the series signal input "Q" into 4 parallel output "D_0", "D_1", "D_2", and "D_3" with two selects S_1, S_0.

- Based on the function of "AND", we may write down the Boolean expression as follows:

$$D_0 = Q, \ if \ S_1 S_0 = 00 \rightarrow D_0 = S_1' S_0' Q$$

$$D_1 = Q, \ if \ S_1 S_0 = 01 \rightarrow D_0 = S_1' S_0 Q$$

$$D_2 = Q, \ if \ S_1 S_0 = 10 \rightarrow D_0 = S_1 S_0' Q$$

$$D_3 = Q, \ if \ S_1 S_0 = 11 \rightarrow D_0 = S_1 S_0 Q \tag{7.8}$$

Truth Table of 1:4 De-multiplexer

Q	S_1	S_0	D_3	D_2	D_1	D_0	
0	0	0				0	
1	0	0				1	$D_0=Q$
0	0	1			0		
1	0	1			1		$D_1=Q$
0	1	0		0			
1	1	0		1			$D_2=Q$
0	1	1	0				
1	1	1	1				$D_3=Q$

- The symbol of 1:4 De-multiplexer is in Figure 7-18,

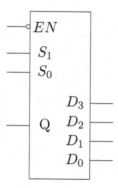

Figure 7-18   The symbol of 1:4 De-multiplexer

Where, for a 1:2^2 De-multiplexer,

- There are 2^2 parallel outputs (D_3, D_2, D_1, D_0). For which , there must be 2 selects (S_1, S_0).

- 'Q" is a series signal input.

- EN=1, enable the De-multiplexer, and it works as specified above.

- EN=0, disable the De-multiplexer, which forces all of the outputs (D_3, D_2, D_1, D_0) to be "0". However, it does not turn off the De-multiplexer.

1:8 De-multiplexer

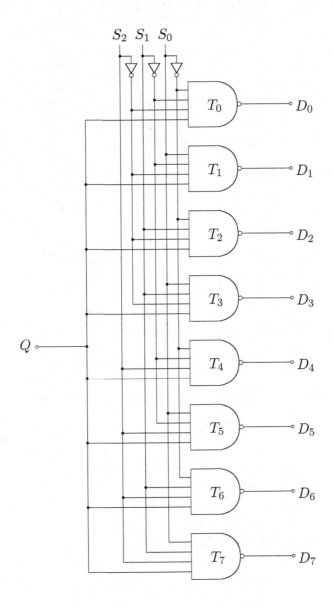

Figure 7-19   1:8 De-Multiplexer

The 1:8 De-multiplexer is shown in Figure 7-19. Based on the function of "AND" we may write down the Truth Table, symbol and Boolean expression as follows.

The Truth Table of 1:8 De-multiplexer is

S_2	S_1	S_0	D_7	D_6	D_5	D_4	D_3	D_2	D_1	D_0	Q
0	0	0								0	0
0	0	0								1	1
0	0	1							0		0
0	0	1							1		1
0	1	0						0			0
0	1	0						1			1
0	1	1					0				0
0	1	1					1				1
1	0	0				0					0
1	0	0				1					1
1	0	1			0						0
1	0	1			1						1
1	1	0		0							0
1	1	0		1							1
1	1	1	0								0
1	1	1	1								1

- The symbol of 1:8 De-multiplexer is in Figure 7-19,

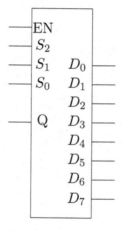

Figure 7-20   Symbol of 1:8 De-multiplexer

- The Boolean expression of the 1:8 De-multiplexer is

$$D_0 = Q, \ if \ S_2 = 0, \ S_1 = 0, \ S_0 = 0, \ \rightarrow \ D_0 = S_2' S_1' S_0' Q$$

$$D_1 = Q, \ if \ S_2 = 0, \ S_1 = 0, \ S_0 = 1, \ \rightarrow \ D_0 = S_2' S_1' S_0 Q$$

$$D_2 = Q, \ if \ S_2 = 0, \ S_1 = 1, \ S_0 = 0, \ \rightarrow \ D_0 = S_2' S_1 S_0' Q$$

$$D_3 = Q, \ if \ S_2 = 0, \ S_1 = 1, \ S_0 = 1, \ \rightarrow \ D_0 = S_2' S_1 S_0 Q$$

$$D_4 = Q, \ if \ S_2 = 1, \ S_1 = 0, \ S_0 = 0, \ \rightarrow \ D_0 = S_2 S_1' S_0' Q$$

$$D_5 = Q, \ if \ S_2 = 1, \ S_1 = 0, \ S_0 = 1, \ \rightarrow \ D_0 = S_2 S_1' S_0 Q$$

$$D_6 = Q, \ if \ S_2 = 1, \ S_1 = 1, \ S_0 = 0, \ \rightarrow \ D_0 = S_2 S_1 S_0' Q$$

$$D_7 = Q, \ if \ S_2 = 1, \ S_1 = 1, \ S_0 = 1, \ \rightarrow \ D_0 = S_2 S_1 S_0 Q \tag{7.9}$$

- Where, for a 1:8 De-multiplexer,

  - There are 2^3 parallel outputs (D_7, D_6, D_5, D_4, D_3, D_2, D_1, D_0). For which there must be 3 selects (S_2, S_1, S_0).

  - The input is a series signal bit Q.

  - EN=1, enable the De-multiplexer, which works as specified above.

  - EN=0, disable the De-multiplexer, which force all of the outputs (D_7, D_6, D_5, D_4, D_3, D_2, D_1, D_0) to be "0". However, it does not turn off the De-multiplexer.

1:16 De-multiplexer

- For a 1:16 De-multiplexer,

  - There are 2^4 parallel outputs (D_{15}, D_{14}, D_{13}, D_{12}, D_{11}, D_{10}, D_9, D_8, D_7, D_6, D_5, D_4, D_3, D_2, D_1, D_0). For which, there must be 4 selects (S_3, S_2, S_1, S_0).

  - The input is a series signal input "Q".

The schematic diagram of the 1:16 De-multiplexer is shown in Figure 7-21.

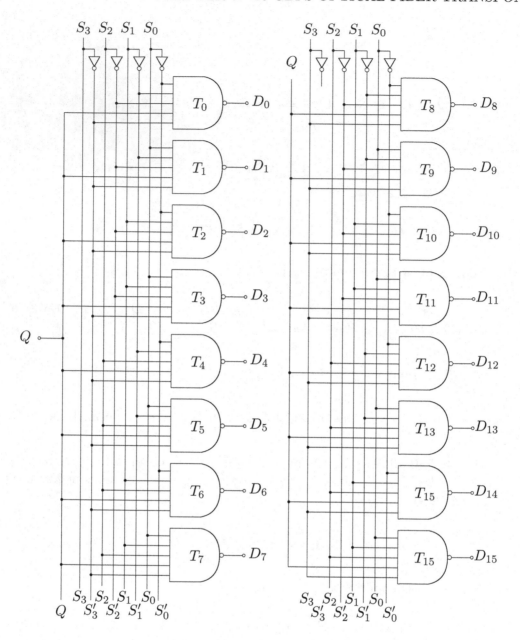

Figure 7-21   1:16 De-Multiplexer

- Truth Table of 1:16 De-multiplexer:
Based on the function of "AND", considering the schematic diagram of 1:16 De-multiplexer in Figure 7-21, we may write down the Truth Table of 1:16 De-multiplexer as follows:

	S_3	S_2	S_1	S_0	D_7	D_6	D_5	D_4	D_3	D_2	D_1	D_0	Q
	0	0	0	0								0	0
0	0	0	0	0								1	1
	0	0	0	1							0		0
1	0	0	0	1							1		1
	0	0	1	0						0			0
2	0	0	1	0						1			1
	0	0	1	1					0				0
3	0	0	1	1					1				1
	0	1	0	0				0					0
4	0	1	0	0				1					1
	0	1	0	1			0						0
5	0	1	0	1			1						1
	0	1	1	0		0							0
6	0	1	1	0		1							1
	0	1	1	1	0								0
7	0	1	1	1	1								1
	1	0	0	0									
8	1	0	0	0									
	1	0	0	1									
9	1	0	0	1									
	1	0	1	0									
10	1	0	1	0									
	1	0	1	1									
11	1	0	1	1									
	1	1	0	0									
12	1	1	0	0									
	1	1	0	1									
13	1	1	0	1									
	1	1	1	0									
14	1	1	1	0									
	1	1	1	1									
15	1	1	1	1									

	S_3	S_2	S_1	S_0	D_{15}	D_{14}	D_{13}	D_{12}	D_{11}	D_{10}	D_9	D_8	Q
	0	0	0	0									
0	0	0	0	0									
	0	0	0	1									
1	0	0	0	1									
	0	0	1	0									
2	0	0	1	0									
	0	0	1	1									
3	0	0	1	1									
	0	1	0	0									
4	0	1	0	0									
	0	1	0	1									
5	0	1	0	1									
	0	1	1	0									
6	0	1	1	0									
	0	1	1	1									
7	0	1	1	1									
	1	0	0	0								0	0
8	1	0	0	0								1	1
	1	0	0	1							0		0
9	1	0	0	1							1		1
	1	0	1	0						0			0
10	1	0	1	0						1			1
	1	0	1	1					0				0
11	1	0	1	1					1				1
	1	1	0	0				0					0
12	1	1	0	0				1					1
	1	1	0	1			0						0
13	1	1	0	1			1						1
	1	1	1	0		0							0
14	1	1	1	0		1							1
	1	1	1	1	0								0
15	1	1	1	1	1								1

- The Boolean expression of the 1:16 De-multiplexer is

D_0	$=Q,$	if	$s_3 s_2 s_1 s_0$	$=0000.$	$\rightarrow$	D_0	$=s_3' s_2' s_1' s_0' Q.$
D_1	$=Q,$	if	$s_3 s_2 s_1 s_0$	$=0001.$	$\rightarrow$	D_1	$=s_3' s_2' s_1' s_0 Q.$
D_2	$=Q,$	if	$s_3 s_2 s_1 s_0$	$=0010.$	$\rightarrow$	D_2	$=s_3' s_2' s_1 s_0' Q.$
D_3	$=Q,$	if	$s_3 s_2 s_1 s_0$	$=0011.$	$\rightarrow$	D_3	$=s_3' s_2' s_1 s_0 Q.$
D_4	$=Q,$	if	$s_3 s_2 s_1 s_0$	$=0100.$	$\rightarrow$	D_4	$=s_3' s_2 s_1' s_0' Q.$
D_4	$=Q,$	if	$s_3 s_2 s_1 s_0$	$=0101.$	$\rightarrow$	D_5	$=s_3' s_2 s_1' s_0 Q.$
D_6	$=Q,$	if	$s_3 s_2 s_1 s_0$	$=0110.$	$\rightarrow$	D_6	$=s_3' s_2 s_1 s_0' Q.$
D_7	$=Q,$	if	$s_3 s_2 s_1 s_0$	$=0111.$	$\rightarrow$	D_7	$=s_3' s_2 s_1 s_0 Q.$
D_8	$=Q,$	if	$s_3 s_2 s_1 s_0$	$=1000.$	$\rightarrow$	D_8	$=s_3 s_2' s_1' s_0' Q.$
D_9	$=Q,$	if	$s_3 s_2 s_1 s_0$	$=1001.$	$\rightarrow$	D_9	$=s_3 s_2' s_1' s_0 Q.$
D_{10}	$=Q$	if	$s_3 s_2 s_1 s_0$	$=1010.$	$\rightarrow$	D_{10}	$=s_3 s_2' s_1 s_0' Q.$
D_{11}	$=Q,$	if	$s_3 s_2 s_1 s_0$	$=1011.$	$\rightarrow$	D_{11}	$=s_3 s_2' s_1 s_0 Q.$
D_{12}	$=Q,$	if	$s_3 s_2 s_1 s_0$	$=1100.$	$\rightarrow$	D_{12}	$=s_3 s_2 s_1' s_0' Q.$
D_{13}	$=Q,$	if	$s_3 s_2 s_1 s_0$	$=1101.$	$\rightarrow$	D_{13}	$=s_3 s_2 s_1' s_0 Q.$
D_{14}	$=Q,$	if	$s_3 s_2 s_1 s_0$	$=1110.$	$\rightarrow$	D_{14}	$=s_3 s_2 s_1 s_0' Q.$
D_{15}	$=Q,$	if	$s_3 s_2 s_1 s_0$	$=1111.$	$\rightarrow$	D_{15}	$=s_3 s_2 s_1 s_0 Q.$

- The abbreviated Truth Table of the 1:16 De-multiplexer is

	Q	EN	S_3	S_2	S_1	S_0	OUT
	Q	0	x	x	x	x	0
0	Q	1	0	0	0	0	D_0
1	Q	1	0	0	0	1	D_1
2	Q	1	0	0	1	0	D_2
3	Q	1	0	0	1	1	D_3
4	Q	1	0	1	0	0	D_4
5	Q	1	0	1	0	1	D_5
6	Q	1	0	1	1	0	D_6
7	Q	1	0	1	1	1	D_7
8	Q	1	1	0	0	0	D_8
9	Q	1	1	0	0	1	D_9
10	Q	1	1	0	1	0	D_{10}
11	Q	1	1	0	1	1	D_{11}
12	Q	1	1	1	0	0	D_{12}
13	Q	1	1	1	0	1	D_{13}
14	Q	1	1	1	1	0	D_{14}
15	Q	1	1	1	1	1	D_{15}

- The symbol of the 1:16 De-multiplexer is in Figure 7-22,

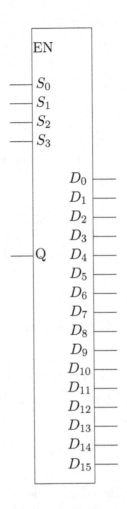

Figure 7-22   Symbol of 1:16 multiplexer

7.2.3   The block diagram of the transponder and FIFO (First-In-First-Out)

The block diagram of the transponder

The available 10Gbps optical fiber transponder is connected to the local station via the network interface processors as shown in Figure 7-23. In which, both of 50km are the optical fiber transmission link with 50km distance. And the network interface processor is connected to the transponder via 300 pin. In Figure 7-23,

- The input is a network interface processor. Which is used for the pick-up/drop-down the data in the local station.

- The output is a network interface processor. Which is used for the pick-up/drop-down the data in the local station as well.

- Each network interface process is connected to a 300 pin insider the 10Gbps optical fiber transponder.

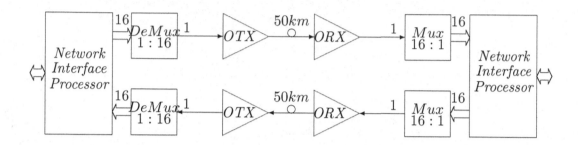

Figure 7-23   Functional Block Diagram of 10Gbps optical fiber transponder

Now, the function of Figure 7-23 can be realized as shown in Figure 7-24. Where,

- The input is the output of an optical fiber transmission link with 50km distance from the last station.

- The output is the input of an optical fibers transmission link with 50km distance to the next station.

- The units between the input and the output are two 10Gbps optical fiber transponders placed in a local station. The top one is transmission from the left side to the right side, the bottom one is transmission from the right side to the left side.

- Two network interface processors are used to pick-up/drop-down the data vis 300 pin.

It is clear that the function of the 10Gbps optical fiber transponder are:

- To amplify the optical power with 19dB gain, so as to give a compensation for the dissipation loss in 50km optical fiber transmission from the last station. In which, the input of the transponder is a very sensitive photo-detector with sensitivity of -17dBm. The output of the transponder is a laser with +2dm output. And the dissipation loss in a 50km optical fiber is around 15dB (0.3dB/km).

- To recovery the clock and data from the photo-detector, then inter-change data (drop-down and pick up data) with the local station via the network interface processors.

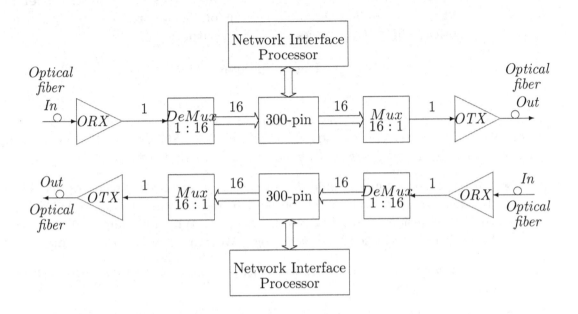

Figure 7-24   Two 10Gbps optical fiber transponder

- More in detail. The data in the optical fiber is "light" and "dark". "light" corespondent to "1" and "dark" corespondent to "0".

  − When the optical data enter the transponder, the **photo-detector** in **ORX** will convert the optical data ("light" and "dark") into electric data ("1" and "0"). Which is a 9.95328Gbps series signal data.

  − **DeMux** is a digital receiver. The function of the DeMux are
    * To recovery the clock and data from the 9.95328Gbps series signal data input.
    * Then the De-multiplexer is used to convert the 9.95328Gbps series signal data into 16 channels of 622.08Mbps parallel data.

  − **300 pin** is used to drop-down the parallel data from the receiver De-Mux to the local station and pick-up the parallel data from the local station to the transmitter Mux via the network interface processors.

— **Mux** is a digital transmitter. The function of the Mux is to convert the 16 channels of 622.08Mbps parallel data into 9.95328Gbps series signal data by multiplexer.

— The 9.95328Gbps series signal data from Mux is sent to the **laser** in **OTX** to modulate the drive current of the **laser**, so as to obtain an optical 9.95328Gbps series signal data output.

Big challenges

Now the big issue is that how to guarantee that the output 9.95328Gbps series optical data correctly corresponds to the input 9.95328Gbps series signal optical data in a transponder? We will settle down this issue by two technologies:

- First, there is a unique 9.95328Gbps clock go through the whole optical fiber transmission link. Which can be used to re-time the output 9.095328Gbps series data inside Mux before sent to the laser. Which is completed inside the Mux with phase-locked loop technology. We will discuss it after then.

- Secondary, the 16:1 conversion in Mux is realized by a FIFO (First-In-First-Out) technology.

Now, we will discuss the FIFO (First-In-First-Out) technology first.

7.2.3.1   FIFO (First-In-First-Out) technology

As shown in Figure 7-25, there are two conversions inside a transponder:

- One is the 1:16 conversion in the digital receiver DeMux with the help of a shift register (RG OUT).

- One is the 16:1 conversion in digital transmitter Mux with the help of a shift register (RG IN) and FIFO technology.

In the first conversion 1:16, the 16 channels of parallel 622.08Mbps data is located in the shift register (RG OUT) in order as shown in Figure 7-25.

After a transmission in a micro-strip circuit board, the 16 channels of parallel 622.08Mbps data is located in a shift register (RG IN) in the same order as in the shift register (RG OUT).

Now, in the second conversion 16:1, if the 16 channels of parallel 622.08Mbps data in RG OUT can be read out into a series signal 9.95328Gbps data by FIFO (FIRST-IN-First-OUT) technology, theoretically, the output of series signal 9.95328Gbps data shall be the same as the input series signal 9.95328Gbps data as shown in Figure 7-25.

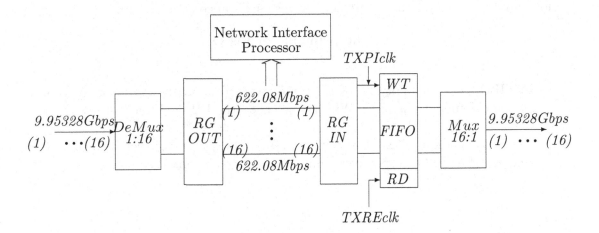

Figure 7-25   First-In-First-OUT technology

FIFO (First-IN-First-OUT) technology

As shown in Figure 7-25, in FIFO unit,

- The first step is "write": namely, "write" the data from the register (RG IN) into FIFO unite by a network clock TXPIclk from the transmission link.

- The second step is "read": namely, "read" the data from the register (RG IN) into Mux (16:1 De-multiplexer) by the reference clock TXREclk from the transmission link according to the First-IN-First-OUT process.

Where, in FIFO process,

- The frequency of two clocks TXPIclk and TXREclk shall be the same.

- However, the tolerance of the time difference between two clocks TXPIclk and TXREclk can be 6.0ns.

In the other words, that two clocks TXPIclk and TXREclk are not necessary to be "time-in line", but the oscillation frequency must be the same. Two clocks "time-in line" is a big issue from the technical point of view.

In fact, in phase-locked loop inside the transponder, the phase-locked process includes two step:

- One is "frequency locked", namely, the frequency of the two clocks is going to be the same.

- Second one is "phase-locked" or "time-in-line", namely, the time of two clocks is going to be the same.

" frequency-locked" is easy to be reached. However, "time-in-line" is very difficult, not only have to spend a lot of time, but also sufferance a big jitters.

The principle of FIFO

In FIFO, every memory, in which the data word is written in first, comes out first when the memory is read. Which is so called the FIFO (First-In-First-Out).

As shown in Figure 7-26, two electric system always are connected to the input and output of a FIFO, respectively: one that writes and one that reads.

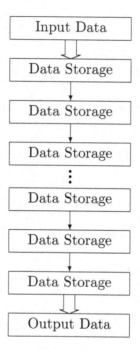

Figure 7-26   First-In-First-OUT Data Flow

Figure 7-27 shows us an example to implement the FIFO. In which,

$$Input\ Data \rightarrow Storage \rightarrow Output\ Data$$

is implemented inside a SRAM. In which,

- "Input Data" is written by using "Write pointer" controlled by "Write Clock".

- "Output Data" is read by using "Read Pointer" controlled by "Read Clock". In which, both of "same frequency" and "same phase" between the "Write Clock" and the "Read Clock" are unnecessary.

— If "same frequency" between the "Write Clock" and the "Read Clock" is required, but "same phase" between them is not required, it is called "the Synchronous FIFO" such as "ACT7881" from Taxes Instruments Inc.

— If "same frequency" is not required, (of cause, "same phase" is not required too), it is called the "Asynchronous FIFO".

- In 10Gbps optical fiber transponder, "same frequency" between the "Write Clock" and the "Read Clock" is required in FIFO. Thus, which is a Synchronous FIFO.

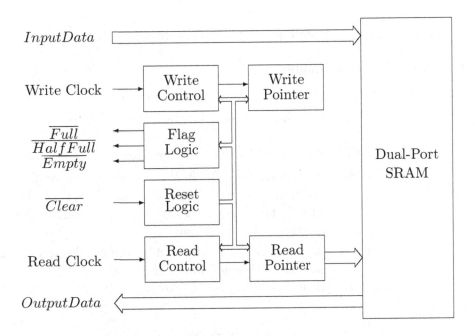

Figure 7-27   Block Diagram of FIFO With Static Memory

SRAM-Static Random Access Memory

SRAM is an abbreviation of "Static Random Access Memory". In fact, it is a bistable latch circuit to store each bit. As an example, take a cell of SRAM to see how to "write" and "read" each bit as shown Figure 7-28. In which,

- Two "M2" forms a flip-flop to store one bit data ("1" or "0").

- Two "M1" forms two latches.

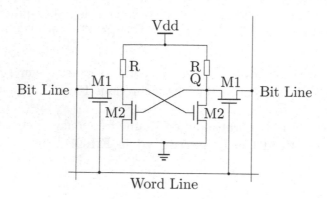

Figure 7-28   Bistable Latching Circuit to Store Each Bit

- "World lines" is "Enable":

  - The latch "M1" is open when the "world line" is "1".
  - The latch "M1" is closed when the "world line" is "0".

  - Two "Bit line": one for "write", one for "Read".
    * "Write" data from the "Bit line" only when "Word line" is "1". Which means that "Enable" forces the "Latch M1" to be opened, and the "Write" data toggle the flip-flop and update the original stored information ("1" or "0") on "Q".
    * "Read" data to the "Bit Line" only when "Word line" is "1". Which means that "Enable" forces the "Latch M1" to be opened, and then "Read" the data on "Q" to the "Bit Line".

The signal of a Synchronous FIFO SN74ACT7881 from Taxes Instruments Inc. is shown in Figure 7-29.

As shown in Figure 7-29,

- After "Reset FIFO",

- "Input data" "D1,D2,D3" is written in order When "Write Enable=1".

- "Output data" "D1,D2,D3" is read in order when "Read Enable=1".

- The "Input data" are written started at the negative edge of the "Write Clock".

- The "Output data" are read started at the negative edge of the "Read Clock".

- Obviously, 'same frequency" between the "Write clock" and the "Read clock is required only in FIFO. And both of them are not "time-in-line". Which means that SN74ACT7881 is a "synchronous FIFO".

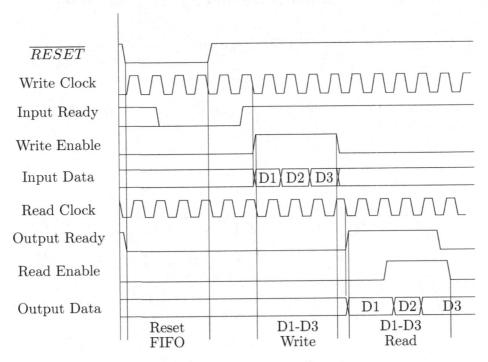

Figure 7-29   Signals in Synchronous FIFO SN74ACT7881
With Multilevel Synchronization of Status Outputs

7.2.4   Register

Register

In fact, **Register** is a set of storage units, which can *"hold"* or *"storage"* the information of binary system ("0" or "1").

- As well known, the flip-flop is a *bi-stable device.*

  – Which can store one bit information ("0" or "1") when it is stable without any trigger.

  – However, it can be *"Toggle"* by any trigger to change from one state to another state, namely, from "0" to "1" or from "1" to "0".

- In this case, "n-flip-flop" can store any "n-bit" information in binary system, which can be controlled by an unique "clock".

- Except *"n-flip-flop"* unit, a register also contain some *"combination gate"*. Where,

— "n-flip-flop" storages the "n-bit" information.

— "Combination gate" controls the transmission of the "n-bit" information at when which should be controlled and how it should be transmitted.

4 bits register

For the sake of the simplification, we take a 4 bits register in Figure 7-30 to explain how it works.

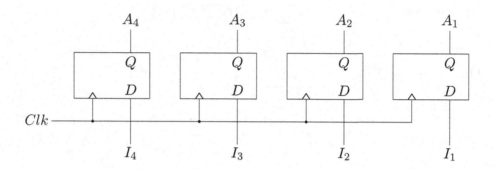

Figure 7-30   4 bits register

As shown in Figure 7-30, the 4 bits register is consist of 4 D-flip-flops with Clk as the common clock.

As well known, the function of D-flip-flop is

- Q=D from the positive edge of Clk $\rightarrow$ 1.

- Q=0 from the negative edge of Clk $\rightarrow$ 0.

Thus, for a 4 bits register in Figure 7-30, we have

- Q=D when Clk=1, and then $A_n = I_n$ (n=1,2,3,4).

- Q=0 when Clk=0 and then A_n is independent from I_n (n=1,2,3,4).

8 bit register

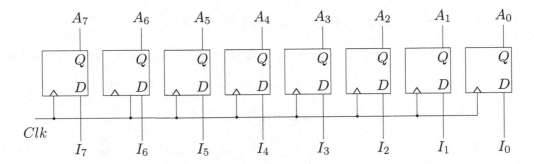

Figure 7-31   8 bits register

For a 8 bits register in Figure 7-31, we have

- Q=D when Clk=1, and then $A_n = I_n$ (n=0,1,2,3,4,5,6,7).

- Q=0 when Clk=0 and then A_n is independent from I_n (n=0,1,2,3,4,5,6,7).

16 bit register

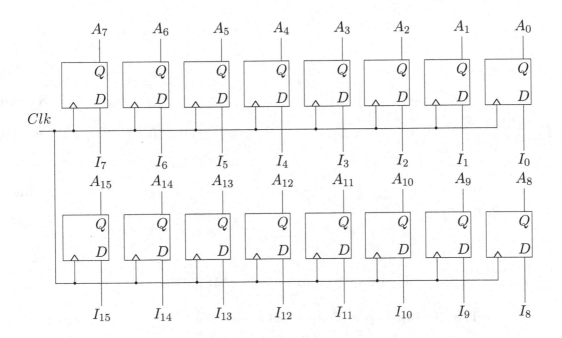

Figure 7-32   16 bits register

For a 16 bits register in Figure 7-32, we have

- Q=D when Clk=1, and then $A_n = I_n$ (n=0,1,2,3,4,5,6,7,8,9,10,11,12,13,14,15).

- Q=0 when Clk=0 and then A_n is independent from I_n (n=0,1,2,3,4,5,6,7,8,9,10,11,12,13,)

D-flip-flop

Each unit in register is a D-flip-flop. To understand the register more in detail, we may recall the D-flip-flop in schematic level.

- How to tell the difference between "flip-flop" and "latch"?

  - The "flip-flop" is sensitive to the transit from "0" to "1" of the clock.
  - The "latch" is sensitive to the duration of "Clk=1" of the clock.

- The flip-flop as a basic unit to storage 1 bit information may form four flip-flops with different combination gate as follows:

  - SR flip-flop (means "set", "reset").
  - D flip-flop (means "Data"), used for register.
  - T flip-flop (means "Toggle"), used for counter.
  - JK flip-flop.

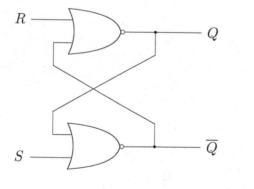

S	R	Q	$\overline{Q}$	
0	0	0	1	(No change)
0	0	0	1	(No change)
1	0	1	0	(Set)
0	1	0	1	(Reset)
1	1	x	x	(Undefined)

Figure 7-33   Storage unit - SR flip-flop

- The **SR flip-flop** in Figure 7-33 is based on the "Set" and "Reset" idea. And the function of **SR latch** gate in Figure 7-34 are

  - "S" may "set": means that Q=1 when S=1, R=0.

- "R" may reset: means that Q=0 when R=1, S=0.

- The function above is hold when "clock" Clk=1. Means that the "latch" is "open".

- "Q" is independent from "S" and "R" when Clk=0. Means that the "latch" is close.

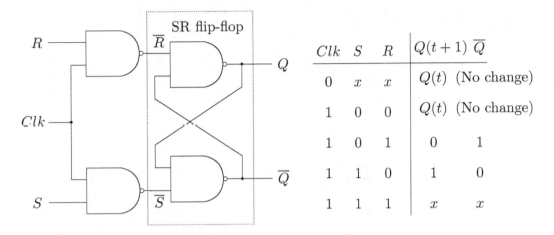

Clk	S	R	$Q(t+1)$	$\overline{Q}$
0	x	x	$Q(t)$ (No change)	
1	0	0	$Q(t)$ (No change)	
1	0	1	0	1
1	1	0	1	0
1	1	1	x	x

Figure 7-34   SR latch gate

- However, the SR-Latch can not over come a big issue, that

    - Q(t) may be "1" or "0" when both of S=1 and R=1 at same time.

- Gated D-Latch in Figure 7-35(a) is a good idea to over come this big issue by using a "negative gate" between "S" and "R", so as R=0 when S=1, or S=0 when R=1. Thus, D-Latch is more useful than the SR-Latch. Where

    - "D" is input data. "Q" is output data.

    - Q=1 when D=1 (means that S=1, R=0).

    - Q=0 when D=0 (means that S=0, R=1) without the issue of S=R=1.

    - **When clock Clk=1, Q follows D start at the positive edge of the clock Clk.**

    - When clock Clk=0, Q is independent from D.

- Now in Figure 7-35(d),

    - Q follows D start at $t = t_1$ (the positive edge of clock Clk).

    - Again, Q follows D start at $t = t_2$ (the positive edge of clock Clk).

    - Again, Q follows D start at $t = t_3$ (the positive edge of clock Clk).

    - Again, Q follows D start at $t = t_4$ (the positive edge of clock Clk).

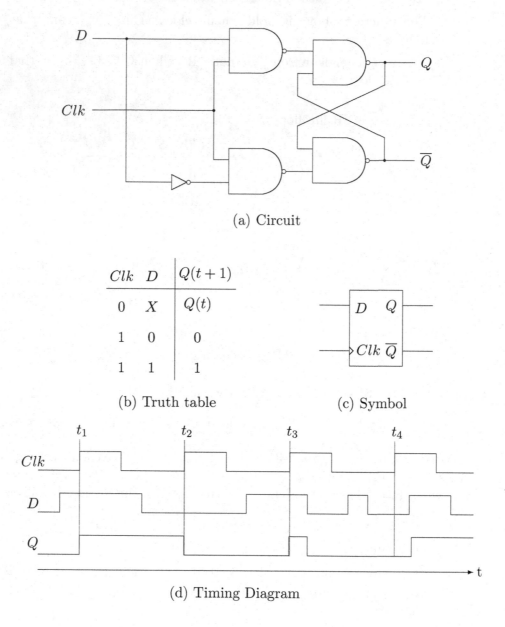

(a) Circuit

Clk	D	$Q(t+1)$
0	X	$Q(t)$
1	0	0
1	1	1

(b) Truth table   (c) Symbol

(d) Timing Diagram

Figure 7-35   D-Latch

- Based on the function of D-Latch above, it is easy to understand the 4 bit register in Figure 7-30 , the 8 bit register in Figure 7-31 and the 16 bit register in Figure 7-32.

- All of those registers will obey the rule above, namely,

  - Q follows D start at the time when Clk transit from "0" to "1" (called positive edge). And then $A_n = I_n$ at positive edge of the clock "Clk".

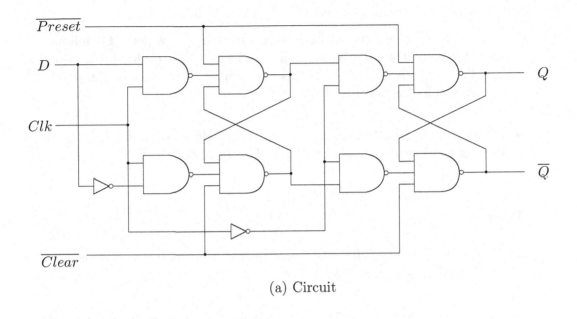

(a) Circuit

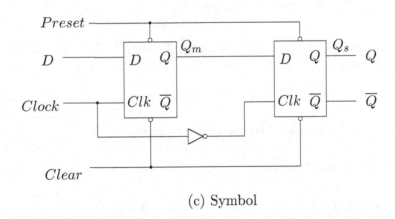

(c) Symbol

Figure 7-36 Master-slave D-flip-flop with "Clear" and "Preset".

- Based on the function of D-latch above, the n-bit register may storage n-bit binary information (n=4,8,16), respectively.

- In this case, the data from "D" may be transfer to "Q" (storage) controlled by the "Clk".

- We may add "Clear" and "Preset" to each register. Where, "Clear" and "Preset" have priority over "Clk". A Master-slave D-flip-flop with "Clear" and "Preset" is shown in Figure 7-36.

- Where the Master-slave D-flip-flop in Figure 7-36 includes two D-flip-flops:

  - The left one is called the master D-flip-flop, in which, "Q" follows "D" started at the positive edge of the clock "Clk".

  - The right one is called the slave D-flip-flop, in which, "Q" follows "D" started at the negative edge of the clock "Clk". (Where $\overline{Clk} = 0$ when Clk=1). Figure 7-36(c) shows the symbol of the circuit, in which, the slave flip-flop is controlled by $\overline{Clk}$.

- Now a timing diagram in Figure 7-37 may help us to understand the operation of the Master-slave D-flip-flop in Figure 7-36.

Where

- In master flip-flop (left), "Q_m" follows "D" start at the positive edge of the clock "Clk" as shown in Figure 3-36.

- In slave flip-flop (right), "$Q = Q_s$" follows "D" start at the negative edge of the clock "Clk" as shown in Figure 3-36.

- Thus, the function of the Master-slave flip-flop is that "Q" follows "D" start at the negative edge of the clock "Clk".

Now, the "clear" are added to the NAND gates.

- When "Clear=0", it forces "Q=0" no mater the "D" and the clock "Clk".

- "Q" is independent from "Clear" when "Clear=1".

Meanwhile, the "Preset" are added to the NAND gates as well.

- When "Preset=0" it forces "Q=1" no mater "D" and clock "Clk".

- "Q" is independent from "Preset" when "Preset=1".

Obviously, both of the "Clear" and the "Preset" are priority over the clock "Clk".

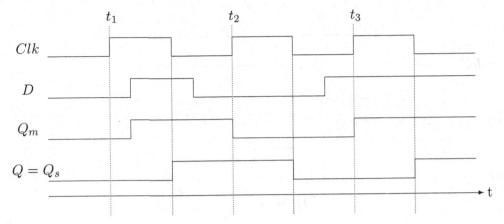

Figure 7-37   Timing Diagram of the Master-slave D-flip-flop

7.2.5   Clock and Synthesizer

We will mention ourself, that 10Gbps optical fiber transponder is used for the long haul optical fiber transmission link. In the transmission link, how to keep all of the 9.9528Gbps series data in order? **The clock. The 9.9528GHz clock**, which controls the 9.9528Gbps series data in order. And the 622.08MHz clock, which controls 622.08Mbps parallel data in order. And the 622.08Mhz clock is from 9.9528GHz clock via 1/16 divider.

Now the 9.9523GHz clock is transmission with the 9.9523Gbps series data and the 9.9523Gbps series data input is very weak when they have been transmitted 50km in the optical fiber link.

- How to recover the very weak 9.9523Gbps series data?

- How to pick up the very weak 9.9523GHz clock from the very weak 9.9523Gbps series data?

In Digital receiver DeMux

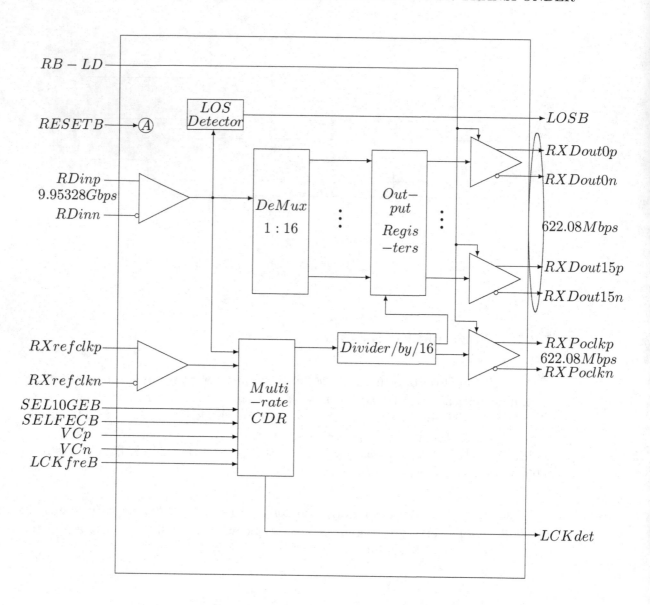

Figure 7-38   De-multiplexer - DeMux

DeMux in Figure 7-38 is a digital receiver. The function of the DeMux are

- To recovery the clock and data from the 9.95328Gbps series signal data input.

- Then the De-multiplexer is used to convert the 9.95328Gbps series signal data into 16 channels of 622.08Mbps parallel data.

Where, it is required, that

- Before 1:16 De-multiplexer, the 9.95328Gbps series data have to be re-time by 9.95328GHz clock. This clock comes from the input 9.95328Gbps series data via the CDR (clock and data recovery).

- After 1:16 De-multiplexer, the 622.08Mbps data has to be re-time by 622.08MHz clock in output register. This clock comes from the 9.95328GHz clock via 1/16 divider.

7.2.6 CDR- Clock and Data Recovery

CDR- Clock and Data Recovery

- How to recover the 9.95328Gbps series signal data in CDR?

By a recovery circuit in CDR as shown in Figure 7-39.

- How to pick up the 9.95328GHz clock from the 9.95328Gbps series signal data in CDR?

By a phase-locked-loop in CRD as shown in Figure 7-40.

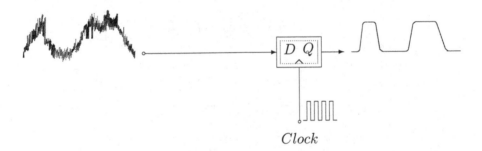

Clock

Figure 7-39   Data-recovery

Now we start from *the idea of the data-recovery* shown in Figure 7-39, the input is a noisy data stream come from a long haul ($40kM \sim 50km$) optical fiber transmission links. Which contains the clock and the data signal with a random noise signal.

- The clock recovery circuit (CRC) is to be used to "re-time" the input data stream by a supper low noise clock so as to reduce the jitter.

- The idea of the clock recovery circuit (CRC) is *to "re-sample" the midpoint of each bit by a D-flip-flop* that is driven by a supper low-noise clock. Which means that

– The output Q=1 if the midpoint of each bit $D \geq D_0$, D_0 is a certain value.

– The output Q=0 if the midpoint of each bit $D < D_0$.

However, in many applications, the clock may not be available independently. For example, for a long haul optical fiber communications link, the optical fiber carries only the random series data stream, providing no separate clock waveform at the receiver end. Therefore, we need an independent supper low jitter 9.9523GHz clock to do the data recovery. Which may be obtained by, for instance, a digital-hybrid phase-locked loop as shown in Figure 7-40.

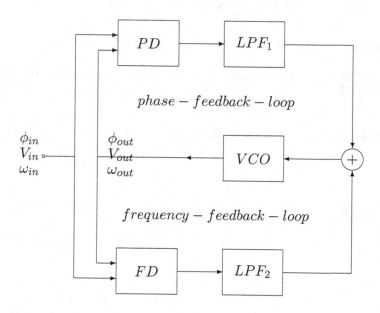

Figure 7-40   A digital-hybrid phase-locked loop

Now we will discuss digital-hybrid phase-locked loop as shown in Figure 7-40. Which is consist of two phase-locked-loops:

- The XOR gate as the phase detector in the digital-phase-locked loop can be used as the phase discriminator.

- The phase/frequency detector in analogy-phase-locked-loop can be used as the frequency discriminator.

Both of them combine together to form a phase/frequency discriminator to control the VCO.

- The output pulses from XOR gate indicate the required direction of phase corrections, these pules are smoothed by a loop filter in digital loop when loop is phase-lock-in.

- When the loop is phase-lock-out, the output of the discriminator in analogy phase-locked-loop will force the frequency of the VCO to track the input frequency until the loop phase-lock-in. In this case, the digital loop will be operation again and the VCO will generate a supper low jitter clock.

Thus, the digital-hybrid phase-locked loop displays a supper low jitter output as well as a good loop stability [91].

A diagram of a digital-hybrid phase-locked loop

A digital-hybrid phase locked loop is shown in Figure 7-40. In which,

- A digital phase-locked loop is phase-aligned the input phase ϕ_{in}, so as to obtain a supper low output jitter.

- An analog phase-locked loop is used to track the input frequency ω_{in} to obtain a wider pull-in range.

Data recovery circuit

Based on the discussion above, the data recovery circuit is shown in Figure 7-41.

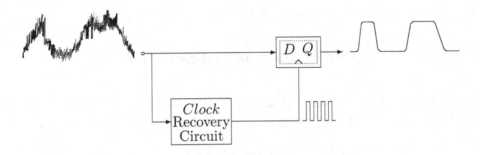

Figure 7-41   Data-recovery in 10Gbit/s optical fiber transponder

Based on the discussion above, we understand that

- CDR (clock and data recovery) generate a clock that is at the same frequency as the incoming data rate at CML series input.

- The clock is applied to a D flip-flop to "re-sample" the midpoint of each bit of the input random data stream.

    – If the input signal "$D \geq D_0$" at the positive edge of the clock, the output of D flip-flop $Q = 1$.

    – If the input signal "$D < D_0$" at the positive edge of the clock, then $Q = 0$.

Since the re-sample is once a bit, the output of D flip-flop will provide the re-timed data stream as shown in Figure 7-41.

The CDR can be forced to lock to RXREFCLK regardless of a valid high-speed series input signal. Thus, if the input high-speed series data stream is stopped for some reason, the VCO in CDR can provide a reference clock as same as the RXREFCLK, inputs to the receiver DeMux.

In digital transmitter - Mux

The Mux in Figure 7-42 is a fully integrated MSA-compliant quadrate SONET transmitter, operating at OC-192 and data rate of 9.95328Gbps with

- Serializer,

- Clock multiplication unit (CMU),

- Loss-of-signal (LOS) detection circuit.

CMU - Clock multiplication unit

There is a phase-locked-loop (PLL) in CMU, which generates the series output clock 9.95328GHz, TSclk, by multiplying the reference clock 622.08MHz, TXrefclk.

Which means that there are two clocks in Mux as shown in Figure 7-42,

- The series output clock (9.9523GHz) TSclk, which is generated by a PLL with multiplying the input reference clock (622.08MHz), TXrefclk (See Table 7-5 and Table 7-6).

- The transmit parallel clock TXPclk, which is the TSclk (9.9523GHz) divided by 16.

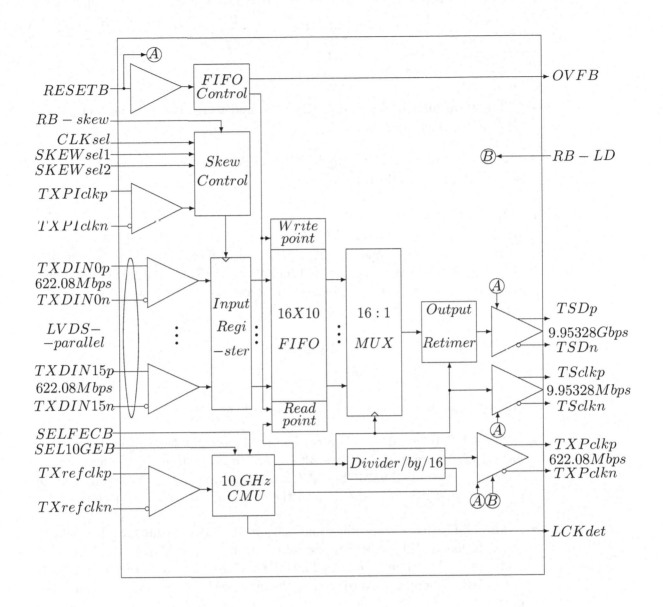

Figure 7-42   Multiplexer-Mux

Table 7-5 Data Rate Selection

Series Data Rate	Parallel Date Rate	Rate Selection	Rate Selection
TSD(Gbps)	TXIN(Mbps)	SELFECB(pin A11)	SEL10GEB(pin C13)
9.953	622.08	1	1

Table 7-6 Reference Clock Frequency Selection

Series Data Rate	Ref. Clock Freq.	Ref. Clock Freq.	Ref. Clock Freq.
TSD(Gbps)	TXREFclk(MHz)	TXPclk(MHz)	REF155ENB(pin A8)
9.953	622.08	622.08	1

The function of these two clocks are

- The transmit parallel clock TXPclk is used to re-time the parallel data at 16:1 multiplexer (16:1 MUX).

- The series output clock (9.9523GHz) TSclk is used to re-time the input series data at the optical transmitter.

- In FIFO (First-In-First-Out), the transmit parallel clock TXPclk is used to "read" the data, the clock TXPIclk is used to "write" the data. Both of them are the same frequency 622.08MHz, but different phase. After the data is clocked into the register, it is fed into 10-word FIFO. The FIFO eliminates difficult system loop timing issues. It accommodates up to 6.0 ns of delay difference between TXPIclk and TXPclk.

Symmetry

In 10Gbps optical fiber transponder, there are two clocks

- The CDR (clock data recovery) in digital receiver DeMux in Figure 7-38 has a PLL generates a clock 9.9523GHz, which is at the same frequency as the incoming data rate 9.9523Gbps at CML series input. The clock is aligned by a PLL so that it sample the data in the center of the data eye pattern.

- The CMU (clock multiplication unit) in digital transmitter Mux in Figure 7-42 has a PLL generates the series output clock 9.9523GHz, TSclk, by multiplying the reference clock 622.08MHz, TXrefclk. The series output clock 9.9523GHz, TSclk, is used to re-time the input series data at the optical transmitter.

Meanwhile, there are two reference clocks in 10Gbps optical fiber transponder,

- The reference clock 622.08MHz in digital transmitter Mux in Figure 7-42, TXrefclk, is used to generates the series output clock 9.9523GHz by CMU.

- The reference clock 622.08MHz in digital receiver DeMux in Figure 7-38, RXrefclk, is used to generates the series clock 9.9523GHz by MULTI-RATE CDR at the time when the input data stream of the digital receiver DeMux in Figure 7-38 is stopped for some reason.

There is a reference clock - RXrefclk 662.08MHz input the digital receiver Mux. The recommended operating condition of the reference conditions are list in Table 7-7.

Table 7-7 The Reference Clock

Parameter	Conditions	Min	Type	Max	Units
Reference clock frequency	-	-	622.08	-	MHz
Stability of reference clock freq.	-	-150	-	+150	ppm
Reference clock duty cycle	40	-	60	-	%
Reference clock rise and fall times	-	-	0.3	-	ns

7.3 Layout of 10 Gb/s optical fiber transponder

7.3.1 Schematic

A block diagram of 10 Gb/s optical fiber transponder

MTP103 Transponder is designed to provide a SONET or SDH compliant electro-optical interface between the SONET or SDH photonic physical layer and the electrical section layer.

Figure 7-43 shows a block diagram for transponder module. The module contains a 10Gb/s optical transmitter (EML TX), a 10 Gb/s optical receiver (RX TIA+LA), a Multiplexer chip (MUX 16:1) and a De-Multiplexer chip (De-MUX 1:16) in the same physical package to provide sixteen $622Mb/s$ LVDS channels on both Input (16 Data In) and Output (16 Data Out) side.

a. 300 pin provide an Adder/Dropper interface to down load any channel of the 16 Data Out to the local station and up load any channels from local station to 16 Data In.

b. Meanwhile, 300 pin also provide a interface to send the alarm signal to the local station and receive the control signal from local station to modify the error from the transponder.

c. Microprocessor Control circuit provide a control signal communication with the local station, which includes a lose signal to tell local station that the photo-receiver can not receive enough signal from the local station.

d. A laser Control Circuit combined with the microprocessor used to control the temperature of Laser in a certain variation range.

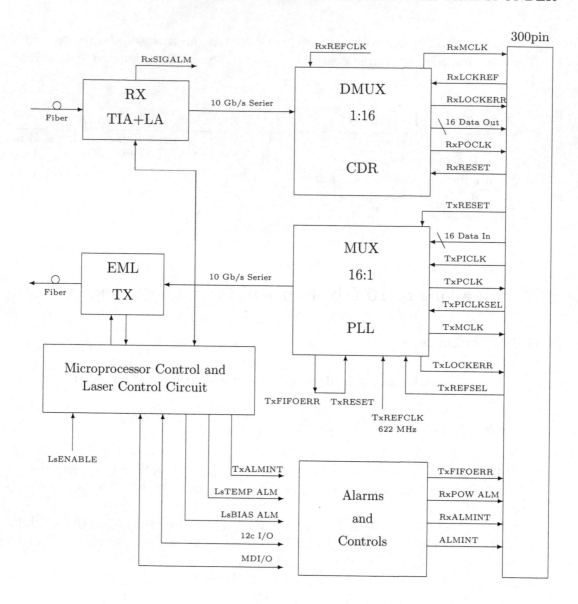

Figure 7-43   MTP103, 10Gbps Transponder Block Diagram

The schematic of OC192 10 Gb/s Transponder

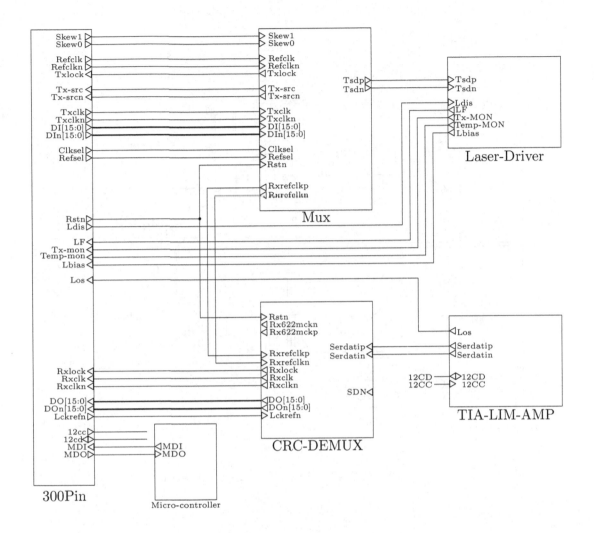

Figure 7-44   The schematic of OC192 10 Gb/s Transponder

The main schematic of 10Gbps transponder is shown in Figure 7-44. In which:

    a. TIA-LIM-AMP is the Photoelectric converter.

    b. Laser-Driver is the electro-optical converter.

    b. Mux is the multiplexer (16:1). The schematic of Mux can be made by the data sheet from the Multiplexer company.

    c. CRC-DEMUX is the de-multiplexer (1:16) with clock recover circuit in it. The schematic of DeMux can be made by the data sheet from the De-multiplexer company.

    d. Micro-controller.

    e. 300pin mainly is the interface of the data adder/dropper.

7.3.2 Layout

300 pin   Mux

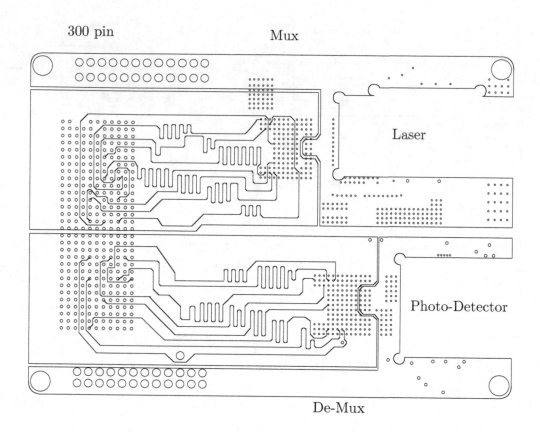

Laser

Photo-Detector

De-Mux

Figure 7-45   Layout of a signal layer

As shown in Figure 7-44, 10Gbps optical fiber transponder is consist of three parts:

a. Laser: which is built in an independent metal box.

b. Photo-detector with amplifier: which is built in another metal box.

c. The 12 layers of printed circuit board: which includes multiplexer (MUX), de-multiplexer (De-MUX) and 300 pin, between them are two set of 16 channels of 622Mbps data stream. Which can be built up on a 12 layers of printed circuit board. In which, one signal layer is shown in Figure 7-45.

Clearly, the one signal layer is a printed circuit board layer with 16 channels, each channel is working at 622Mbps data transmission.

a. First, the photo-detector converts the 10 Gbps optical data stream from the optical fiber into 10Gbps electrical data stream.

b. Then the 10 Gbps electrical data stream is divided into 16 channels of

622Mbps data stream via De-multiplexer.

c. The 16 channels of 622Mbps data stream are connected to 300 pin via the microstrip lines.

d. Where, as the data adder/dropper, the 300 pin download any channels of 16 data stream to local station and up load any channel data stream from the local station, which become a new 16 channels of data stream, sent to Multiplexer via the microstrip lines.

e. The new 16 channels of data stream are converted into 10 Gbps electrical data stream via Multiplexer, and send to laser via the microstrip line.

f. The laser converts the 10Gbps electrical data stream into 10Gbps optical data stream sent to the optical fiber.

In all of the processing above, the processing from (b) to (e) is implemented in the 12 layer of printed circuit board. Therefore, the design of a 10Gbps optical transponder is concentrated on the 12 layers of printed circuit board design and the Laser and the photo-detector design are two special designs independently.

12 layers of PCB design

Now the 12 layers of printed circuit board design is supported by the program "Orcad", in which,

a. The program "Capture" is used to draw the circuits diagram, which includes Multiplexer, De-multiplexer, 300 pin, processor, alarm circuit, signal loss circuit and power supply etc.

b. The program "Layout" converts all of the circuit diagram from "Capture" into 12 layer of PCB (printed circuit board). In which the Multiplexer, De-multiplexer and 300 pin etc are the structure diagrams on the PCB, respectively.

The signal layer of PCB design

The signal layer of PCB design is the **mainly design and key design**, if you didn't pay more attention onto the PCB design especially the signal layer design, the test result of the 10 Gbos transponder design will be fail.

a. The key point of this design is the path length of each channel in 16 channels must be equal. The reason of this point is that there are two processing inside the PCB, one is that 10 Gbps data stream is converted into 16 channels of 622Mbps data stream via De-multiplexer; the second one is that the 16 channels of 622Mbps data stream are converted into 10Gbps data stream. Each bit in these two 10Gbps data stream shall be in a "same properly position". If the path length of each channel of 16 is not the same, the "same properly position" in the second 10 Gbps data stream will be destroyed, which will results in a big bit error rate.

b. As shown in Figure 7-45, each path of 16 channels must be smooth without any angle less than 90 degree. Otherwise, the path with small angle will cause some wave reflection of the data stream. Which will interrupt the data stream transmission in this path.

c. Each layer is a printed circuit board, the bottom of each layer is a whole metal as the ground. Therefore, the 16 channel paths is inside a sandwich. Which prevent any electrical interruption from the other layer.

d. The top layer and the bottom layer are loaded the power supply, processor, signal loss circuit and alarm circuit etc. In each of these two layer, the whole plan is divided into digital signal region and analog signal region, so as to prevent the 10 Gbps data signal interrupted by the power supply and other analog signal. Meanwhile the path length of the 10 Gbps data stream shall be as short as possible.

e. The connection between 12 layer is implemented by "Via", which is a gold-plated hole go through 12 layer. Some of the "Vis" are used as the connection between each layer. Most "Via" are used as ground. Therefore, 12 layer is constructed as a lot of ground to protect the available circuits and the channel paths with a sandwich structure. .

Chapter 8

Introduce to Quantum communication

8.1 Introduction

8.1.1 The conception of the quantum communications

In 2015, NIST (National Institute of Standards and Technology) held a workshop on "Cybersecurity in a Post-Quantum World," which was attended by over 140 people, and "action at a distance" in quantum mechanics has been verified by scientists. The scientists in NIST have been realized several pairs of entangled photons. One of each photon pair is moved to A and another one is moved to B. A and B are separated. And the measurement of each photon pair verified the "action at a distance" in quantum mechanics is truly investigated.

"Action at a distance" is the concept that an object can be moved, changed, or affected without being physically touched (as in mechanical) by another object. It is nonlocal interaction of objects that are separated in space. For instance, if a pair of particles is generated in such a way that their total spin is known to be zero, and one particle is found to have clockwise spin on a certain axis, the the spin of the other particle, measured on the same axis, will be found to be counterclockwise spin, as to expected due to their entanglement. In the entangled particles, such a measurement will be on the entangled system as the whole. It thus appears that one particle of an entangled pair **"knows"** what measurement has been performed on the other, as well as measurement result. Which is so called **"action at a distance"**. For instance, although one particle is on the earth, another one is on the moon, one of an entangled pair **"knows"** what measurement has been performed on the other, as well as the measurement results with no time. "Action at a distance" is very surprising, that no any normal communication mechanism between the particle pair and no time to be taken as one "knows" the measurement has been performed on the other. This phenomena provide us an idea to realized a quantum communication in the space with almost no time.

To implement the quantum communication:

One way is that, after a pair of entangled photons is generated, keeping one of the entangled photons in your hand and sending another one of the entangled photons into another location along the based infrastructure of optical fiber communications link.

Another way is keeping one of the entangled photon pair in your hand, sending another one of them on the satellite to form the satellite quantum communications.

8.1.2   The conception of Quantum Cryptography

"Action at a distance" in quantum mechanics also is interested for the Cryptography.

The origins of quantum cryptography can be traced to the work of Wiesner, who proposed that if single-quantum states could be stored for long periods of time they could be used as counterfeit-proof money. Wiesner eventually published his idea in 1983, but they were of largely academic interest owing to impractically of isolating a quantum state from the environment for long time periods. However, Bennett and Brassard realized that instead of using single quantum for information storage, they could be used for information transmission. In 1984 they published the first quantum cryptography protocol now known as "BB84". A further advance in theoretical quantum cryptography took place in 1991 when Ekert proposed that Einstein Podolsky Rosen Paradox(EPR) entangle two-particle states could be used to implement a quantum cryptography protocol whose security was based on Bell's inequalities. Also in 1991, Bennett and collaborators demonstrated that Quantum Key Distribution (QKD) was potentially practical by constructing a working prototype system for the BB84 protocol, using polarization photons.

In 1992 Bennett published a "minimal" QKD (Quantum Key Distribution) scheme ("B92") and proposed that it could be implemented using single-photo interference with photons propagating for long distance over optical fibers. Since then, other QKD protocols have been published and experimental teams in UK, Switzerland and the USA have developed optical fiber-based prototype QKD systems. The aim of these is the conceptual feasibility of QKD, rather than to produce the definitive system, or to address a particular cryptographic application. Thus we can expect that the experiences with the current generation of systems will lead to implements towards demonstrating the practical feasibility of QKD as well as a definition of the applications where it could be used.

The remainder of this chapter is organized as follows:

- The significance of QKD.

- The B92 QKD protocol.

- The immunity of QKD to eavesdropping.

- Some of the practical issues that arise in implementing QKD.

In the last three decades, public cryptography has become an indispensable component of our global communication digital infrastructure. These networks support a plethora of applications that are important to our economy, our security, and our way of life, such as mobile phones, internet commerce, social networks, and cloud computing. In such a connected world, the ability of individuals, businesses and governments to communicate securely is of the utmost importance.

Many of our most crucial communication protocols relay principally on three cryptographic functionalities: public key encryption, digital signatures, and key exchange. Currently, these functionalities are primarily implemented using Diffie-Hellman key exchange, the RSA cryptosystem, and elliptic curve cryptosystems. The security of these depends on the difficulty of certain number theoretic issue such as Integer Factorization or the Discrete Log issue over various teams.

In 1994, **Peter Shor** of Bell Laboratories showed that quantum computers, a new technology, which can leverage the physical properties of matter and energy to perform calculation, and then which can efficiently solve each of these problems. Thus a powerful quantum computer will put many forms of modern communication, from key exchange to encryption to digital authentication. It is reconveyed that the quantum computers could be utilized to solve certain problems faster than classical computers, which has inspired great interest in quantum computing.

A large international community has emerged to address the issue of information security in a quantum computing future, in the hope that our public key infrastructure may remain intact by utilizing new quantum-resistant primitives. In the academic world, this new science bears the name "**Post-Quantum Cryptography**". This is an active area of research, with its own conference series, PQCrypto, which started in 2006. It has received substantial support from national funding agencies, most notably in Europe and Japan, through the EU projects PQCryto and SAFE-crypto, and the CREST Crypto-Math project in Japan.

These efforts have led to advances in fundamental research, paving the way for the deployment of post-quantum cryptosystems in the real world. In the past few years, industry and standards organizations have started their own activities in the field: since 2013, the European Telecommunications Standards Institute (ETSI) has held three "Quantum-Safe Crytography" workshops, and in 2015, NIST (National Institute of Standards and Technology) held a workshop on "Cybersecurity in a Post-Quantum World," which was attended by over 140 people.

Research into the feasibility of building large-scale quantum computers began in earnest after Peter Shor's 1994 discovery of a polynomial-time quantum algorithm for integer factorization. it was unclear whether quantum computing would be fundamentally scale technology. Many leading experts suggested that quantum state were too fragile and subject to the accumulation of error for large-scale quantum computation ever to be realized. This situation changed in the late 1990s with the development of quantum error correcting codes and threshold theorems.

This threshold theorems show that if the error rate per logical operation ("quantum gate") in a quantum computer can be brought below a fixed threshold, then arbitrarily long quantum computations can be carried out in a reliable and fault-tolerant manner by incorporating error-correction steps throughout the execution of the quantum computation.

Over the years, experimentalists have gradually developed improved hardware with ever lower error rates per quantum gate. Simultaneously, theorists have developed new quantum error correction procedures yielding higher fault-tolerance thresholds. Recently, some experiments that are nominally below the highest theoretical fault-tolerance thresholds (around 1%). This is a significant milestone, which has spurred increased investment from both government and industry. However, it is clear that substantial long-term efforts are needed to move from present day laboratory demonstration involving one to ten qubits up to large-scale quantum computers involving thousands of logical qubits encoded in perhaps hundreds of thousands of physical qubits.

8.2   The principle of quantum entanglement

8.2.1   The phenomena of quantum entanglement

Quantum entanglement is a physical phenomenon that occurs when pairs or groups of particles are generated or interact in such a way that the quantum state of each particle cannot be described independently, instead, a quantum state must be described for the system as a whole. Measurement of physical properties such as position, momentum, polarization, etc., performed on one entangled are found to be appropriately correction.

For example, if a pair of particles is generated in such a way that their total spin is known to be zero, and one particle is found to have clockwise spin on a certain axis, the spin of the other particle, measured on the same axis, will be found to be counterclockwise spin. as to expected due to their entanglement. In the entangled particles, such a measurement will be on the entangled system as the whole. It thus appears that one particle of an entangled pair "**knows**" what measurement has been performed on the other, as well as measurement result. Which is so called "**action at a distance**". That is , it is the nonlocal interaction of two particles that are separated in space, for instance, although one particle is on the earth, another one is on the moon, one of an entangled pair "**knows**" what measurement has been performed on the other, as well as the measurement results, with no time.

Now, assuming that when a particle with zero spin decays, two particles with spin are generated caused by the even and are separated along an axis with opposite direction to A and B, respectively. And found that if the spin of one particle at A is, say, clockwise spin, the other one at B is counterclockwise spin definitely. If one particle at A is, say, counterclockwise spin, the other one at B is clockwise spin definitely. Especially,

When you measure the spin of two particles at A and B at the same time, respectively, you may find that they are violation of Bell's inequality.

When you measure one of the two particles, the other one seems "knows" the measurement and the measurement result, although they are separated a large distance in space and no any communication mechanism between them.

Which means although they may be separated a large distance in space, they are not independent to each other. This phenomena is called that the two particles are in anti-correlation , or say, that two particles are "**entangled**".

Basic conception of entanglement

Assuming that the decay of a π pion with zero spin caused an electron and a positive electron to be generated [92]. This two particles are moving along an axis with opposite direction to A and B, respectively. In which, Alice at A fund that the polarization of the electron at A is up-spin. Bob at B found that the polarization of the positive electron at B is down-spin, or, Alice at A found that the polarization of the electron at A is down-spin. Bob at B found that the polarization of the positive electron at B is up-spin. So as the superposition of two particles forms a zero spin state. The system obeys the momentum conservation law. Before measurement, two particles forms a entangled state $|\psi >$, which is the superposition of two product states as

$$|\psi >= \frac{1}{\sqrt{2}}\Big(|\uparrow> \otimes|\downarrow> -|\downarrow> \otimes|\uparrow> \Big)$$ (8.1)

Where $|\uparrow>$ indicates that the polarization state of the electron is up-spin, $|\downarrow>$ indicates that the polarization state of the positive electron is down-spin.

The first term $|\uparrow> \otimes|\downarrow>$ means that the spin of the electron is "up-spin" if and only if the positive electron is "down-spin".

The second term $|\downarrow> \otimes|\uparrow>$ means that the spin of electron is "down-spin" if and only if the positive electron is "up-spin".

The superposition of two term is the state of the whole entangled system. However, no one can predict which spin will be for each particle, provide that to measure each particle, respectively. Quantum mechanics can not predict which term will be presented, instead, quantum mechanics can predict that the probability of each term is 50%.

Creation of quantum entangled state

In the early years, the way to create quantum entangled state is by atom correlation as shown in Figure 8-1,

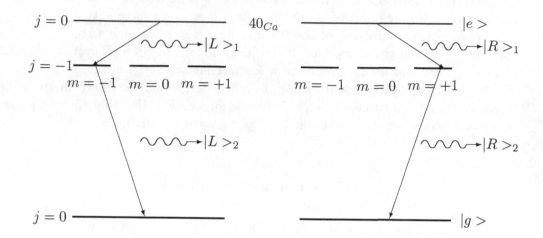

Figure 8-1   Creation of entangled particles from atom

Where the atom 40_{C_a} with zero spin decays from the excited level $|e>$, causes two particles to be created, in which:

The state of the left one with left-circular spin ($m = -1$) is noted as $|L>_2$,

The state of the right one with right-circular spin ($m = +1$) is noted as $|R>_2$.

Both of $|L>_2$ and $|R>_2$ are entangled to each other, forms a new system with zero spin.

The wavelength of the photons (noted as $|L>_2$ and $|R>_2$) is 551.3nm. The wavelength of the photons (noted as $|L>_1$ and $|R>_1$)is 442.7nm.

j is the angular momentum, m is the magnetic quantum number.

Magnetic quantum number ($m = -1, m = 0, m = +1$) specifies the orientation in space of an orbit of a given energy. This number divides the sub-shell into individual orbits which hold the electrons.

Angular momentum j is the rotational analog of linear momentum. It is conserved quantity, the angular momentum of a system remains constant unless acted on by external torque.

Nowadays, one way to create the quantum entanglement is by using the incidence of a laser beam onto beta-barium borate crystal, a nonlinear crystal as shown in Figure 8-2. In this case, most of the photons will go through the crystal and only a few photons will cause pairs of entangled photons because of the spontaneous emission. Where, the photons with vertical polarization are distributed in the vertical polarization hone. The photons with horizontal polarization are distributed in horizontal polarization hone. And the entangled photons are distributed in the intersection of the vertical polarization hone and the horizontal polarization pone.

Now, a lot of scientists have an interest in the quantum entanglement and the quantum entanglement were created not only in photons, electrons but also in molecules, in Buckminsterfullerene (for instance, C_{60}), even in entangling macroscopic diamond at room temperature. For example, as shown in Figure 8-2, an entangled photons pair can be created by spontaneous parametric down-conversion (SPDC). In which, a strong laser beam is incident directly at a nonlinear crystal, BBO (beta-barium borate) crystal. Most of the photons continue straight through the crystal. However, occasionally, some of the photons undergo spontaneous down-conversion with type II polarization correction, and the resultant corrected photon pairs have trajectories that are constrained to be within two cones, whose axes are symmetrically arranged relative to the pump beam. Also, due to the conservation of energy, the two photons are always symmetrically located within the cones, relative to the pump beam. Importantly, the trajectories of the photon pairs may exist simultaneous in the two lines where the cones intersect. This result in entanglement of photon pairs whose polarization are perpendicular. [93], [94]

Spontaneous parametric down-conversion (SPDC) is simulated by random vacuum fluctuations, and hence the photon pairs are created at random times. The conversion is very low, on the order of 1 pair every 10^{12} incoming photons. However, if one of the pair (the signal) is detected at any time then its partner (the "idler") is known to be presented.

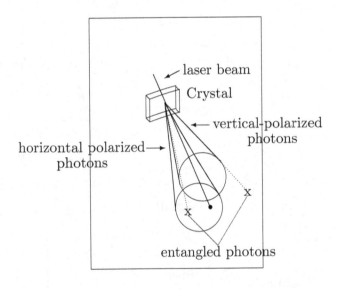

Figure 8-2   Creation of entangled photons from crystal

The spontaneous parametric down-conversion of a nonlinear $\chi^{(2)}$ crystal in Figure 8-3(a) obeys the energy conservation and momentum conservation as shown in Figure 8-3(b). In which,

a. A pump light of frequency

$$\nu_{pump} = h^{-1}(E_e - E_g) \tag{8.2}$$

is incident, which induces an electron transiting from ground state (E_g) into excited state (E_e) and results in a light absorption as shown in Figure 8-3(b). The process of light absorption will excite the atoms from ground state into the excited state. This is called the pumping transition.

b. After then, an atom in excited state decay by spontaneous emission to the ground state, releasing two photons with frequency ν_s and ν_i, respectively, because of the nonlinearity of the nonlinear χ^2 crystal.

The input light and two output lights obey the energy conservation as follows,

$$h\nu_{pump} = h\nu_s + h\nu_i \tag{8.3}$$

Which leads to

$$\omega_{pump} = \omega_s + \omega_i \tag{8.4}$$

as indicated in Figure 8-3(b). Meanwhile, the input light and two output lights obey the momentum conservation as indicated in Figure 8-3(b),

$$\vec{k}_{pump} = \vec{k}_s + \vec{k}_i \tag{8.5}$$

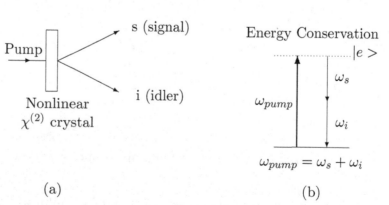

Figure 8-3   (a) Spontaneous parametric down-conversion in crystal
(b) Momentum conservation and energy conservation in crystal

These several years, the focus of the quantum entanglement research turns to consider the application of the quantum entanglement, such as in communication, computer and cryptography. A lot of outstanding research lead to consider how to use the "action in distance" of quantum entanglement for the communication. However, physicists still consider that the basic mechanism of the quantum entanglement is not so clear.

8.2.2 EPR Paradox and Bell's inequality

Since physicists still consider that the basic mechanism of the quantum entanglement is not so clear, we need a theoretical explanation and some experience to support the "action in space" of the quantum entanglement.

The Einstein Podolsky Rosen (EPR) Paradox

In 1935, Einstein Podolsky Rosen Paradox(EPR) published an important paper in which they claim that the whole formation of quantum mechanics can not be complete. Which means that there must be exist some elements of reality that are not described by quantum mechanics. They concluded that there must be a more complete description of physical reality involving some hidden variable that can characterize the state of affairs in the world in more detail than the quantum mechanical state. This conclusion leads to paradoxical results.

Now we will describe the point of view in the paper of Einstein Podolsky Rosen Paradox(EPR) argument as follows.

Considering two particles with spin-(1/2) A and B, both of them form a system. After some time, A and B are separated completely and can not affect each other. When Alice (investigator) measures a certain component of the spin value of A, based on the angular momentum conservation law, Einstein Podolsky Rosen Paradox(EPR) argument may expect the spin value of B. And then Einstein Podolsky Rosen Paradox(EPR) argument consider that, based on the reality criterion, they may confirm that 8 components of B spin should have a determined values, Which are the real physical components, and they are existed before measurement. However, Quantum mechanics doesn't allow to determinate 8 components of a spin simultaneously. Therefore, which can not be considered that it provide a complete description of the physical existence. If you conform that quantum mechanics is complete and perfect, you have to consider that the measurement on A will affect the state of B. Which will lead to conform the fact of "action in distance" of the entangled quantum.

Obviously, the real judgment of Einstein Podolsky Rosen Paradox(EPR) argument is based on an assumption of "locality", namely, if two system A and B are separated without any affect each other, then any measurement on A can not affect on the behavior of the state of B. Therefore, this real physical investigation based on the locality is called "local realism".

Bohmian mechanics — Hidden variables

When the EPR paper was published there already existed a hidden variable theory of quantum mechanics, which is called Bohmian mechanics completed form in 1952.

The hidden variable theory of Bohmian mechanics though that the behavior of an electron should have a "quantum potential" except the classic behaviors such as electromagnetic potential. The "quantum potential" is actually a kind of wave. Which obeyed the Schrodinger equation and spreads out from the electron. However, the effect of the quantum potential is independent of its intensity but is depend on the shape of the wave. Therefore, the quantum potential can reach to the end of the universal without any attenuation.

In Bohmian mechanics, electron is a classical article, but has a "quantum potential". An electron can spread a quantum potential to inquire information around him-self. So as he understand all of the information around him-self. When an electron is incident into a double slit, his quantum potential will find the exist of the double slit, so as to lead him-self to change his behavior according to the interference model. If one slit is closed, his quantum potential will find the change, so as to lead him-self to change his behavior. Consequently, if some one try to measure the particular position of an electron, the interaction between the measurement equipment and the electron, and his quantum potential will happen invisibly.

Bohmian mechanics greatly satisfies observation, however, the mathematic formula is very complicated. Meanwhile, Bohmian mechanics has to abandon the locality, means that all of the cause and effect should be hold in a particular area, without any transient action or propagation in transcendence of time and space. However, the quantum potential may spread the information required by article instantaneously.

Bell's inequality

John Stewart Bell was born in northern Ireland. He majored in the experience physics. However, he is extremely interested in the theoretical physics, especially in quantum-mechanics.

In 1964, Bell proved a powerful inequality $|P_{xz} - P_{zy}| \leq 1 + P_{xy}$, named Bell's inequality. Under the assumptions of locality and reality from Einstein Podolsky Rosen Paradox(EPR), Bell's inequality set up a strictly limitation for the measurement results of the quantum correlation between two separated articles if we measure two separated particles simultaneously. The conventional point of view of quantum mechanics requires that the correlation between two separated system will be beyond the allowance from the locality realism. Bell's inequality provide a criterion between the quantum uncertainty and hidden theory based on the local realism from Einstein, Podolsky and Rosen (EPR). Till now, all of the experiences indicated that quantum mechanics is correct, and the hidden theory based on the locality realism is false.

The violence of Bell's inequality implies that prediction of local realism from Einstein Podolsky Rosen Paradox(EPR) can not match the quantum mechanics. And all of the experience results is in good agreement with the prediction from quantum mechanics, which means that the correlation between two articles is far away over the prediction from the hidden theory of local realism. Consequently, physicists refuse to accepted the explanation from hidden theory of local realism. If they would not accepted the explanation from quantum mechanics, they have to consider that which is a cause and effect of super-luminal effect.

Some researchers have interpreted this result as showing that quantum mechanics is telling us that nature is non-local, that particles can affect each other across great distances in a time, which is too brief to have been due to ordinary causal interaction. Others object to this interpretation, and the problem is still open and hotly debated among both physicists and philosophers. It has motivated a wide range of research from the most fundamental quantum mechanical experiments through foundations of probability theory of stochastic causality as well as the metaphysics of free will.

Proof of Bell's inequality

In the developing of the local realism from Einstein Podolsky Rosen Paradox (EPR), Bell consider an experience idea: Assuming that an article with zero spin decays from the excited level $|e>$, causes two particles with sin-(1/2) A and B to be created. And A and B separated along the opposite direction in a large distance. Then measure the spin of each article, respectively in a three dimensional coordinate. For each article, the measurement result is noted as "+" if the measurement result is "up-spin", or noted as "-" if the measurement is "down-spin".

Now, assuming that an article with zero angular momentum decays cause two photons A and B are created and separated in opposite direction. According to the conservation of angular momentum, the polarization direction of one photon should be the same as the polarization direction of other one. Which can be conformed by setting a polarizer on a plane perpendicular to the moving direction of each photon, to see the polarization state of each photon. If A go through the polarizer and B will be sure to go through the polarizer, the correlation of two photons is 100%. However, If the setting is that two polarizer are perpendicular each other, then photon B will be stopped if A go through the polarizer. Which is 100% anti-correlation between A and B. The test results are list in Table 8-1.

Table 8-1

parallel θ^0	First pair	Second pair	Third pair	Forth pair	$\cdots$	n pair
Alice	+	-	-	+	$\cdots$	
Bob	-	+	+	-	$\cdots$	
Correlation	+1	+1	+1	+1	$\cdots$	$/n = +$
Orthogonal $\theta = 90^0$	First pair	Second pair	Third pair	Forth pair	$\cdots$	n pair
Alice	+	-	+	-	$\cdots$	
Bob	-	-	+	+	$\cdots$	
Correlation	+1	-1	-1	+1	$\cdots$	$/n = 0$

In which,

a. *correlation* = 100% if the polarization of A is parallel to that of B.

b. *correlation* = 50% if the polarization of A is perpendicular to that of B.

If the polarization of A is neither parallel nor perpendicular to the polarization of B, instead, the intersection angle of two polarization Y_A and Y_B is θ as shown in Figure 8-4, the probability of coincidence of A and B will be variable with the angle θ.

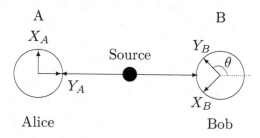

Figure 8-4   Observation to spin of articles by two observers

Now, The measurement is going at A and B independently by Alice at A and Bob at B, respectively.

If the measurement result for Alice is up-spin and is down-spin for Bob, we say that the measurement result of A is coincident with the result of B, noted as +1. Vice verse.

If the measurement results are up-spin both for Alice and Bob or are down-spin both for Alice and Bob, we say that the measurement result of A is not coincident with the result of B, noted as −1.

In this case, when we measure the entanglement of two articles,

The test result of the entanglement is coincident for ever, if Y_A is parallel to Y_B. Which is totally correction between A and B.

The probability of coincidence of A and B is 50% if Y_A is perpendicular to Y_B.

The test results are list in Table 8-2.

Table 8-2 In XYZ coordinate system

A_x	A_y	A_z	B_x	B_y	B_z	probability of +1
+	+	+	-	-	-	N_1
+	+	-	-	-	+	N_2
+	-	+	-	+	-	N_3
+	-	-	-	+	+	N_4
-	+	+	+	-	-	N_5
-	+	-	+	-	+	N_6
-	-	+	+	+	-	N_7
-	-	-	+	+	+	N_8

In which "+" used for up-spin, "-" used for down-spin.

Definition:

P_{xy} is the probability when article A is in x direction and B is in y direction.

P_{zy} is the probability when article A is in z direction and B is in y direction.

P_{xz} is the probability when article A is in x direction and B is in z direction.

Then we have, from Table 8-2

$$P_{xy} = -N_1 - N_2 + N_3 + N_4 + N_5 + N_6 - N_7 - N_8 \qquad (8.6)$$

$$P_{zy} = -N_1 + N_2 + N_3 - N_4 - N_5 + N_6 + N_7 - N_8 \qquad (8.7)$$

$$P_{xz} = -N_1 + N_2 - N_3 + N_4 + N_5 - N_6 + N_7 - N_8 \qquad (8.8)$$

Consider that the sum of all N_n, ($n = 1, 2, 3, 4, 5, 6, 7, 8$), are 100%, namely,

$$N_1 + N_2 + N_3 + N_4 + N_5 + N_6 + N_7 + N_8 = 1 \qquad (8.9)$$

we have, from (8.7),

$$\begin{aligned} P_{zy} &= -N_1 + N_2 + N_3 - N_4 - N_5 + N_6 + N_7 - N_8 \\ &+ 1 - (N_1 + N_2 + N_3 + N_4 + N_5 + N_6 + N_7 + N_8) \\ &= 1 - 2N_1 - 2N_4 - 2N_5 - 2N_8 \qquad (8.10) \end{aligned}$$

Similarly, we have

$$P_{xz} = 1 - 2N_1 - 2N_3 - 2N_6 - 2N_8 \qquad (8.11)$$

And then, we have

$$\begin{aligned} |P_{xz} - P_{zy}| &= |-2N_3 + 2N_4 + 2N_5 - 2N_6| \\ &= 2|(N_4 + N_5) - (N_3 + N_6)| \\ &\leq 2[|(N_4 + N_5) + |(N_3 + N_6)|] \qquad (8.12) \end{aligned}$$

Now, consider that all N_n are positive, we have

$$2[|(N_4 + N_5) + |(N_3 + N_6)|] = (N_3 + N_4 + N_5 + N_6) + (N_3 + N_4 + N_5 + N_6)$$

$$= (N_3 + N_4 + N_5 + N_6) + (1 - N_1 - N_2 - N_7 - N_8)$$
$$= 1 + P_{xy} \tag{8.13}$$

And then, we have

$$|P_{xz} - P_{zy}| \leq 1 + P_{xy} \tag{8.14}$$

This is Bell's inequality.

From the process of the proof, Bell' inequality is obtained based on the assumption of one dimensional hidden theory and local realism. Now,

If the measurement results matches the Bell's inequality, implicates that it matches the hidden theory, consequently, the quantum mechanics is not completed.

However, if the measurement results is violation of Bell's inequality, the prediction of quantum mechanics is correct or the measurement results is favorable for quantum mechanics.

Therefore, Bell's inequality has been used as the criterion to tell us which one is correct, the hidden theory or quantum mechanics?

The initial explanation of Bell's inequality

Bell's inequality provides us the measurement results according to the local realism. Obviously, Bell discovered that Bell's inequality not only can not match the prediction of quantum mechanics, but also can not match the measurement results. Therefore, Bell's inequality refuse the explanation that local realism belongs to quantum mechanics. Therefore, Bell published his paper' name "On the Einstein Podolsky Rosen Paradox" in ≪Physics≫ 1964,[95]. The conclusion of this paper is:

To determinate an individual measurement results, some paper required critically to inserted some external parameters into quantum mechanics, and required that this action should be no any impact to the prediction of the quantum mechanics. For this theory (hidden variable theory), it is necessary to have a mechanism that the alternate of the setting value in the first measurement device, would give the impact to the reading value of the second measurement device inspire that how far is the distance that two devices are separated. Beside this, the signal involved in this mechanism shall be propagation intransitively from device A to device B. All of those can not match the Lorentz invariance.

On the other hand, according to the local realism, the setting value of the first device couldn't affect the reading value of the second device when the separated distance between two devices is far enough. Therefore, the explanation of Bell above can be expressed as : "if the hidden variable theory doesn't alternate the prediction of quantum mechanics, it is necessary to be violence of Bell's inequality". Therefore, if the hidden variable theory is violence of inequality, the final explanation is: **any domain hidden variable theory can not cover all of the prediction of quantum mechanics.**

Bell's inequality experiments

Bell test experiment or Bell's inequality experiment is a very difficult task since the condition to realize the experiment is very critical. Therefore, the experiment was developed till 70s 20 century. In this 30 years from 1972, most of scientists take the twin photons as the testing elements, since scientists consider that it is better to take the polarization state as the test element.

1882, Alain Aspect leading a team in France, realized the first Bell test experiment in critical condition. The test results totally matches the prediction of quantum mechanics, instead, the difference between the test results and the prediction of the local hidden variable theory is by 5 standard variance. After then, a lot of physicists repeater the Aspect's test experiment, and step by step, the experiment model is close to close to the conceive of the initial EPR by Einstein.

1998, the scientists in Ambrosian university, Austria, make two photons separated by 400m, the difference between the test results and the local hidden variable theory reaches to 30 standard variance.

2003, Pittman and Franson reported that two photons created by two independent sources separated, and the test results is violence of Bell's inequality.

Now we would like to introduce some typical experiments as follows.

Aspect's experiment 1982

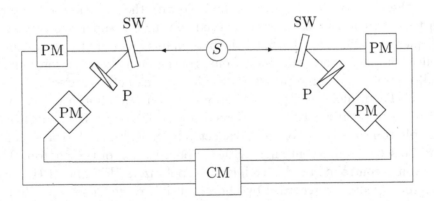

Figure 8-5   Scheme of Aspect experiment

The Bell's inequality experiment by Alain Aspect etc. in Figure 8-5, published in ≪Physical Review Letters≫, Vol. 39, pp.1804, 1982, is the most efficient one.

The scheme of the Aspect's experiment is shown in Figure 8-5. Where:
"S" is "Source".
"SW" is Acoustic optical switch device".
"P" is "Polarizer".
"CM" is "Coincidence Monitor".

This experiment is to measure the polarization of two photons, which are induced by the atom 40_{C_a} decay from the excited level $|e>$ as shown in Figure 8-3(b) and then separated in oppositive direction simultaneously.

a. "Source": The source "S" is a single Ca atom-beam. Two laser beams irradiate perpendicularly onto the Ca atom beam, so as the Ca atom transits from ground to exited level $|e>$, this Ca atom-beam decay from the excited level induce two photons with correlation, one is at $\lambda = 551.3nm$, another one is at $\lambda = 442.7nm$. Where, two laser-beam are focused to a interaction region, which is a cylinder with $1mm$ length and $60\mu m$ diameter. The typical density of the Ca atom-beam in this cylinder is 3×10 atoms/cm. The density is so low that which is low enough to prevent the $442.7nm$ resonance light to be intercepted. The first laser beam is provided by a single-mode krypton ion laser with resonance wavelength $\lambda_k = 406.7nm$. The second laser-beam is provided by a single-mode dye laser with the resonance wavelength $\lambda_D = 581nm$. The polarization of these two laser-beam are parallel each other. Each with $40mW$ power. The typical correlation is $4 \times 10^7/s$.

b. The polarizers "PI" and "PII" are heap chip polarizers. Each one is consist of 10 piece of plan grass, which inclines to a Brewster angle. Insert a line polarizer in front of the 10 piece of plan grass.

c. Setting an acoustic–optical switch device "SW" $6m$ away from the source both left side and right side, respectively. The principle of the acoustic-optical switch is that the refractive of the water is variable with the variation of the pressure slightly. In this switch, a reverse sensor is used to set up a super-acoustic stand-wave with frequency of $25MHz$. Then, let the photon incident onto the switch in almost totally reflection, so as the conversion from transparency to reflection happen once every half period of the super acoustic stand-wave (namely the frequency is $50MHz$).

d. Then, along the path, whatever the output photon of the switch "SW" go through to polarizer PI or is reflected to the polarizer "PII", each polarizer will let the photon go through or stop, respectively. Since the polarization of each polarizer can be set with different angle with the polarization of the photon. The fate of each photon is controlled by the setting of the polarizers "PI" and "PII", respectively and the test results are recorded by photo-multiplier, which set up behind the polarizer "PI" and "PII", respectively.

This experience is used to record the fate of each photon by photo-multiplier to evaluate the correlation of a pair of photons. The essential characteristic of this experience is that when the photon is propagation along a path, it can alternate its following path arbitrarily in a very fast way so that the signal doesn't get enough time to go over from left side to right side although the signal is propagation with light speed.

To judgment the wish from two great scientists Einstein and Bell, Aspect etc. work very hard for 8 year, until 1982, he force the time that each polarizer holds certain polarization from $60ns$ falling down to $10ns$. Which is shorter than the time ($\Delta t = 40ns$) that the photon propagation from left polarizer "PI" to right polarizer "PI" (with distance of $L = 13m$) and the photon emission life (about $5ns$). Which realize to alternate the polarization of the polarizer during the photon flight. And then it satisfies the Bell's local condition.

According to Aspect etc., the experiment has been last 12000 second typically. This period contain three steps:

a. First step: the experiment arrange is described above.

b. The second step: took away all of the polarizers "PI" and "PII" in both side.

c. The third step: took away only one polarizer "PI" or "PII" in each side.

Which can correct the system error from the test results.

In this experiment, according to Bell's inequality, if the fact is realism, the value of the function (about the individual test result from four polarization angle A_1, A_2 B_1 B_2 of four polarizers, respectively) shall be in the range between -2.0 to 2.0. However, all test results indicated that it is violence of Bell's inequality. Meanwhile, the value of the function is totally match the prediction of quantum mechanics (using the wave function to describe the photon).

Therefore, the reality is non-locality, and can be described by quantum mechanics; meanwhile quantum mechanics is non-locality. In fact, if the the reality is non-locality, it is still violence of Bell's inequality although the distance between A and B is very very far away (even in terms of light years).

Innsbruck experiment:

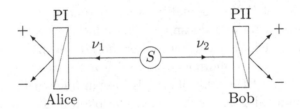

Figure 8-6   Scheme of Innsbruck experiment

Another typical experiment as shown in Figure 8-6 is "Innsbruck Experiment" realized by the scientists in Innsbruck university, Austria.

a. First, the distance between two test stations A and B is above $400m$.

b. The polarizer is connected to a computer in each station, and each polarizer can alternate the switch randomly from channel + to channel − or vice reverse.

c. The propagation material between polarizer and source is optical fiber. And the source produces a twin photon.

In the process of the experiment, a twin photon separated from the light source and propagation along opposite direction, respectively. And two polarizers "PI" and "PII" and computers in two test station A and B record and process all of the data at "+" "-", respectively.

It worth to point out that the observer Bob do not know the polarization operation by observer Alice each time since the operation time is limited in $1.3\mu s$ each time.

The output data of two channel "+" and "-" has been recorded in computer with an atom clock. The polarization of each polarizers can be set by computer arbitrarily in very short time. In this case, the photons get no time to transmission any massage between two stations.

After comparison of the data obtained in station A and station B one by one, the scientists in Innsbruck experiment team assertion: as long as an operation of one polarizer acts, the twin photon will alternate simultaneously in this way that if the polarization at A is "up-spin", the polarization of B would be "down-spin" without any delay, vise reverse. This conforms that the twin photon is totally correlation, or say, this is non-locality.

The reported: *the final conclusion of the experiment is: the test results is very advantageously coincidence with the prediction of quantum mechanics, and is indisputably violence of Bell's inequality.*

Effect:

From the discussion above, with strict separation of relativistic measurements are all violence of Bell's inequality. Which conform that the idea from Einstein to describe the twin photon doesn't work. Which is because that Einstein deal with the correlation of EPF twin photon as the normal photons, and then the correlation properties of two photons separated from source is ignored. However, the real situation is that a EPF twin photon is inseparable entity. so it is not reasonable to "provide" the localism for each individual photon. In some sense, the relationship in the space and in the time between twin photon is a directly provence of inseparable property of quantum mechanics. No wander, the EPR paper cause Bell shock, since multiple particle system will lead to pure quantum effect, which is not so clear before EPR paper published.

Nowadays, the quantum effect based on "non-locality" has launched the quantum massage research in full flourish, such as:

a. Quantum Key Distribution — QKD.

b. Quantum data compression,

    c. Quantum teleportation.

    d. Quantum computer.

In the process of evaluation of Bell's inequality for 30 years, quantum mechanics correctness again experienced a severe test of high tech specifications. The conclusion of the test is: all of the test till now do not support an accusation against quantum mechanics. On the other words, **the localism hidden variable theory can not replace quantum mechanics**. Bell also conformed that **"any localism hidden variable theory can not reappear the total statistic prediction of quantum mechanics"** (**Bell theorem**).

    In summery,

    a. Bell's inequality and its verification has far-reaching scientific significance, which turn the philosophy argument about quantum mechanics into an operational evaluation.

    b. The process of Bell's inequality experiments and its test results is also far-reaching scientific significance. Which leads human peep into the magical beauty in information science.

    Nobel prize winner Yuebifuxun though that "Bell's inequality and Bell theorem is the most important advance in physics".

    Philosopher Stapu thought that "Bell theorem is the most profound scientific discovery, it not only verified the completeness of quantum mechanics but also have a positive and lasting impact to expend people's thinking and vision ".

8.3 Quantum Key Distribution — QKD

8.3.1 BB84 protocol and QKD

In 1984 Bennett and Brasard realized that the single quanta can be used for information transmission. They published the first quantum cryptography protocol now known as "BB84" protocol, in which they mentioned quantum key distribution (QKD) protocol. Which is the first one to describe how to use the polarized photons to do the information transmission.

    A further advance in theoretical quantum cryptography took place in 1991 when Ekert proposed that Einstein-Podolsky-Rosen (EPR) entangled two-particle states could be used to implement a quantum cryptography protocol whose security was based on Bell's inequalities.

    Also in 1991, Bennett and collaborators demonstrated that QKD was potentially practical by constructing a working prototype, using polarized photons.

    In 1992, Bennett published a "minimal" QKD scheme ("B92") and proposed that it could be implemented using single-photon interference with photons propagating for long distance over optical fibers.

    In BB84, when we use the polarized photons as the information carrier to do the information transmission, we need, as shown in Figure 8-7:

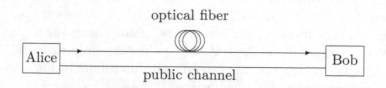

Figure 8-7   Transmission of Quantum states

a. A sender, known as Alice.

b. A receiver, known as Bob.

c. A quantum transmission channel, could be optical fiber.

d. A public channel (for instance, radio or internet) for communication between Alice and Bob. Of cause, a public channel can not guarantee the security of the information transmission, the transmitting information could be intercepted by the third person, the eavesdropper, known as Eve. Which has been considered in BB84 protocol.

The safety of the BB84 protocol is based on the property of the quantum mechanism: two non-orthogonal quantum states could not be thoroughly identified. In this case, BB84 utility two pair of quantum states,

a. One pair is in linear basis "+": where the horizontal polarization (0^0) is noted as $| \rightarrow>$, the vertical polarization is noted as $| \uparrow>$.

b. Another pair is in diagonal basis "x": where 45^0 polarization is noted as $| \nearrow>$, 135^0 polarization is noted as $| \searrow>$.

These two pair of polarizations are non-orthogonal to each other and then it is not possible to be thoroughly identified. For instance, if you choose basis "+" to measure $| \uparrow>$, the probability to get the answer of $| \uparrow>$ is 100%. However, if you choose basis "+" to measure $| \nearrow>$, the probability to get the answer of $| \uparrow>$ is 50%, and the original information is lost. On the other words when you get final answer of $| \uparrow>$ state, we could not make decision whether the original state is $| \uparrow>$ or $| \nearrow>$. Means that two non-orthogonal states could not be thoroughly identified. Where basis "+" and basis "x" are indicated in Table 8-3.

	Basis	$|0>$	$|1>$
Table 8-3	+	$| \uparrow>$	$| \rightarrow>$
	x	$| \nearrow>$	$| \searrow>$

Which means that the classical bit is "0" or "1", the quantum bit (qubit) is $| \uparrow>$ or $| \rightarrow>$ in basis "+", in which, $| \uparrow>$ is equivalent to "0", $| \rightarrow>$ is equivalent to "1".

Meanwhile, the classical bit is "0" or "1", the quantum bit (qubit) is $| \nearrow>$ or $| \searrow>$ in basis "x", in which, $| \nearrow>$ is equivalent to "0", $| \searrow>$ is equivalent to "1".

Now we will discuss the BB84 protocol as follows:

BB84 QKD protocol:

The first step—Quantum communication phase:

Alice randomly create a bit (0 or 1), then randomly choose a basis to convert to a quantum bit (qubit) as indicated in Table 8-3, in which:

a. if Alice choose basis "+", the classical "0" is equivalent to $| \uparrow >$, "1" is equivalent to $| \rightarrow >$;

b. if Alice choose basis "x", the classical "0" is equivalent to $| \nearrow >$, "1" is equivalent to $| \searrow >$.

After Alice randomly choose a quantum state in a basis, she send the photon to Bob via the quantum channel. Alice repeaters this process for several times.

Bob doesn't know that which basis has been chosen by Alice, he might choose a basis ("+" or "x") to measure the received quantum state. Bob measures each quantum state what he has received and recodes the result of the quantum state and the basis what he has chosen every time.

Public discussion phase:

After then, Alice broadcast her bases of measurements. Bob broadcast his bases of measurements.

Alice and Bob discard all events where they use different bases for a signal (almost 50%) and remains the other evens, then take the sharded data sequence as their cryptography key as listed in Table 8-4.

Table 8-4

Alice's random bit	0	1	1	0	1	0
Alice's basis what she chosen	+	+	x	+	x	x
qubit what Alice sent	$\|\uparrow>$	$\|\rightarrow>$	$\|\searrow>$	$\|\uparrow>$	$\|\searrow>$	$\|\nearrow>$
Bob's basis what he chosen	+	x	x	x	+	x
Bob's qubit what he measured	$\|\uparrow>$	$\|\nearrow>$	$\|\searrow>$	$\|\rightarrow>$	$\|\searrow>$	$\|\nearrow>$
Shared data sequence	0		1			0

The eavesdropper — Eve? :

To check is there any intercept from Eve (named from eavesdropper), Alice and Bob have to take some of the cryptography key to see is there any error in it. From Eve side, Eve has to measure the polarization state of each photon, which will result in the error of the cryptography key. Now

a. If Eve choose a basis (which is the same as the Alice's basis) to measure the polarization of each photon, Bob could not find any error.

b. However, if Eve's basis is the same as Bob's basis, Eve's measurement would result in the error when Bob measures. Which means that Eve has 50% probability to choose Bob's basis to measure each photon's polarization, which will alter the photon's polarization, meanwhile, Bob has 50% probability receiving the massage which is different from Alice. Therefore, the error probability caused by Eve will be 25%.

The threshold value is that if the number of the error evens is over p, say 11%, the cryptography key abort. Otherwise, they process to the next step.

The second step:

Alice and Bob each convert the polarization data of all remaining data into binary string called a raw key by, for example, mapping a vertical or 45^0 photon to "0" and a horizontal or 135^0 photon to "1". When they find the interception from Eve, they can perform classical postprocessing such as error correction and amplification to create a final key, in which the number of the sharded data sequence is less than the original one.

BB91 QKD protocol:

In 1991, BB91 QKD protocol based on two "entangled" particles" is published in a paper by Arthur Eckert [96]. This means that although two particles are separated by large distance in space, they are not independent of each other. Suppose two entangled particles are photons, if one of the particles is measured according to the rectilinear basis and found to have a vertical polarization, then the other particle will also be found to have a vertical polarization if it is measured according to the rectilinear basis. If, however, the second particle is measured according to the circular basis, it may be found to have either left-circular or right-circular polarization.

In this 1991 paper, Ekert [96] suggested that the security of the two-qubit protocol is based on Bell's inequality. The Bell's inequality demonstrates that some correlations predicated by quantum mechanics cannot be reproduced by the local theory.

To do this, Alice and Bob have to collect enough data to test Bell's inequality.

Now, the steps of the BB91 QKD protocol of two entangled photons are as follows:

a. Alice create entangled pairs of polarization photons, keeping one particle for her-self and sending the other particle of each pair to Bob.

b. Alice randomly measures the polarization of each particle she kept according to the rectilinear or circular basis. She records each measured basis and polarization.

c. Bob randomly measures each particle he received according to the rectilinear or circular basis. He records each measured basis and polarization.

d. Alice and Bob tell each other via public channel that which measured basis were used, and they kept the data from all particle pairs at which they chose the same measured basis.

e. They convert the remaining data to a string of bits using a convention way, such as: left-circular=0, right-circular=1; horizontal=0, vertical=1.

The difference between BB84 protocol and BB91 protocol:

One important difference between the BB84 and BB91 protocol is that:

With BB84, the key created by Alice and Bob must be stored classically until it is used. Therefore, although the key was completely secure when it was created, its continued security over time is only as great as the security of its storage.

Using the BB91 protocol, Alice and Bob could potentially store the prepared entangled particles and then measure them and create the key just before they going to use it, so as to eliminating the insecure stored key issue to minimal.

Information Reconciliation and Privacy Amplification:

Information Reconciliation: The next step is Information Reconciliation and Privacy Amplification.

Information Reconciliation is a way to correct the error so as to guarantee that the cryptography key in Bob hand is identical with the cryptography key in Alice hand. The process of the the error correction is completed via the public channel. In this case, it is possible that Eve might intercepted some of the cryptography key. However, the intercepted by Eve or channel noise will cause the error of the key. In this case, the error will be aborted. And the cryptography key will be shorter than original one after "Information Reconciliation".

Privacy Amplification: Privacy Amplification is a way to eliminate or abort the part what Eve has intercepted. The interception by Eve might happen at the time during the cryptography key is transmission or the procession of "Information Reconciliation" is going.

The procession of "Privacy Amplification" is to create a new cryptography key from the remaining data in both of Alice's and Bob's hand. Therefore, the new cryptography key will be shorter than before, and the probability of intercepted by Eve will be smaller and smaller than before.

8.4   Quantum optical fiber telecommunications

Quantum communication refers to the new type of communication using quantum entanglement information transfer. Quantum communication is mainly related to:

    a. Quantum cryptography communication.

    b. Quantum teleportation.

    c. Quantum dense coding.

This discipline has been gradually moving from theory to practice and consequently moving to the application.

Quantum communication is an interdisciplinary of quantum mechanics and information theory, which features with high efficiency and confidential. Which is different from the classical communication, however, the development of quantum communication has to step on the classical communication.

	Classical communication	Quantum communication
Signal	"0" or "1"	"$\mid \uparrow>$" or "$\mid \downarrow>$"
Type	electrical signal	polarization state of photon
Type	integer number	statistic probability
correlation	non-correlation	correlation or entanglement
Feather	non-confidential	high efficiency and confidential

In classical communication, the basic signal is "0" and "1", which is an electric signal. In quantum communication, the basic signal is the photon polarization state $\mid \uparrow>$ or $\mid \rightarrow>$, in which, $\mid \uparrow>$ is equivalent to "0", $\mid \rightarrow>$ is equivalent to "1". Which is a photon polarization state signal "spin-up", "spin-down", respectively. On the other word, which is the information of the photon but not the photon itself. Therefore, the photon polarization state signal $\mid \uparrow>$ or $\mid \rightarrow>$ can be separated with the photon it-self. But the photon polarization state signal $\mid \uparrow>$ and $\mid \rightarrow>$ are entanglement and then are correlated, which means that if the polarization state of the article A is $\mid \uparrow>$, the polarization of the article B must be $\mid \rightarrow>$, vise reverse.

Classical communication

The classical communication is based on the digital communication. In classical communication,

    a. The basic signal is "0" and "1". Which can be electric signal ("1" is "on" and "0" is "off"), also can be optical signal ("1" is "light" and "0" is "dark").

    b. After "0" and "1" are created in the computer, they are stored in the memory units and then are picked up to be encoded and computation, then restored in the memory units. In this case, "0" or "1" is held in the memory units.

    c. Nowaday, to realize the classical telecommunications in a long distance, the signal "0" or "1" has to be optical signal "0" or "1", and to be transferred on the all optical fiber communications networks (Internet) or satellite communications networks. In this case, it can not be avoided that the classical massage can be eavesdropped by the eavesdropper from the Internet or other way since the eavesdropping has no any affect to the signal "0" or "1".

Quantum communication

The quantum communication is based on the quantum mechanics. In quantum digital communication,

a. The basic signal is the photon polarization state $|0>$ or $|1>$, which correspondent to "0" or "1", respectively. Where, $|0>$ correspondent to the state with "spin-up" and $|1>$ correspondent to the state with "spin-down".

b. After $|0>$ and $|1>$ are created in the quantum computer, the one of them is held in the quantum computer A, the other one has to be transferred to the other place B. And $|0>$ and $|1>$ are not independent, but are entanglement or correlated.

c. $|0>$ or $|1>$ can be measured by polarizer so as to be transferred into "0" or "1", respectively. However, the measurement of $|0>$ will impact to $|1>$ immediately, vise reverse. Therefore, any eavesdropping will be found immediately by A and B. Therefore, quantum communication features with high efficiency and confidential.

In summery

Quantum communication is based on the quantum mechanics, which is total different from the classical communication although the realization of the quantum communication has step on the classical communication networks. Therefore, the creation of the entangle articles, the storage of the polarization state of photon, the transfer of the polarization state, the measurement of the polarization state and the quantum encoding, all are based on quantum mechanics and the information theory.

Difference between quantum mechanics and classical mechanics

Quantum communication is based on the quantum mechanics. quantum mechanics is total different from the classical mechanics. In classical mechanics, the state of a article is defined by its position, momentum. In quantum mechanics, however, the state of a article is defined by its state vector (or polarization state vector). In simplification, state vector is a statistic probability. That means that the state of the article is described by probability, which means that the state of the article can be this, or can be that, but not necessary to be which one. Is that the moving speed of the article is too fast to be investigated? No, quantum mechanics think that the uncertainness of the article is the real state but the information of the article we obtained is not completed.

For instance, the state of article can be noted as a vector, such as direction

"↑", "↓", "←", "→". Consequently, we defined the "direction" as a physic quantity. However, some polarization state is a "circle", "circle" state is a real state as the "vector", merely the direction is uncertain. If we measure the direction of the "circle" state, It will be the same probability into another state. In the same reason, when scientist preparation several articles with the same state, and measure the position of each article for several times, we will find that every test result is different from the others. Which is in normal since the state of any article is random essentially.

8.4.1 Quantum key distribution (QKD) based on the optical fiber infrastructure

When we discuss that "how to realize quantum communication?", we'd better to say "**how quantum key distribution (QKD) works?**"

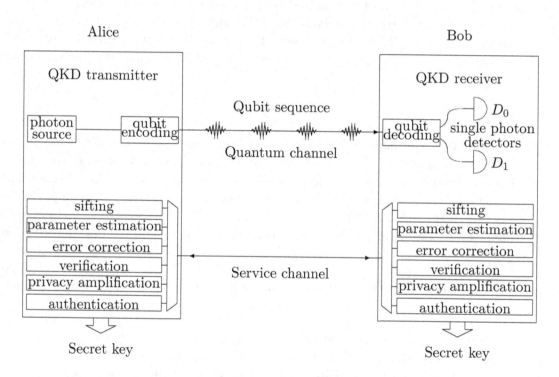

Figure 8-8   A typical prepare-and-measurement QKD setup

Quantum key distribution (QKD) is a process that uses an authenticated communication channel together with a quantum communication channel in order to establish a secret key. There are several different protocols of implementing quantum distribution, all of which require both a quantum channel and an authenticated classical channel as shown in Figure 8-8. The quantum channel is used to send quantum state of light, the authenticated classical channel is used for the sender, Alice, and the recipient, Bob, to compare certain measurement related to these quantum states and perform certain post-processing steps to distil a correct and secret key. The quantum channel uses optical fibers or space/satellite links to send photons (quantum states of light) between Alice and Bob, whereas the classical channel could be a simple (authenticated) telephone line that Alice and Bob use to talk to each other. Interestingly, both of these can be public. Quantum channel necessary shows Alice and Bob when an eavesdropper (Eve) has been listening in, and it is a fact of the QKD protocols hat the classical channel could be broadcast publicly without promising security.

As shown in Figure 8-8, quantum key distribution begins by Alice deciding to distribute some cryptography key to Bob. Both Alice and Bob have the specialized optical equipment necessary for establishing the quantum channel, as well as access to a classical channel where they can communicate with one another. Alice uses a light source to send a steam of photons (quantum states) one-at-a-time. Each photon can be thought of as one bit of information. As each photon is sent, she randomly choose to prepare it in one of two "basis" ("+" and "×"). Basis can be described as a perspective from which a photon is measured.

As the recipient, Bob needs to record values for each photon he receives via the quantum channel. To do this, he must, like Alice, make a measurement of each one, and he therefore also chooses one of the two possible "basis" and records which one he measured in. These choices are random and do not require any information about the bases that Alice chose when she was sending each bit. Afterward, Alice and Bob then communicate over the classical channel to compare which basis each bit was measured in at each end of the quantum channel. Sometimes Alice and Bob will randomly choose the same basis, and these are the bits for which they will get the same value for the photon (which is useful, so they will keep this bit as part of the key). When Alice and Bob measure the photon using different bases, they throw this bit away and do not use it in the final key.

After each bit has been sent and received, Alice and Bob can speak publicly about which basis they used to measure each photon, and this can provide enough information for each of them to generate key from the received quantum states, but not enough information for an adversary to reconstruct the key. Thus, an eavesdropper will not be able to discover the transmission key for two important reasons:

a. Firstly, the adversary cannot directly observe the photon without changing them, therefore being detected and having these bits discarded by Alice and Bob.

b. Secondly, the adversary cannot directly observe the photon through observing the measurement of Alice and Bob, either, since Alice and Bob do not disclose the final measurement result for each quantum state. Rather, they only disclose which basis they used to measure it. By this time, it is too late for the adversary to measure the photon, because it has already been received by Bob, so knowing the basis that Alice used is not useful. It is well-established using information theoretic proofs that the measurement information is inadequate for an adversary to use to reconstruct the generated key.

Authenticating the QKD channel:

An important consideration in quantum key distribution is how to authenticate the classical communication channel to ensure that two people communicating are actually Alice and Bob. The authenticated classical communication may be realized a few different ways:

a. The most secure method for authentication does not require any computational assumptions and use a small amount of random secret data that is initial data without sacrificing randomness or secrecy. Subsequent communication using QKD normally uses part of the generated key authenticating subsequent QKD sessions and part of the keying material for encryption.

b. If initially Alice and Bob do not share an authentication key, a natural alternative is to use public key signatures to authenticate their initial QKD session. Provided that the public key signature is not broken by future algorithmic advances, even if the public key signature is later broken. Subsequent QKD session between Alice and Bob may be authenticated using a portion of previous QKD keys, so that they only need to relay on the short-term security of the public key signature once.

QKD protocols and their implementations

Several QKD protocols have been successfully implemented in academic and commercial research labs, transmitting keys over long distances through either fibers or free space. These protocols fall into categories, and while theoretically identical in their analysis, are differentiated experimentally in part by how eavesdropping is detected:

1. **Prepare-and-measure QKD** allows legitimate parties to detect eavesdroppers by comparing the amount of error they might expect in their communications to the actual error of their measurements. This technique relies upon the fact that an adversary intercepting a quantum state must measure it, and in this adversary attempting to guess the original state to forward it to the recipient, they will introduce a percentage of identifiable error.

2. **Entanglement-based QKD** allows legitimate parties to detect eavesdroppers by virtue of the fact that if the sender and recipient each have a photon the two of which are related by quantum-mechanical entanglement-interception or measurement by an adversary will change the two-photon system in a way that the legitimate parties can readily detect.

Some implementations of QKD protocol are list as follows:

a. One example of a QKD protocol is the BB84 protocol [BB84], which was the first protocol proposed for quantum key distribution and remains one of the most widely implemented protocols inside of commercial products and used within research labs. QKD based on the BB84 protocol has been demonstrated *over 100km in length* in both optical fiber and free space, reaching transmission speeds on the order of one megabit per second over short distances ([SMA07, LUC13]). Optical fiber based QKD products have already been commercially deployed and are in use today to distribute quantum safe keys in real networks.

b. The SARG protocol [SARG04] is similar to BB84 and has been used over a period of 21 months (from Oct. 2014) of continuous operation on the international SwissQuantum networks [STU11].

c. Another protocol which aim to have convenient implementations while still enabling long-distance and high transmission rate QKD are the Different-Phase-Shift protocol [WAN12] and the Coherent OneWay protocol [STU09], which both have *exceeded 250 km transmission distance* in optical fiber.

d. There is also Continuous Variable QKD protocol that is the only QKD protocol that does not require single-photon detectors. This protocol relies upon homodyne detection and continuous encoding of quantum state [GRO02, QL1].

e. In addition to the above commonly used protocols, there is on-going reach into protocols that aim to reduce the security assumptions of the actual implementation of the QKD devices. For example, the measurement device independent (MDI) protocol allows for the complete removal of the single photon detectors and measurement devices from the security consideration [BHM96, Ina02, LCQ12]. That means that users of the QKD system do not need to trust the distribution of their measurement devices.

f. Another possibility exists in the Ekert protocol [EKE91], which proposes the use of quantum entanglement to implement QKD with complete device-independent security. Once implemented, this would ultimately mean that trust assumptions of the QKD system implementation by a manufacturer could be reduced to a minimum.

Quantum key distribution penetration testing and security research ('quantum hacking') is an active of academic and commercial work, and has identified some implementation specific vulnerabilities in QKD systems, leading to improved system designs.

QKD in networks

Quantum key distribution is intransically a point-to-point technology, but has been demonstrated in a routed network topology over multiuser optical fiber networks [SEC09, CHE10, STU10, STU11], to secure data transmission such as encrypted telephone communication and video-conference throughout all nodes in a network. Therefore, the point-to-point nature of QKD can be implemented in such a way as to secure communications throughout a multiuser network. These networks are currently being explored through industrial and academic research into optical fiber networks. Additionally, current work on free apace QKD links are also progressing toward the ultimate goal of using a satellite as a trusted node to enable free space quantum key distribution around the globe.

Optical fiber quantum key distribution can be implemented on existing optical fiber infrastructures, but the quantum channel can not pass through optical amplifiers. The maximum distance over which QKD photons can travel is limited because of the absorbtion of the signal that occurs over long distance transmission in optical fiber. Classical signals in the optical infrastructure use repeater notes throughout the network to amplify and propagate the signal. However, due to "*no cloning*" theorem of quantum information, there are challenge in developing a repeater system for a QKD network. The present solution to this problem is to concatenate multiple QKD system to let keys propagate via intermediate notes, which requires that the inter notes must all be trusted to some extent.

While routing QKD using trusted notes is one solution to the distance limitations and point-to-point nature is of quantum key distribution, current research is exploring quantum repeater architectures that exploit something known as quantum entanglement in order to extend the range of QKD links beyond 400km.

Another way of overcoming distance related challenges to implement QKD is to send the signals through free space rather than optical fiber, as signals are diffracted less rapidly through the medium of air than they are through the medium of optical fiber. There is a trade-off with this approach; it is a more difficult engineering problem to protect against noise from atmospheric fluctuations. Several international research terms are currently working to develop satellites for use in quantum key distribution. These systems would have the benefit of not only being able to receive point-to-point signals over distance of a few hundred km from the ground to low earth orbit [ELS12], but furthermore, a network of these satellite could act as trusted intermediary nodes capable of transmitting free space links all around the world. While this is an area of active research, satellite based quantum key distribution has yet to be demonstrated.

Current limitations of quantum key distribution, in general, are its higher costs for dedicated hardware, its transmission distance of a few hundred kilometers, and its key generation rate which decreases with increasing distance between sender and receiver. However, for special applications for which strong security conditions must be met, QKD will likely become an increasingly attractive option in the upcoming years as research extends the distances over which quantum key distribution can be performed.

8.4.2   The conception of Quantum Teleportation

To describe the Conception of Quantum Teleportation, we'd better to recall the entanglement by "Schrodinger's cat". When a cat is inside the toxic closed box for a while, you ask "is the cat alive or died?" The answer from the quantum physics of view is that "it is either alive or died". People will say, it is easy to know that "is the cat alive or died?" by open the door of the box. However, according to the quantum physics of view, the state of "alive" or "died" is the results of human observation, namely, it is the results of macroscopic interference so as the cat become "alive" or "died". It is not the real situation before the door opened. Similarly, the microscopic particles is still remain in the superposition state of two states "alive" and "died" before non-interference, or say, which is either "0" and "1" before non-interference.

1993, the American scientist C.H. Bennett proposed the conception of quantum teleportation, which means that the quantum communication is to transmit the quantum state, which carry information.

After than, 6 scientists from different country proposed a scheme to realize "Teleportation" by using quantum channel with classical channel. In this case, an unknown quantum state at A is transmitted to another place B, then make another particle at B onto the prepared quantum state. And the original particle still remain at A. Which means that the "transmission signal" from A to B is "quantum state" only but not the photon it-self. However, the information has been transmitted from A to B by the preparation of another particle to the quantum state at B. Which is so called "Teleportation". This is an important idea, according to this idea, a large information can be transmitted from A to B with three features:

1. Security:

a. First, the the quantum cryptography key is random and used only once. Although it has been eavesdropped, Eve is unable to get the correct key, and is unable to break the information.

b. Secondary, in the process of quantum communication between Alice and Bob, two entangled articles, one in Alice hand and another one in Bob hand, any observation and interference any particle will cause the decay of the article. Therefore, the part of information obtained by Eve is a "broken" information, but not the original information.

2. High efficiency:

The unknown quantum state is in entanglement before testing, which represents two state. For instance, one quantum state represents two number "0" and "1". Seven quantum states represents $2^7 = 128$ states and "2×128" number. Therefore, the quantum communication efficiency is two times of classical communication efficiency or more.

3. Instant transmission:

In quantum communication, the information transmission is instant transmission since the original information photon is still in Alice hand, and Bob get the information is by the preparation of another particle to the quantum state. Which can be implemented instantly. Which is equal to say that the speed of the information transmission is almost 10,000 times of light velocity "c".

The quantum communication can be implemented on the available optical fiber infrastructure with **no optical amplifier** in the transmission line. Which is so called *point-to-point transmission*. We may ask "why **no optical amplifier?**" in the quantum transmission system?" The one of the reasons is because that the noise from amplified spontaneous emission (ASE) intrinsic to amplification can severely impact the qubit error rate of the quantum channel. The spectral components of the ASE noise that are in band with the quantum signal reduce the fidelity of the received qubits.

To implement multi-point of information transmission, we need a quantum repeater to store the polarization state of two entangled articles as long as possible. This repeater can be used to communication with each point-to-point branch to form a star quantum networks.

In this scheme, the entangled state of non-locality plays important key role. Quantum mechanics is a non-locality theory. This point has been proved by the experiment results of the Bell's inequality. And then, the quantum mechanics displays a lot of counter-intuitive effect. In quantum mechanics, scientist may create two article state with entanglement correlation. The entanglement correlation can not be explanation classically. Quantum entanglement means that two or more articles are in non-locality correlation. Quantum teleportation is very important not only for physicist understanding and revealing the mysterious rule in nature, but also for realizing high capacity information transmission and quantum cryptography communication by using quantum state as the information carrier.

8.4.3   The implement of Quantum Teleportation

Quantum Teleportation

Quantum communication is based on quantum entanglement to implement the "Teleportation". "Teleportation" is one kind of disembodied and "complete" information transfer by quantum channel and classical channel.

However, from physical point of view, we may imaging the process of "Teleportation" as follows:

a. Alice extracts all information of the original article,

b. Alice send all of the information to Bob by quantum channel and classical channel,

c. According to the received information, Bob select the basic unit of the same original to create a perfect replica of the original.

However, **"uncertainty principle of quantum mechanics"** does not allow to accurately extract all the original information, and the replica can not be perfect. Therefore, physicists consider that seems like the "Teleportation" is a fantasy for a long time. Therefore, we really need the experimental implement of the "Teleportation".

The discussion above mostly belongs to the theoretical discussion. The implementation and application of quantum entanglement is a very difficult technology since quantum state is a random and instant event. Meanwhile,

How to create the entangled particles from Alice?.

How to transfer the quantum state?

How to receive the quantum state in Bob?

How to create the quantum repeater with the storage technology of quantum state?

Each of them is a new technology for the scientists and engineers in the world.

The implement of Quantum Teleportation

United States

Since the American scientist C.H. Bennett in IBM proposed quantum communication theory in 1993, National Science Foundation and Defense Advanced Research Projects Agency in United State has been investigated deeply the quantum communication.

European Union

European Union focus international forces working on quantum communication from 1999. The number of research projects are more than 12.

According to the ≪New scientists≫, a scientist group from Italy and Austria reported that they have identified a number of photons bounced back to Earth from the satellite $155km$ away over the earth as shown in Figure 8-9. Which Achieves a major breakthrough of **space quantum information transmission** for the first time in the world.

The research team in University of Padua, leaded by Paul·Veronica Meath and Kai Shaer·Ba Boli, use binoculars with diameter $1.5m$ at Laser Ranging Observatory named "Matera" in Italy emitted photons to a satellite away $1500km$ over earth named "Ajisai" from Japan, and then required the Japanese satellite to bounce back the photons to the original emitted place. This experiment displayed that the quantum encoded communication may be realized by the satellite communication, which can not be eavesdropped anymore.

In June 2007, a research team from Austria, Britain and Germany, hit the maximum quantum communication distance of 144 kilometers in space. Over which, It is very difficult, since atmosphere will interference the vulnerable quantum entanglement state. Therefore, based on the satellite experiment above, the research team consider a new idea that the photon can be emitted from the satellite. Since the atmosphere will be getting thinner and thinner as height increasing. The photon traveling on the space several thousand km is equivalent to the photon traveling $8km$ on the ground.

To verify that it is observable that the photon bounce backed from the satellite, the research use binoculars with diameter $1.5m$ at Laser Ranging Observatory named "Matera" in Italy emitted a normal laser beam to the "Ajisai" satellite. The "Ajisai" satellite as shown in Figure 8-9 is consist of 318 mirrors. The laser beam bounce back from the 318 mirrors to the original emitted place, the Observatory.

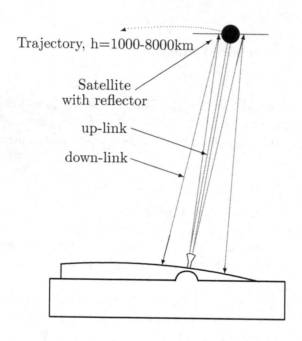

Figure 8-9   Quantum satellite communications

Japan

Japan's Ministry of Posts take quantum communication as a strategic project of the 21st century.

China

From the 1980s China began working in the field of quantum communications. Recently, the quantum research group of Chinese University of Science and Technology made outstanding achievements.

In 2003, the researchers from Korea, China and Canada presented a propose about the "**decoy state quantum key**" theoretical scheme, which expects to be a good solution for the issue that the rate of quantum communication security is limited by the communication distance.

"decoy state quantum key"

In the summer of 2006, the research group of jianwei Pan in China, Los Alamos National Laboratory in USA, the Joint Study Group of University of Munich and University of Vienna in Europe, independently realized the "**decoy state quantum key**" scheme.

By the end of 2008, the research team of jianwei Pan in China successfully developed the prototype of quantum communication network based on the "**decoy state quantum key**" scheme. Successfully built up the quantum telephone network with three nodes in Hefei, China for the first time in the world. Meanwhile, European joint experimental team announced that the team built up the quantum telecommunication network too.

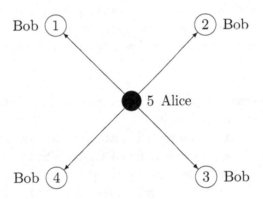

Figure 8-10   Quantum communication star network

In September 2009, the research team of jianwei Pan in China successfully built up all pass quantum communication star network with 5 nodes. Two customers from any node can communication independently. The all pass quantum communication star network with 5 nodes can be realized as shown in Figure 8-10 in principle. In which,

a. The start network is consist of 5 point-to-point optical fiber branches (5 - 1,

5 - 2, 5 - 3, 5 - 4) without any optical amplifier. Which is because that quantum entanglement state can not pass any optical amplifier as we mentioned in this subsection. Therefore, the distance of each fiber branch is limited in $20km$, so as to hold the quality of the quantum entanglement state in each fiber branch.

b. The node 5 is a quantum repeater controlled by Alice, which is used to create 5 pairs of entangled photon, one photon of each pair is sent to a node n ($n = 1, 2, 3, 4$) at Bob. Therefore, any two customers can communication independently by five fiber branches via the repeater 5.

c. The process of quantum communication on each fiber branch just like the process of quantum communication between Alice and Bob.

d. In implementation of all pass start fiber network, we have to face several technologies challenge such as the low temperature storage of quantum entangled state, "**decoy state quantum key**" technology and etc.

This all pass fiber start network can cover a medium-sized cities in China. The service includes quantum Internet communications and quantum telephone communications.

Cold Atoms Storage Technology

By using "**Cold Atoms Storage Technology**", the research team of jianwei Pan in China realized quantum entanglement swapping with storage and readout functions, create quantum entanglement between two clod atom system connected by $300m$ optical fiber. This quantum entanglement between two cold atom system can be readout and can be converted into photon entanglement for further transmission and quantum operations. This experiment realized "quantum repeater", which is extremely important step to realize long distance quantum communications in the world.

2010, the research team of Chinese Science and Technology university and Tsinghua University achieve an innovation that the experience of the space quantum communication reaches to a record of $16km$ from the record of several hundred meters. After then, the research team of Chinese Science and Technology university and Shanghai Research Institute reach to a new record of $97km$ across a Chinese lake. In this experiment, they create the same qubit at a new place. This experiment transmitted 1100 pairs photons to a distance of 97km away. Which means that the satellite quantum communication is feasible.

Due to the feature of high efficiency and confidentiality of quantum information, scientists consider the quantum communication will become the next generation of IT technology. Based on the research of quantum entanglement, correction and storage technology have obtained an extremely important progress by China, Chinese Academy of Science Started a strategic and pilot science and technology project, try to lunch the first "quantum communication satellite" in the world by 2015.

The research team is consist of Professor jianwei Pan, Chengzhi Peng, Yuxiang Chen in Chinese Science and Technology university and Jianyu Wang in Shanghai Institute of Technical Physics of Chinese Academy of Science, Yongmei Huang in Optoelectronic Technology Institute of Chinese Academy of Science. The research team has realized the quantum teleportation and quantum entanglement distribution in the space away about $100km$ by October 2011 over the Qinghai Lake. It is proved experimentally that whether the quantum teleportation from earth to satellite or quantum entanglement two channels distribution from satellite to earth are both feasible. This experiment set up a technical basement not only for the large area quantum communication but also for the examination of quantum mechanics.

The transmission distance is $97km$ over a high loss space (Chinese lake), means that the transmission distance should be more than $1000km$ in the low loss space. Which means that the quantum satellite communication over all world is feasible.

Quantum storage and Quantum repeater

For a classical communication repeater, the repeater is used to shape and amplify pulse signal after transmitted a long distance (such as 40km). For a quantum repeater, the scientist proposed a theoretical repeater, which can be realized by a combination of *quantum storage technology, quantum entanglement switching technology and quantum purification technology.* So as to realize large scale and long distance quantum communications. However, the experiment of quantum storage is still a difficult issue in the world. For instance, a research team from China (Luming Duan), Austria and America have proposed another scheme of quantum repeater. Which is easy to be realized experimentally. However, after some experimental operation, it is found that the entanglement state is too sensitive to the channel length jitter in this scheme and the bit error rate is excessive growth and then it can not be realize in the practical long distance quantum communications.

To overcome this difficult issue, Pan Jianwei, Chen Zengbing and Zhao Bo proposed a theoretical scheme of quantum repeater with the quantum storage function, which is not sensitive to the channel length jitter, low bit error rate and high efficiency. Meanwhile, the co-operation and development of Pan Jianwei (China) with the scientists from Austria and America realized experimentally the basic unit of quantum repeater after achieved photon - quantum entanglement, and quantum teleportation step by step.

In October 2013, important progress was achieved in high-dimensional quantum information storage, which is obtained by Key Laboratory of Quantum Information of Chinese Academy of Science leaded by academician Guangcan Guo. The research team of this laboratory leaded by professor Baosen Shi realized the storage and release of a single photon pulse in a cold atoms system, which is spatial structure photon pulse and carries orbital angular momentum. The report has been published in ≪nature · communications≫. In which it is proved:

a. The storage of high-dimensional quantum state is feasible.

This team prepares two cold radicals via two magnetic optical traps (MOT), using one of them to prepare a marked single photon in terms of nonlinear process. And then make this photon carry certain angular momentum in terms of passing spiral phase plate. And this photon has a special spatial structure. After then, this photon is stored in another cold radicals as the storage media. The experiment displays that the orbital angular momentum carried by the single photon is able to be stored with high-fidelity.

b. Meanwhile, in terms of the Sagnac interferometer obtained with carefully designing, the team has verified that the storage of superposition of single photon orbital angular momentum is feasible, in which, quantum tomography technology and interferometry technology have been used. The superposition of states is that quantum information differs from classical information. These experiments displays that the quantum repeater with storage of high-dimensional quantum state is feasible.

Milestone events of quantum communication implementation

1. 1993, C. H. Bennett proposed the conception of quantum teleportation. After than, 6 scientists from different country proposed a scheme to realize "Teleportation" by using quantum channel and classical channel based on quantum entanglement. In this case, the one quantum state of two entangle article - an unknown quantum state, at A (Alice) is transmitted to B (Bob), then make other particle at B onto the prepared quantum state, the original article still leave at A. The basic idea of this scheme is that separated the original information at A (Alice) into classical information and quantum information. Both of them are sent to B (Bob) via classical channel and quantum channel, respectively. The classical information is obtained by certain measurement, quantum information is the information without extracted by Alice. After received both of classical information and quantum information at B, a replica of the original article is able to be prepared completely at B (Bob). In this transmission, the transferred object is the quantum state, but not the original article it-self. The sender (Alice) can even be ignorant of this quantum state. In this scheme, quantum state can be used as the carrier of information, complete the large volume of information transmitting through the quantum state transfer.

2. In 1997, Anton Zeilinger and jianwei Pan in Innsbruck University, Austria, publish a paper in ≪Nature≫ to report that the research group realized a single photon teleportation for the first time in the world. Which is the first time in the world to send one quantum state of photon from Alice (at A) to Bob (at B) experimentally. In this experiment, the transmitted object is the quantum state (as

the quantum information) only. but not the photon it-self, which should be as the information carrier. However, the quality of quantum entanglement will be getting worse as increasing transmission distance because of the environmental Noise. An then, how to get high quality of quantum entanglement is a key issue in the world.

3. The research team in University of Padua, leaded by Paul·Veronica Meath and Kai Shaer·Ba Boli, use binoculars with diameter $1.5m$ at Laser Ranging Observatory named "Matera" in Italy emitted photons to a satellite away $1500km$ over earth named "Ajisai" from Japan, and then required the Japanese satellite to bounce back the photons to the original emitted place. This experiment displayed that the quantum encoded communication may be realized by the satellite communication, which can not be eavesdropped anymore.

In June 2007, a research team from Austria, Britain and Germany, hit the maximum quantum communication distance of 144 kilometers in space. Over which, It is very difficult, since atmosphere will interference the vulnerable quantum entanglement state. Therefore, based on the satellite experiment above, the research team consider a new idea that the photon can be emitted from the satellite. Since the atmosphere will be getting thinner and thinner as the height increasing. The photon traveling on the space several km is equivalent to the photon traveling $8km$ on the ground.

To verify that it is observable that the photon bounce backed from the satellite, the research use binoculars with diameter $1.5m$ at Laser Ranging Observatory named "Matera" in Italy emitted a normal laser beam to the "Ajisai" satellite. The "Ajisai" satellite is consist of 318 mirrors. The laser beam bounce back from the 318 mirrors to the original emitted place, the Observatory.

4. In 2012, the research team of Chinese scientist jianwei Pan achieved $97km$ kilometers quantum teleportation and entanglement distribution in China. Which has set up a basic technology for lunching quantum communications satellite to the space. The transmission distance is $97km$ over a high loss space (Chinese lake), means that the transmission distance should be more than $1000km$ in the low loss space. Which means that the quantum satellite communication over all world is feasible.

8.4.4 Argument

After the introduction of quantum teleportation implementation above, now you may have your point of view about the argument between Einstein and Bell. A lot of physicist still can not believe that the quantum communication on optical fiber is feasible, they consider that the quantum state can not carrier information to realize quantum teleportation. Therefore, we would like to introduce some more information about this argument.

On the magazine ≪Nature ≫ by 2008,08,14, 5 scientists from Swiss reported their newest research results. In which the scientists pointed out that atom, electron and microscopic materials in the universe always display some abnormal strange behavior, which sometimes completely contrary with the classical scientific law. For instance, an object may display in two or more places simultaneously. Two entangled articles may spin with opposite direction to each other. Perhaps these phenomena can be explained by quantum physic only. From quantum physic point of view, an even occurs in one object, may inference another objet, no matter the second objects is far away from the first object. This is so called "quantum entanglement".

Einstein can not accepted "quantum entanglement" behavior, he call this behavior is "distant ghosts behavior". According to the explaination from quantum mechanics, two entangled articles can perception and impact each other's status no matter how far apart the two particles. Scientists try to exam whether this magical property is true or not and what is the reason behind for several tens years. In fact, we may explain this phenomena with visualization method, that two entangled objects release certain unknown article or other high speed signal, so as to impact the partner. Previously, the fact, that there are hidden signals, has been verified in the area of classical physics. After then, people dispel the suspicion about hidden signals. However, there is a magic possibility has to be verified, namely, the transmission speed of the unknown signals is possible to be beyond light speed.

To verify this possibility, scientist from Swiss star to study the entangled articles experimentally. First, separated the photon pair, and send them to the received stations along optical fiber provided by Swiss telecommunications company. The distance between two receiver is about $18km$. The photons will be measured with detector along the travel path so as the scientists can determine the "color" from the start to the end. Finally, the receiver confirm that every article pair is still in entanglement after separated and traveling to the receiver. After analyzed one photon, the scientist can correctly determine the behavior of the other photon. Any hidden signal traveling from one receiver to another receiver only spend one million trillion of a second, which transmission speed guarantee that the receiver can correctly detected the photon. From which, it is reasonable to speculate that the transmission speed of any hidden signal is 10000 times of light speed at least.

Einstein not only can not accept the idea about "quantum entanglement", but also conform that there is no object that its moving speed is beyond light speed according to the "Relativity" of Einstein published in 1905. Einstein consider: there is no object that its moving speed is beyond light speed. Light speed is a nature constance: the light speed is the same for any investigator. When object is accelerating, the weight of the object will be increased simultaneously , and the accelerating of an object needs energy to support. When the wight of the object is increasing the energy caused by the accelerating will be increasing either. After calculation, Einstein think that the wight of the object will become infinity when the moving speed of the object is going to be light speed, and then the energy to support this moving speed is infinity either. Therefore, it is impossible that the moving speed is beyond light speed.

The experimental conclusion from scientists is different from the point of view of Einstein and is helpful for us to understand the quantum communication step by step although quantum teleportation still is an issue in physicist.

8.5 Quantum satellite communications

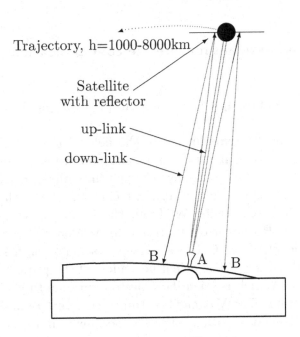

Figure 8-11   Quantum satellite communications

The conception of quantum satellite communication

Till now, most of scientists consider that quantum satellite communication is a new project, which is a theoretical project before its implementation is realized step by step.

Now, Figure 8-11 can help us to understand the conception of quantum satellite communication based on the fact that

a. The quantum state can not pass any optical amplifier as we mentioned before.

b. However, the laser beam can be bounced back from the satellite as long as the height of the satellite is lower than for example several thousand km over the earth, since the atmosphere will be getting thinner and thinner as height increasing, the photon traveling on the space several thousand km is equivalent to the photon traveling $8km$ on the ground.

Therefore,

a. The laser beam from the ground A (Alice) can be bounced back from the low height satellite to the ground B (Bob) to form a satellite communications.

b. Ground station B (Bob) is one station of the quantum optical fiber communication link.

In this case, the quantum satellite communications can be used to connect to the quantum optical fiber communication link from one area to another area on the earth, regardless of the distance between them.

Implementation of quantum satellite communications

Austria and China are stepping up cooperation on quantum physics.

The optical ground station for experiments in the field of quantum communication opened at the Institute for Quantum Optics and Quantum Communication (IQOQI) in Vienna, which will pave the way for quantum physical experiments in outer space. At their joint press conference, Karlheinz T?chterle, the Austrian Minister for Science and Research, Heinz W. Engl, the Rector of the University of Vienna, Helmut Denk, the President of the Austrian Academy of Sciences, and Hejun Yin, the Vice -President of the Chinese Academy of Sciences, inspected the "Vienna Quantum Space Test Link". The Scientific leaders of the project are Anton Zeilinger and Jian Wei Pan. Which is scientific milestone in quantum physics and in Austro -Chinese cooperation. The Vienna Quantum Space Test Link plays a key role in the context of the planned quantum physics experiments in outer space. By enabling the development of the necessary innovative technologies and infrastructures, it marks an important step toward technologies of the future.

The prepared distribution of entangled light particles enables the creation of unconditionally secure cryptographic keys. Currently, fiber optics technology is limited to applications over comparably short distances. Intercontinental quantum communication is therefore restricted to optical terminals on satellites and ground stations. Quantum key relay protocols, on the other hand, enable and guarantee the secure distribution of cryptographic keys between two stations on Earth, regardless of the distance between them. In the joint project QUESS (Quantum Experiments on Space Scale), the European partners will prepare the ground station while the

Chinese partners will provide the satellite, which will be launched within the next five years. A source for entangled photons will be on board the satellite, enabling the testing of quantum communication to satellites using individual photons. The Vienna group will be responsible for scientific coordination of the ground stations in Europe.

Strategically important partnership: "In the context of Austrian research activities , quantum physics plays a very prominent role and already has a highly successful track record," said the Austrian Minister for Science and Research, Karlheinz T?chterle, at the press conference last Friday. The eleven ERC grants and the four Wittgenstein award winners give a clear reflection of Austria's excellent achievements, and the nine START awards in the field of quantum physics clearly underline the fact that Austria also focuses on promoting talented young scientists. in Austria, quantum physics enjoys a broad institutional basis. Close cooperation between universities and the Academy of Sciences also contributes towards strengthening its position. "The quantum space project enables Austrian scientists from the Vienna IQOQI to further develop their internationally acknowledged role as leading players in their field, while also confirming most impressively their attractiveness as scientific partners for China," emphasized the Minister.

"The University of Vienna is a leading player in the excellent field of quantum physics. Scientists from the University of Vienna such as Anton Zeilinger, Markus Arndt, Markus Aspelmeyer and Frank Verstraete stand for top scientific achievements that enjoy a worldwide reputation," underlined the Rector of the University of Vienna, Heinz W. Engl. "The University of Vienna attaches utmost importance to promoting both the international perspective and the international dimension of research and teaching. Cooperation with Chinese partners contributes to strengthening the worldwide network built by the University of Vienna."

"The Vienna Quantum Space Test Link, contractually agreed upon in 2011, is an outstanding example of the successful and long-term cooperation between the Chinese Academy of Sciences and the Austrian Academy of Sciences, involving participation by the University of Vienna," stressed the President of the Austrian Academy of Sciences, Helmut Denk.

In his concluding statement, Anton Zeilinger underlined that "new opportunities are being opened up by this combination of expert knowledge enjoyed by the Vienna scientists and their Chinese partners. The development of unconditionally secure, multi-channel communication represents a very special challenge in quantum physics , and this project enables us to make an important step towards meeting that challenge. "

The goal of the project is to test quantum entanglement over a distance of thousands of kilometers and to pave the way for intercontinental quantum data exchange.

This article refers to the address: http://blog.sciencenet.cn/blog-327614-688559.html text from Web of Science Dan Leibo passengers .

The newest information

Based on the Chinese information, China successful launched a quantum satellite named "Mo Zi Hao" by 2016,08,16.

8.6 Quantum communications networks

In the past, most data is secured by encrypting it with algorithms based on computation-complexity. The security of these algorithms relies on the assumption that the minimum time it would take to decipher the encrypted message without access to the key known given by the state-of-the-art technology. The security assurance provided by computationally complex algorithms is that the information will no longer be relevant by the time and adversary could complete the operations required to decipher the encrypted message.

A new approach to the key distribution problem, however, may offer a potential solution and eventual realization of unconditional security. Which is known the quantum key distribution (QKD).

The security principle of QKD is based the quantum physics or based on the quantum entanglement as we mentioned in section 8.2. Base on the phenomena of quantum entanglement, QKD (quantum key distribution) has the potential to continuously distribute security keys over communication networks, including free-space and fiber optical links, without compromising the security of the key or relying on unproven computational and mathematical assumptions. The security of QKD is achieved by encoding random data in the quantum state of individual photons, transmitting the photons over a quantum channel, and perform a protocol BB84, for example, to distill a final, shared secret key. One of the important security principles of BB84 is the detection of eavesdropping. In BB84, Alice in transmitter chooses between two "conjugate" bases for encoding her qubit. The non-orthogonality of these quantum states makes it impossible for Eve (an eavesdropper) to measure the qubit values with perfect fidelity. When Eve makes a measurement on a qubit that is in superposition of states, the quantum system is reduced to only one of the two possible states by Eve's detector. This perturbation to the original state can be used to detect the presence of Eve on a quantum channel who is attempting to actively measure and re-transmitting the quantum bit sequence. The actions of Eve on the quantum channel will induce error and make her presence known to the communicating parties, Alice and Bob, through an increase in the quantum channel bit error rate.

QKD has several unique security advantages:

a. *Quantum bits used in the key can not be recorded by Eve.* The photons that passively extracted and measured by Eve can no be received by Bob and do not become part of the key. Meanwhile, Eve's attempt to actively measure and retransmit bits to Bob will increase Bob's error rate and expose her tempering attempts.

b. *More powerful computing technologies can not help Eve guess the key.* QKD does not use mathematical complexity to protect the information exchanged between Alice and Bob. As the result, more powerful computing technologies, including quantum computing, do not help Eve obtain the key.

c. *Alice and Bob can exchange a secure key in Eve's presence.* Post-processing steps such as privacy amplification theoretic estimate error in the quantum channel to Eve's tampering. Alice and Bob use information theoretic estimates to determine the size of their key to ensure its security. In BB84, a secure key can be established for quantum bit error rate (QBER) approximately less than 11%.[97]

d. *Security is based on the known laws of physics.* The security of QKD is based on the fundamental assumptions of quantum mechanics including superposition and the no-cloning theorem. For Eve to mount a successful and undetectable attack on an ideal QKD system, she must demonstrate that these assumptions can be violated.

8.6.1 The development of quantum communications in western

According to the report from Robert J. Runser et al. (http://blog.sciencenet.cn/blog-327614-688559.html), since the early QKD demonstration of the 1990s, many research groups through the world have begun experimental investigations to show the applicability of quantum communication over the distance in free space and dedicated-fiber links.

a. In 2002, the Los Alamos group achieved terrstrial free-space QKD through a 10km air mass at ground lavel in daylight, demonstrated the facibility of QKD from a ground station to a low earth orbiting satellite.[98]

b. Today the longest distance free space link that has been demonstrated to distribute keys was **over 144km** at night using an entanglement-based QKD scheme. [99]

c. Fiber QKD has also made a tremendous amount of progress. In 2003, a group from Mitsubishi [100] reported a fiber distance over 87km. This was quickly exceeded in 2003 by a Toshiba group [101] which achieved 122km. To data the world record in fiber QKD is held by Los Alamos which demonstrated a distance of **185km** working in collaboration with the super-conducting transition Edge Sensor (TES) detector group at NIST-Boulder and Albion College. [102]

Practical constraints in QKD system deployment, such as optical loss in the transmission medium and precision tolerances of the components, appear to limit the current generation of QKD system to about **200km**.

Breakthroughs in on-demand single-photon sources (which can be obtained by reducing the optical pulse source to a single-photon/ns level sent by Alice.) [103],[104], noiseless single-photon detectors [105], [85], and novel optical fiber type [106] could improve the reach of these system in the future. It may also be possible to take advantage of new protocols to extend the transmission distance such as the *decoy state protocol* [107]. By using the combination of these innovations, QKD system may be able to move beyond metropolitan area fiber distance in the near future.

Although point-to-point quantum communications are important, multi-user quantum networking will require that Alice can route her quantum communications to other parties. In this case, Alice and Bob require a conventional network connection to exchange classical information to complete the QKD protocol and a means to transmit their encrypted information. For these reason, it is important to investigate the compatibility of quantum communications with conventional optical channels on the same fiber infrastructure, and the technical challenges associated upgrading an existing optical network to support quantum communications service.

8.6.2 Quantum communications networks

A paper "Progress toward quantum communication networks: opportunities and challenges" provided by Robert Runser etc. [1] investigate the ability of optical fiber network to support both of optical traffic and single-photon quantum communications signal on a shared infrastructure in terms of the DWDM technology. In which:

a. The effect of Ramman scattering from conventional channels on the quantum bit error rate of a QKD system is analyzed.

b. Additionally, the potential impact and mitigation strategies of other transmission impairments such as four wave mixing, cross-phase modulation, and noise from mid-span optical amplifiers are discussed.

c. Review recent trends toward the development of automated and integrated QKD systems which are important steps toward reliable and manufacturable quantum communications systems.

---

[1]rrunser@ieee.org

Bibliography

[1] Cisco Systems.Inc. USA, "Introduction to DWDM Technology", June 4,2001. pp.1-1.

[2] Cisco Systems.Inc. USA, "Introduction to DWDM Technology", June 4,2001. pp.1-8.

[3] ITU-T G.694.1 "Spectral grids for WDM applications DWDM frequency grid" ITU-T website (http://www.itu.int/rec/T-REC-G.694.1/en).

[4] "DWDM ITU Table, 100MHz spacing" telecommengineering.co(http: www.telecomengineering.com/downloads/DWDM

[5] (http://www.infinera.com/products/ils2.html).

[6] Cisco Systems.Inc. USA, "Introduction to DWDM Technology", June 4,2001. pp.1-10.

[7] Cisco Systems.Inc. USA, "Introduction to DWDM Technology", June 4,2001. pp.1-13.

[8] Cisco Systems.Inc. USA, "Introduction to DWDM Technology", June 4,2001. pp.1-11.

[9] Cisco Systems.Inc. USA, "Introduction to DWDM Technology", June 4,2001. pp.3-5.

[10] Cisco Systems.Inc. USA, "Introduction to DWDM Technology", June 4,2001. pp.3-6.

[11] Cisco Systems.Inc. USA, "Introduction to DWDM Technology", June 4,2001. pp.3-15.

[12] Phillip. "Optical Multiplexing and Demultiplexing", "Optical Telecommunicaion networks", Spring 2012.

[13] I.P. Kaminow, "FSK with direct detection in optical multiple-access FDM networks", IEEE J. Selec. Areas Commun., vol. 8, pp. 1005-1014, 1990.

[14] J. Stonc and L. W. Stutz, "Pigtailed high-fineses tunable fiber Fabry-Perot interferometers with large, medium and small free spectral ranges," Electron. Lett. vol. 23, pp. 781-783, 1987.

[15] C. Dragone, C. A. Edwards, and R. C. Kister, "Integrated optical N x N multiplexer on silicon," IEEE Photon Technol. Lett., vol. 3, pp. 896-899, 1991.

[16] M. Zimgibl, C. Dragone, and C. H. Joyner, "Demontration of a 15 x 15 array wavelength multipler on InP," IEEE Phonton Technol. Lett. vol. 4, pp. 1250-1253, 1992.

[17] A. Yi-Yan, R. J. Deri, M. Seto, and R. J. Hawkins, "GaAs/GaAlAs asymmmetric Mach-Zehnder demultiplexer with reduced polarization dependence," IEEE Photon. Technol. Lett., vol. 1, pp. 83-85, 1989.

[18] H. A. Haus and Y. Lai, "Narrow band optical channel-dropping filters," J. Lightwave Technol. vol. 10, pp. 57-62, 1992.

[19] R. C. Alferness, L. L. Buhl, U. Koren, B. I. Miller, M. Young, T. L. Koch, C. A. Burrus, and G. Raybon, "Broadly Tunable InGaAsP/InP buried rib waveguide vertical coupler filter," Appl. Phys. Lett. vol. 60 pp. 980-982, 1992.

[20] H. Sakata, "Sidelobe suppression in grating-assisted wavelength-selectivity couplers," Opt. Lett., vol. 17, pp. 463-465, 1992.

[21] C. Bornholdt, F. Kappe, R. Muller, H.-P. Nolting, F. Reier, R. Stenzel, H. Venghaus, and C. M. Weinert, "Meander coupler. A novel wavelength division multiplexer/demultiplexer," Appl. Phys. Lett., vol. 57, pp. 2517-2519, 1990.

[22] H. Venghaus, C. Bornholdt, F. Kappe, H.-P. Nolting, and C. M. Weinert,, "Meander-type wavelength demultiplexer with weighted coupling," Appl. Phys. Lett., vol. 61, pp. 2018-2020, 1992.

[23] M. Kuznetsov, "Cascaded Coupler Mash-Zehnder Channel Dropping Filters for Wavelength-Division-Multiplezed Optical Systems" Journal of Lightwave Technology, vol. 12, nO. 2, fEBRUARY 1994, PP. 226-231.

[24] P. S. Cross and H. Kogelnik, "Sidelobe suppression in corrugated waveguide filter," Opt. Lett.. vol. 1, pp. 43-45, 1977.

[25] M. Kuznetsov, "Cascaded Coupler Mach-Zehnder Channel Droping Filters for Wavelength-Division-Multiplexed Optical Systems", Journal of Lightwave, Vol.12, No.2, February 1994, ppp.227.

[26] M. Kuznetsov, "Cascaded Coupler Mach-Zehnder Channel Droping Filters for Wavelength-Division-Multiplexed Optical Systems", Journal of Lightwave, Vol.12, No.2, February 1994, ppp.227.

[27] M. Kuznetsov, "Cascaded Coupler Mach-Zehnder Channel Droping Filters for Wavelength-Division-Multiplexed Optical Systems", Journal of Lightwave, Vol.12, No.2, February 1994, ppp.228.

[28] M. Kuznetsov, "Cascaded Coupler Mach-Zehnder Channel Droping Filters for Wavelength-Division-Multiplexed Optical Systems", Journal of Lightwave, Vol.12, No.2, February 1994, ppp.228.

[29] M. Kuznetsov, "Cascaded Coupler Mach-Zehnder Channel Droping Filters for Wavelength-Division-Multiplexed Optical Systems", Journal of Lightwave, Vol.12, No.2, February 1994, ppp.228.

[30] M. Kuznetsov, "Cascaded Coupler Mach-Zehnder Channel Droping Filters for Wavelength-Division-Multiplexed Optical Systems", Journal of Lightwave, Vol.12, No.2, February 1994, ppp.229.

[31] M. Kuznetsov, "Cascaded Coupler Mach-Zehnder Channel Droping Filters for Wavelength-Division-Multiplexed Optical Systems", Journal of Lightwave, Vol.12, No.2, February 1994, ppp.229.

[32] M. Kuznetsov, "Cascaded Coupler Mach-Zehnder Channel Droping Filters for Wavelength-Division-Multiplexed Optical Systems", Journal of Lightwave, Vol.12, No.2, February 1994, ppp.230.

[33] Manfred Kreuzer, IBM, Germany, "Strain Measurement with Fiber Bragg Grating Sensors". pp. 4.

[34] Amnon Yariv, Canifolia Institude of Technology, "Optical Electronics" 3rd Edition, 1985 CBS College Publishing. ISDN 0-03-070289-5. pp. 416-418.

[35] Manfred Kreuzer, IBM, Germany, "Strain Measurement with Fiber Bragg Grating Sensors". pp. 3.

[36] Manfred Kreuzer, IBM, Germany, "Strain Measurement with Fiber Bragg Grating Sensors". pp. 3.

[37] Erdogan, Turan (Audust 1997) "Fiber Grating Spectra". Journal of lightwave Technology 15(8): 1277-1294.

[38] Manfred Kreuzer, " Strain Measurement with Fiber Bragg Grating Sensors" HBM Department Germany.

[39] Milorad Cvijetic, Ivan B. Djordjevic, "Advanced optical communication system and networks" 2013 ARTECH HOUSE. Figure 2.20.

[40] M. K. Smit, "New Focusing and dispersive plannar component based on an optical phase array," Electron. Lett., 24(7), pp.385-386, 1988.

[41] H. Takahashi, S. Suzuki, K. Kato, and I. Nishi, "Arrayed-waveguide grating for wavelength division multi/demultiplexer with nanometer resolution," Electron. Lett., 26(2), pp.87-88, 1990.

[42] C. Dragone, " An NxN optical multiplexer using a plannar arrangement of two star couplers," IEE photon. Technol. Lett., 3, pp.812-815, 1991.

[43] C. Dragone, C. A. Edwards, and R. C. Kistter, "Integrated optical NxN multiplexer on silicon," IEEE photon. Technol. Lett., 3 pp. 896-899.

[44] Y. Tachikawa, Y. Inone, M. Kawachi, H. Takahashi, and K. Inoue, "" Arrayed-waveguide grating add-drop multiplexer with loop-back optical paths," Eletron. Lett., 29(24), pp.2133-2134, 1993.

[45] O. Ishida, H. Takashi, S. Suzuki, and Y. Inoue, " Multichannel frequency-selective grating multiplexer with fold-back optical paths," IEEE photon. Technol. Lett., 6(10), pp.1219-1221, 1994.

[46] Xaveer J. M. Leijtens, Berndt Kuhlow and Meint K. Smit, "Arrayed Waveguide Grating", Figure 4.1 (a), pp. 127.

[47] Xaveer J. M. Leijtens, Berndt Kuhlow and Meint K. Smit, "Arrayed Waveguide Grating", Figure 4.1 (b), pp. 127.

[48] Xaveer J. M. Leijtens, Berndt Kuhlow and Meint K. Smit, "Arrayed Waveguide Grating", Figure 4.2, pp. 130.

[49] Xaveer J. M. Leijtens, Berndt Kuhlow and Meint K. Smit, "Arrayed Waveguide Grating", Figure 4.2, pp. 130.

[50] Xaveer J. M. Leijtens, Berndt Kuhlow and Meint K. Smit, "Arrayed Waveguide Grating", Figure 4.6, pp. 135.

[51] Milorad Cvijetic, Ivan B. Djordjevic, "Advanced optical communication system and networks" 2013 ARTECH HOUSE. Figure 1.11.

[52] Han-xiong Lian, "Analytical Thchnology in Electromagnetic Field Theory in RF, Wireless and Optical Fiber Communications" 2009 iUniverse, Printed in the United State of America.

[53] Han-xiong Lian and Jia men Zhang, " A novel General Qunersi-steady-state solution of the Stimulated Raman Scattering in the optical fiber". Sino-Japenese electromagnetic field theory and optical fiber proceding, pp.34-39, 1987,5, Nan Jing, China.

[54] Jia men Zhang and Han-xiong Lain, " The theoretical analysis of the CW optical fiber ocsillator", Communications in China, No.1, 1990.

[55] Han-xiong Lian and Jia men Zhang, "The nonlinearity in singla mode fiber and the trasient solution of the trasient coupling wave equations", Journal of China Institute of Communications, Vol. 10, No. 1 jan. 1989, pp. 1-7 China.

[56] Jia men Zhang and Han-xiong Lian, "The analysis of transient process of the Stimulated Raman Scattering in single mode fiber", Journal of Beijing Institute of Posts and Telecommunations, No. 1, 1989, pp. 35-41.

[57] Milorad Cvijetic, Ivan B. Djordjevic, "Advanced optical communication system and networks" 2013 ARTECH HOUSE. Figure 2.28.

[58] Milorad Cvijetic, Ivan B. Djordjevic, "Advanced optical communication system and networks" 2013 ARTECH HOUSE. Figure 2.29.

[59] Milorad Cvijetic, Ivan B. Djordjevic, "Advanced optical communication system and networks" 2013 ARTECH HOUSE. Figure 2.29.

[60] Milorad Cvijetic, Ivan B. Djordjevic, "Advanced optical communication system and networks" 2013 ARTECH HOUSE. Figure 2.28.

[61] Milorad Cvijetic, Ivan B. Djordjevic, "Advanced optical communication system and networks" 2013 ARTECH HOUSE.

[62] Marc Niklés, Luc Th evenaz, and philippe A. Robert, "Brillouin Gain Spectrum Charaterization in Single-Mode Optical Fiber", Journal of Lightwave Technology, Vol. 15, No. 10. October 1997.

[63] Dr. Sawsan Abdul-Majid, P. Eng, "Optical Communications Networking and Optical Communication Systems", University of Ottawa.

[64] Ivan P. Kminow, Tingye Li, and Alan E. Willner, "Optical Fiber Telecommunication VA: Components and Subsystem" AP (Academic Press) pp.288, 2008.

[65] D. Marcuse, Theory of Dielectric Optical Waveguides, New York: Academic Press, 1974.

[66] D. Gloge, "Dispersion in weakingly guiding fibers," Applied Optics, Vol. 10, pp.2252-2258.

[67] G. Keiser, Optical fiber communications, McGraw-Hill, 1983.

[68] D. marcuse, Loss Analysis of Single Mode Fiber Splices, The Bell System Technical Journal. May-June 1977 pp.703-718.

[69] S. R. Nagel, "Optical fiber - The expanding medium," IEEE Commun. Mag. Vol. 25, no. 4 pp.33-44, 1987.

[70] J. Gowar, "Optical Communication systems", Prentice Hall, 1984.

[71] J. M. Senior, "Optical Fiber Communications", Prentice Hall, 1985.

[72] E. E. Basch, ed., "Optical Fiber Transmissions", Sams/McMillan, 1986.

[73] Govind P. Agrawal, "Fiber-Optic Communication System" The Institute of Optics, University of Rochester, NY WILEY-INTERSCIENCE, A John Wiley and Sons, Inc., Publication. Third Edition. pp.42.

[74] Govind P. Agrawal, "Fiber-Optic Communication System" The Institute of Optics, University of Rochester, NY WILEY-INTERSCIENCE, A John Wiley and Sons, Inc., Publication. Third Edition. pp.40.

[75] Ammon Yariv, "Optical Electronics" 3rd Edition, CBS College Publishing, New York, 1985, pp. 471. ISDN 0-03-070289-5.

[76] Amnon Yariv, Canifolia Institude of Technology, "Optical Electronics" 3rd Edition, 1985 CBS College Publishing. ISDN 0-03-070289-5. pp.480.

[77] C.H. Henry, "Theory of the linewidth of semiconductor lasers," IEEE, Quantum Eletron, 18(2),259. 1982.

[78] G. Grobkopf, R. Ludwig, R. Molt. E. Patzak, R. Schnabel, H. G. Weber, "Semiconductor Laser Amplifier".

[79] Moustafa Ahmed and Ali El-Lafi, "Analysis of small-signal intensity modulation of semiconductor laser taking account of gain suppression". Journal of Physics, Vol. 71, No.1 July 2008, pp99-115.

[80] Fabry, C., and A. Perot, "Theorie at applications d'une nouvell Method de spectroscopie interferentielle," Ann. Chim, Phys. Vol. 16, pp.115, 1899.

[81] Govind P. Agrawal, "Fiber-Optics communication system" Third Edition, 2002 John Wiley and Sons, Inc. pp.96.

[82] Govind P. Agrawal, "Fiber-Optics communication system" Third Edition, 2002 John Wiley and Sons, Inc. pp100.

[83] D.J. Blumenthal, "Lecture 5:Single mode laser design", pp.2.

[84] B.S. Robibson, A.J. Kerman, E.A. Dauler, R.J. Barron, D.O. Caplan, et al., Opt. Lett., vol. 31, pp.444-446, 2006.

[85] B.S. Robibson, A.J. Kerman, E.A. Dauler, R.J. Barron, D.O. Caplan, et al., Opt. Lett., vol. 31, pp.444-446, 2006.

[86] D.J. Blumenthal, "Lecture 5:Single mode laser design", pp.12.

[87] D.J. Blumenthal, "Lecture 5:Single mode laser design", pp.12.

[88] Amnon Yariv, Canifolia Institude of Technology, "Optical Electronics" 3rd Edition, 1985 CBS College Publishing. ISDN 0-03-070289-5. pp. 91.

[89] Mckay, K.G., "Avalanche breakdown in cilicon," Phys. Rev., vol. 94, p.877, 1954.

[90] Milorad Cvijetic, Ivan B. Djordjevic, "Advanced optical communication system and networks" 2013 ARTECH HOUSE. Figure 2.31.

[91] Han-xiong Lian, Ying-xiu Ye, Wenlei Lian, "Supper Low Noise PLL Oscillator and Low Jitter Synthesizer. Theory and Design." iUniverse Inc. 2014, pp.333-336.

[92] Griffiths, David J., "Introduction to Quantum Mechanics" (2nd ed.) Prentice Hall, 2004, ISDN 0-13-111892 7.

[93] P. Kwait; et al. (1995). "New High-Intensity Source of Polarization-Entangled photon Pairs". *Phys. Rev. Lett.* 75(24):pp.4337-4341.

[94] Anton Zeilinger (12 Octorber 2010). "The super-source and closing the communication loophole". *Dance of the photons: From Einstein to Quantum Teleportation*, Farrar, Straus and Giroux. ISDN 978-1-4299-6379-4.

[95] Bell, John, "On the Einstein Podolsky Rosen Paradox", *Physics* 13, pp. 195-200, Nov. 1964.

[96] Ekert, A. K. "Quantum cryptography based on Bell's theorem", Physical Review letters, 1991. 67(6), 661-663.

[97] C. E. Shannon, Bell System Technical Jounrnal, Vol. 28, pp. 656-715 (1949).

[98] R. J. Hughes, J.E. Nordholt, D. Derkacs, C.G. Peterson, *New Jounal of physics*, Vol.4, pp. 43.1-43-14(2002).

[99] R. Ursin, F. Tiefenbacher, T. Schmitt-Manderbach, H. Weier, et al., *arXiv: quant-ph/0606182*, V2, Jul 2006.

[100] T. Hasegawa, T.Nisioka, H. Ishizuka, J. Abe, M. Matsui, S. Takeuchi, *CLEO/QELS*, Baltimors, MD, 2003.

[101] C. Gobby, Z.L. Yuan, A. J. Shields, *Appl. Phys. Lett.*, vol. 84, pp.3762-3764 (2004).

[102] P.A. Hiskett, D. Rosenburg, C.G Peterson, R.J. Hughes, Nam, et al., *New Journal of Physics*, vol.8, 193, 2006.

[103] J.R. Rabeau, F. Jelezko, A. Stacey, B.C. Gibson, et al., *LEOS Summer Topical Meetings*, p.15, July 17-19, 2006.

[104] J. Vuckovie, D. Fattal, C. Santori, G.S. Solomon, Y.Yamamoto, *arXiv:quant-ph/0307025*, V 1, 3 Jul 2003.

[105] D. Rosenberg, S. Nam, P.A. Hiskett, C.G. Peterson, R.J. Hughes, et al., *Appl. Phys. Lett.*, vol. 88 021108, 2006.

[106] G. Humbert, J. Knight, G. Bouwmans, P. Russel, et al., *Optics Express*, vol.12, pp. 1477-1484, 2004.

[107] W. -Y. Hwang, *Phys. Rev. Lett.*, vol.91, pp. 05790, 2003.

[108] S. Fasel, N. Gisin, G. Ribordy, V. Scarani, H. Zbinden, Phys. Rev. Lett., vol.89 107901, 2002.

Printed in the United States
By Bookmasters